Ceramic Matrix Composites

Edited by
Walter Krenkel

Related Titles

Riedel, R., Chen, I.-W. (eds.)

Ceramics Science and Technology

Volume 1: Structures

2008
ISBN: 978-3-527-31155-2

Riedel, R., Chen, I.-W. (eds.)

Ceramics Science and Technology

Volume 2: Properties

2008
ISBN: 978-3-527-31156-9

Ewsuk, KG

Characterization, Design and Processing of Nanosize Powders and Nanostructured Materials

2006
ISBN: 978-0-470-08033-7

de With, G.

Structure, Deformation, and Integrity of Materials

Volume I: Fundamentals and Elasticity /
Volume II: Plasticity, Visco-elasticity, and Fracture

2006
ISBN: 978-3-527-31426-3

Krenkel, W., Naslain, R., Schneider, H. (eds.)

High Temperature Ceramic Matrix Composites

2001
ISBN: 978-3-527-60562-0

Kainer, K. U. (ed.)

Metal Matrix Composites

Custom-made Materials for Automotive and Aerospace Engineering

2006
ISBN: 978-3-527-31360-0

Scheffler, M., Colombo, P. (eds.)

Cellular Ceramics

Structure, Manufacturing, Properties and Applications

2005
ISBN: 978-3-527-31320-4

Ceramic Matrix Composites

Fiber Reinforced Ceramics and their Applications

Edited by
Walter Krenkel

WILEY-VCH Verlag GmbH & Co. KGaA

The Editor

Prof. Dr.-Ing. Walter Krenkel
University of Bayreuth
Ceramic Materials Engineering
Ludwig-Thoma-Strasse 36 b
95447 Bayreuth
Germany

■ All books published by Wiley-VCH are carefully produced. Nevertheless, authors, editors, and publisher do not warrant the information contained in these books, including this book, to be free of errors. Readers are advised to keep in mind that statements, data, illustrations, procedural details or other items may inadvertently be inaccurate.

Library of Congress Card No.:
applied for

British Library Cataloguing-in-Publication Data
A catalogue record for this book is available from the British Library.

Bibliographic information published by the Deutsche Nationalbibliothek
Die Deutsche Nationalbibliothek lists this publication in the Deutsche Nationalbibliografie; detailed bibliographic data are available in the Internet at <http://dnb.d-nb.de>.

© 2008 WILEY-VCH Verlag GmbH & Co. KGaA, Weinheim

All rights reserved (including those of translation into other languages). No part of this book may be reproduced in any form – by photoprinting, microfilm, or any other means – nor transmitted or translated into a machine language without written permission from the publishers. Registered names, trademarks, etc. used in this book, even when not specifically marked as such, are not to be considered unprotected by law.

Typesetting SNP Best-set, Hong Kong
Printing betz-druck GmbH, Darmstadt
Binding Litges & Dopf GmbH, Heppenheim
Cover Design WMX-Design, Bruno Winkler, Heidelberg

Printed in the Federal Republic of Germany
Printed on acid-free paper

ISBN: 978-3-527-31361-7

Foreword

Ceramic Matrix Composites (CMCs) are *non-brittle* refractory materials designed for applications in severe environments (often combining high temperatures, high stress levels and corrosive atmospheres). Compared to other structural materials (such as steels, aluminium or titanium alloys, as well as nickel-based superalloys or monolithic ceramics), they are relatively new, still mostly at the development stage but with a few well-established and promising applications in different high technology domains. We will first summarize here the main features of this new class of materials, then show what impact they have (or could have) on the development of different high tech fields and mention some important historical milestones.

The high strength potential of CMCs is directly related to the use of high strength, high modulus ceramic *fiber reinforcements* of small diameter (typically of the order of 10 µm). Covalent non-oxide fibers, such as carbon or silicon carbide based fibers, are those which presently display the best mechanical properties at high temperature (particularly in terms of creep resistance), but they are oxidation-prone. In this field, the development in Japan of the *SiC-based fiber* family, from the pioneering work of S. Yajima in the mid-1970s, which exhibits better oxidation resistance than carbon fibers, has been an important milestone. Comparatively, refractory *oxide fibers* (such as alumina and alumina-based fibers) display, by their chemical nature, excellent oxidation resistance, good mechanical behaviour at room temperature but creep even at moderate temperatures. As a result, carbon and SiC-based fibers presently are the most commonly used reinforcements in CMCs with a view to applications at high temperatures (say 1200–1800 °C). Small diameter ceramic fibers are extremely fragile and should be properly embedded in a refractory *ceramic matrix* (either oxide or non-oxide), primarily to protect the fibers and to permit load transfer from the matrix to the fibers. Fibers, with a volume fraction of the order of 40–50 %, are still the costly (but key) constituent of CMCs. *Nanoreinforcements*, such as carbon nanotubes or SiC nanofibers, are not presently used in CMCs to a significant extent, owing to handling difficulty, health considerations and cost.

Another key feature of CMCs, compared to polymer or metal matrix composites, is the fact that they are *inverse composites*, which is to say that under load it is the brittle matrix which fails first (in terms of failure strains: $\varepsilon_m^R < \varepsilon_f^R$) and at very low

Ceramic Matrix Composites. Edited by Walter Krenkel
Copyright © 2008 WILEY-VCH Verlag GmbH & Co. KGaA, Weinheim
ISBN: 978-3-527-31361-7

strain, typically ≈ 0.1%. Hence, matrix cracks should be arrested or/and deflected at the fiber–matrix (FM) interface to avoid the early failure of the fibers and thus a brittle failure of the composite. This is achieved through a weakening of the FM-bonding, usually by introducing a thin layer (typically, 50 to 200 nm) of a weak material at the fiber surface, acting as a *mechanical fuse* and referred to as the *interphase*. The most commonly used interphase materials are those with a layered crystal structure (the layers being roughly parallel to the fiber surface and weakly bonded to one another to promote crack deflection). Examples are pyrocarbon (PyC) or hexagonal boron nitride (BN). Historically, they have been formed either (1) insitu in SiC (Nicalon)/glass-ceramic composites by decomposition of the fibers or/and FM interactions during composite high temperature (HT) processing, or (2) deposited by chemical vapour infiltration (CVI) from gaseous precursors, in C/SiC and SiC (Nicalon)/SiC composites. Both these processes were developed in the 1980s. When the FM bonding is weak enough, CMCs behave as *elastic damageable non-linear materials*, which is to say that beyond the proportional limit, the brittle matrix undergoes multiple microcracking under load, the cracks being deflected within (or near) the interphases, the fibers being partly (or totally) debonded and exposed to the atmosphere before the ultimate failure which commonly occurs at a strain of the order of 0.5 to 1.5%. All these damaging phenomena take place with energy absorption and are responsible for the high toughness of the materials (a very uncommon feature for ceramics). An important milestone in this field was the pioneering work of A. Kelly and his coworkers in the early 1970s.

Improving the *oxidation resistance* of non-oxide CMCs is another important issue, particularly with a view to long duration exposures at high temperatures. In C/SiC and SiC/SiC composites, the weak points are the fibers themselves and the interphase. A first strategy is to use well-crystallized pure SiC fibers, which display good oxidation resistance, rather than SiC-based fibers (which usually contain free carbon and are poorly crystallized) or carbon fibers. In this field, the development of the so-called *stoichiometric* SiC fibers in Japan in the late 1990s was an important step. A second approach is to replace the commonly used PyC-interphase (which undergoes oxidation at temperatures as low as ≈ 500 °C) by *BN-interphase* (which displays a better oxidation resistance, at least in dry oxidizing atmospheres). A third possibility is to use *self-healing* coatings (single layer or multilayer), such as a SiC-coating containing a layer of a boron-bearing compound (such as B_4C) or a ternary Si–B–C mixture. The first role of the coating is to close the open residual porosity of the composite to impede the in-depth diffusion of oxygen. SiC-based coatings do undergo microcracking under load but oxygen diffusing along the microcracks would react with the microcrack wall to form a SiO_2–B_2O_3 healing phase. Finally, the best oxidation protection of SiC-based composites is achieved by extending the concept of crack-healing to the matrix itself which is now a *multilayered matrix* with layers of SiC, layers of sealant formers and layers of mechanical fuses, resulting in a lifetime under load exceeding 1000 hours at 1000–1200 °C. Finally, the use of oxidation-prone interphase can be avoided by utilizing a porous (and hence, relatively weak) matrix, but this approach exposes the fibers to the environment and, thus, it is better suited to oxide/oxide composites.

Processing considerations constitute another important point, the main requirement being that fiber degradation should be absolutely avoided. Hence, low temperature/pressure processing techniques are often favored. This is actually the case for the *CVI-process* and the *PIP-process* (Polymer Impregnation and Pyrolysis) where the matrix precursor is gaseous or liquid, respectively. Both are pressureless techniques involving temperatures of the order of 900–1200 °C. Further, the starting material can be a multidirectional fiber preform, e.g. a 3D-fiber architecture. These techniques yield near net shape parts (which can be of large size and complex shape) but with a relatively high residual porosity (10–15 %). In this field, an important milestone was the transfer of the CVI-process to the plant level in the 1980s for the volume production of C/SiC and SiC/SiC composites. In the so-called *RMI-process* (Reactive Melt Infiltration), the matrix is formed in situ by chemical reaction between a liquid precursor and a preconsolidated fiber preform. For SiC-matrix composites, the matrix precursor is liquid silicon (or a silicon alloy) and the fiber preform is consolidated (e.g. by PIP) with carbon, the former reacting with the latter to form the SiC-matrix. RMI is also a pressureless technique (it is conducted under vacuum). It yields near net shape composites with a low residual porosity, but it involves relatively high temperatures (1400–1600 °C for liquid silicon) with a risk of fiber degradation (unless thick fiber coatings are used), and the matrix usually contains unreacted precursor (such as free silicon). Finally, CMCs can also be fabricated according to *ceramic processing routes*. In the so-called Slurry Infiltration/High Pressure Sintering technique (SI-HPS), the reinforcement is impregnated with a suspension of matrix powder (usually a sol for oxide/oxide or a slurry for non-oxide matrix composites). After drying, the material is densified by sintering at high pressure. For non-oxide covalent ceramic powders, such as SiC-powders that display a poor sintering ability, sintering aids (such as oxide mixture forming eutectics) should be added to the slurry, the sintering conditions ($T = 1800$ °C, $P = 10$–50 MPa for SiC) remaining harsh. As a result, only very stable fibers, such as the stoichiometric SiC fibers prepared at high temperatures can be used. This technique yields composites with almost no residual porosity, high crystallinity and high thermal stability, but it is not suited to the volume production of large parts with complex shapes.

CMCs are expected to have a serious impact on the development of new technologies, as suggested by a few successful current applications. Significant *weight saving* is achieved when heavy superalloys are replaced by high strength and tough C/SiC or SiC/SiC composites in aerojet or rocket engines. Fighters are already equipped with CMC engine nozzles and could have, in the future, CMC combustion chambers. *Lifetimes* of parts working at high temperatures are improved by replacing metal alloys by CMCs. A good example is given by CMC braking systems (C/C for aircraft and C/C–SiC for cars) which exhibit longer lifetimes than their steel counterparts and better wear and friction properties at high temperatures. The use of C/C brakes, first on military fighters and then on civil jumbojet aircraft, on the basis of weight saving, braking performance and safety considerations, as well as that of C/C–SiC brakes on Formula 1 racing cars and sport cars, constitute other important milestones. CMCs can considerably extend the *temperature domain* of use of structural ceramics in many fields, such as jet engines and gas turbines (with higher yields and the possibility of reducing (or even suppressing) cooling requirement), heat exchangers and high temperature chemical engineering. Another promising new field of application could be the use of SiC/SiC composites in high temperature *nuclear reactors* (fission and fusion) for power generation, on the basis of their refractoriness, high temperature mechanical properties (creep resistance), compatibility with neutrons and low residual radioactivity after prolonged exposure to radiation.

It thus appears that CMCs – compared to metal alloys and monolithic structural ceramics – constitute a new class of materials which are well suited to applications in harsh environments. However, they are still very new and will undoubtedly require an intense effort of research. Present applications in different demanding fields suggest that they could have a bright future in the development of high technologies.

Honorary Professor, Bordeaux 1 University   *R. Naslain*

Contents

Foreword *V*
Preface *XVII*
List of Contributors *XIX*

1   **Fibers for Ceramic Matrix Composites** *1*
         Bernd Clauß
1.1   Introduction *1*
1.2   Fibers as Reinforcement in Ceramics *1*
1.3   Structure and Properties of Fibers *2*
1.3.1   Fiber Structure *2*
1.3.2   Structure Formation *3*
1.3.3   Structure Parameters and Fiber Properties *4*
1.4   Inorganic Fibers *7*
1.4.1   Production Processes *7*
1.4.1.1   Indirect Fiber Production *7*
1.4.1.2   Direct Fiber Production *7*
1.4.2   Properties of Commercial Products *9*
1.4.2.1   Comparison of Oxide and Non-oxide Ceramic Fibers *9*
1.4.2.2   Oxide Ceramic Filament Fibers *10*
1.4.2.3   Non-oxide Ceramic Filament Fibers *11*
1.5   Carbon Fibers *12*
1.5.1   Production Processes *15*
1.5.1.1   Carbon Fibers from PAN Precursors *15*
1.5.1.2   Carbon Fibers from Pitch Precursors *17*
1.5.1.3   Carbon Fibers from Regenerated Cellulose *17*
1.5.2   Commercial Products *18*
         Acknowledgments *19*

2   **Textile Reinforcement Structures** *21*
         Thomas Gries, Jan Stüve, and Tim Grundmann
2.1   Introduction *21*
2.1.1   Definition for the Differentiation of Two-Dimensional and Three-Dimensional Textile Structures *23*
2.1.2   Yarn Structures *23*

Ceramic Matrix Composites. Edited by Walter Krenkel
Copyright © 2008 WILEY-VCH Verlag GmbH & Co. KGaA, Weinheim
ISBN: 978-3-527-31361-7

2.2	Two-Dimensional Textiles	*24*
2.2.1	Nonwovens	*24*
2.2.2	Woven Fabrics	*25*
2.2.3	Braids	*27*
2.2.4	Knitted Fabrics	*28*
2.2.5	Non-crimp Fabrics	*29*
2.3	Three-Dimensional Textiles	*30*
2.3.1	Three-Dimensional Woven Structures	*30*
2.3.2	Braids	*32*
2.3.2.1	Overbraided Structures	*32*
2.3.2.2	Three-Dimensional Braided Structures	*34*
2.3.3	Three-Dimensional Knits	*37*
2.3.3.1	Multilayer Weft-Knits	*37*
2.3.3.2	Spacer Warp-Knits	*37*
2.4	Preforming	*38*
2.4.1	One-Step/Multi-Step Preforming	*38*
2.4.2	Cutting	*39*
2.4.3	Handling and Draping	*39*
2.4.4	Joining Technologies	*40*
2.5	Textile Testing	*41*
2.5.1	Tensile Strength	*41*
2.5.2	Bending Stiffness	*41*
2.5.3	Filament Damage	*42*
2.5.4	Drapability	*42*
2.5.5	Quality Management	*42*
2.6	Conclusions	*43*
2.6.1	Processability of Brittle Fibers	*43*
2.6.2	Infiltration of the Textile Structure	*43*
2.6.3	Mechanical Properties of the Final CMC Structure	*44*
2.6.4	Productivity and Production Process Complexity	*44*
2.7	Summary and Outlook	*44*
	Acknowledgments	*45*
3	**Interfaces and Interphases** *49*	
	Jacques Lamon	
3.1	Introduction	*49*
3.2	Role of Interfacial Domain in CMCs	*50*
3.3	Mechanism of Deviation of Transverse Cracks	*52*
3.4	Phenomena Associated to Deviation of Matrix Cracks	*53*
3.5	Tailoring Fiber/Matrix Interfaces. Influence on Mechanical Properties and Behavior	*55*
3.6	Various Concepts of Weak Interfaces/Interphases	*59*
3.7	Interfacial Properties	*61*
3.8	Interface Control	*64*
3.9	Conclusions	*66*

4	**Carbon/Carbons and Their Industrial Applications** *69*	
	Roland Weiß	
4.1	Introduction *69*	
4.2	Manufacturing of C/Cs *69*	
4.2.1	Carbon Fiber Reinforcements *71*	
4.2.2	Matrix Systems *73*	
4.2.2.1	Thermosetting Resins as Matrix Precursors *73*	
4.2.2.2	Thermoplastics as Matrix Precursors *74*	
4.2.2.3	Gas Phase Derived Carbon Matrices *75*	
4.2.3	Redensification/Recarbonization Cycles *79*	
4.2.4	Final Heat Treatment (HTT) *80*	
4.3	Industrial Applications of C/Cs *82*	
4.3.1	Oxidation Protection of C/Cs *83*	
4.3.1.1	Bulk Protection Systems for C/Cs *83*	
4.3.1.2	Outer Multilayer Coatings *88*	
4.3.1.3	Outer Glass Sealing Layers *90*	
4.3.2	Industrial Applications of C/Cs *92*	
4.3.2.1	C/Cs for High Temperature Furnaces *97*	
4.3.2.2	Application for Thermal Treatments of Metals *102*	
4.3.2.3	Application of C/C in the Solar Energy Market *105*	

5	**Melt Infiltration Process** *113*	
	Bernhard Heidenreich	
5.1	Introduction *113*	
5.2	Processing *114*	
5.2.1	Build-up of Fiber Protection and Fiber/Matrix Interface *115*	
5.2.2	Manufacture of Fiber Reinforced Green Bodies *117*	
5.2.3	Build-up of a Porous, Fiber Reinforced Preform *118*	
5.2.4	Si Infiltration and Build-up of SiC Matrix *119*	
5.3	Properties *121*	
5.3.1	Material Composition *127*	
5.3.2	Mechanical Properties *128*	
5.3.3	CTE and Thermal Conductivity *130*	
5.3.4	Frictional Properties *131*	
5.4	Applications *131*	
5.4.1	Space Applications *131*	
5.4.2	Short-term Aeronautics *133*	
5.4.3	Long-term Aeronautics and Power Generation *133*	
5.4.4	Friction Systems *134*	
5.4.5	Low-Expansion Structures *135*	
5.4.6	Further Applications *136*	
5.5	Summary *137*	

6 Chemical Vapor Infiltration Processes for Ceramic Matrix Composites: Manufacturing, Properties, Applications 141
Martin Leuchs

6.1 Introduction 141
6.2 CVI Manufacturing Process for CMCs 143
6.2.1 Isothermal-Isobaric Infiltration 144
6.2.2 Gradient Infiltration 145
6.2.3 Discussion of the Two CVI-processes 146
6.3 Properties of CVI Derived CMCs 146
6.3.1 General Remarks 146
6.3.2 Mechanical Properties 148
6.3.2.1 Fracture Mechanism and Toughness 148
6.3.2.2 Stress-Strain Behavior 149
6.3.2.3 Dynamic Loads 151
6.3.2.4 High Temperature Properties and Corrosion 151
6.3.2.5 Thermal and Electrical Properties 153
6.4 Applications and Main Developments 153
6.4.1 Hot Structures in Space 153
6.4.2 Gas Turbines 155
6.4.3 Material for Fusion Reactors 156
6.4.4 Components for Journal Bearings 156
6.5 Outlook 161

7 The PIP-process: Precursor Properties and Applications 165
Günter Motz, Stephan Schmidt, and Steffen Beyer

7.1 Si-based Precursors 165
7.1.1 Introduction 165
7.1.2 Precursor Systems and Properties 166
7.1.3 Cross-Linking Behavior of Precursors 167
7.1.4 Pyrolysis Behavior of Precursors 169
7.1.5 Commercial Available Non-oxide Precursors 171
7.2 The Polymer Impregnation and Pyrolysis Process (PIP) 171
7.2.1 Introduction 171
7.2.2 Manufacturing Technology 173
7.2.2.1 Preform Manufacturing 173
7.2.2.2 Manufacturing of CMC 175
7.3 Applications of the PIP-process 180
7.3.1 Launcher Propulsion 180
7.3.2 Satellite Propulsion 182
7.4 Summary 184

8 Oxide/Oxide Composites with Fiber Coatings 187
George Jefferson, Kristin A. Keller, Randall S. Hay, and Ronald J. Kerans

8.1 Introduction 187
8.2 Applications 189

8.3	CMC Fiber-Matrix Interfaces	189
8.3.1	Interface Control	190
8.3.2	Fiber Coating Methods	191
8.3.3	CMC Processing	194
8.3.4	Fiber-Matrix Interfaces	195
8.3.4.1	Weak Oxides	195
8.3.4.2	Porous Coatings and Fugitive Coatings	197
8.3.4.3	Other Coatings	198
8.4	Summary and Future Work	198

9   All-Oxide Ceramic Matrix Composites with Porous Matrices   205
Martin Schmücker and Peter Mechnich

9.1	Introduction	205
9.1.1	Oxide Ceramic Fibers	206
9.1.2	"Classical" CMC Concepts	207
9.2	Porous Oxide/Oxide CMCs without Fiber/Matrix Interphase	208
9.2.1	Materials and CMC Manufacturing	210
9.2.2	Mechanical Properties	214
9.2.3	Thermal Stability	218
9.2.4	Other Properties	220
9.3	Oxide/Oxide CMCs with Protective Coatings	223
9.4	Applications of Porous Oxide/Oxide CMCs	226

10   Microstructural Modeling and Thermomechanical Properties   231
Dietmar Koch

10.1	Introduction	231
10.2	General Concepts of CMC Design, Resulting Properties, and Modeling	232
10.2.1	Weak Interface Composites WIC	232
10.2.2	Weak Matrix Composites WMC	237
10.2.3	Assessment of Properties of WIC and WMC	238
10.2.4	Modeling of the Mechanical Behavior of WMC	238
10.2.5	Concluding Remarks	243
10.3	Mechanical Properties of CMC	244
10.3.1	General Mechanical Behavior	244
10.3.2	High Temperature Properties	246
10.3.3	Fatigue	251
10.3.4	Concluding Remarks	255
	Acknowledgment	256

11   Non-destructive Testing Techniques for CMC Materials   261
Jan Marcel Hausherr and Walter Krenkel

11.1	Introduction	261
11.2	Optical and Haptic Inspection Analysis	263
11.3	Ultrasonic Analysis	262

11.3.1 Physical Principle and Technical Implementation *263*
11.3.2 Transmission Analysis *264*
11.3.3 Echo-Pulse Analysis *265*
11.3.4 Methods and Technical Implementation *266*
11.3.5 Ultrasonic Analysis of CMC *267*
11.4 Thermography *268*
11.4.1 Thermal Imaging (Infrared Photography) *269*
11.4.2 Lockin Thermography *271*
11.4.3 Ultrasonic Induced Thermography *272*
11.4.4 Damage Detection Using Thermography *272*
11.5 Radiography (X-Ray Analysis) *273*
11.5.1 Detection of X-Rays *273*
11.5.1.1 X-Ray Film (Photographic Plates) *274*
11.5.1.2 X-Ray Image Intensifier *274*
11.5.1.3 Solid State Arrays *275*
11.5.1.4 Gas Ionization Detectors (Geiger Counter) *275*
11.5.2 Application of Radiography for C/SiC Composites *275*
11.5.3 Limitations and Disadvantages of Radiography *277*
11.6 X-Ray Computed Tomography *277*
11.6.1 Functional Principle of CT *277*
11.6.2 Computed Tomography for Defect Detection *279*
11.6.3 Micro-structural CT-Analysis *280*
11.6.4 Process Accompanying CT-Analysis *282*
11.7 Conclusions *283*

12 Machining Aspects for the Drilling of C/C-SiC Materials *287*
Klaus Weinert and Tim Jansen
12.1 Introduction *287*
12.2 Analysis of Machining Task *288*
12.3 Determination of Optimization Potentials *290*
12.3.1 Tool *290*
12.3.2 Parameters *294*
12.3.3 Basic Conditions *294*
12.4 Process Strategies *295*
12.5 Conclusions *300*

13 Advanced Joining and Integration Technologies for Ceramic Matrix Composite Systems *303*
Mrityunjay Singh and Rajiv Asthana
13.1 Introduction *303*
13.2 Need for Joining and Integration Technologies *304*
13.3 Joint Design, Analysis, and Testing Issue *304*
13.3.1 Wettability *305*
13.3.2 Surface Roughness *306*
13.3.3 Joint Design and Stress State *306*

13.3.4	Residual Stress, Joint Strength, and Joint Stability	307
13.4	Joining and Integration of CMC–Metal Systems	309
13.5	Joining and Integration of CMC–CMC Systems	314
13.6	Application in Subcomponents	318
13.7	Repair of Composite Systems	321
13.8	Concluding Remarks and Future Directions	322
	Acknowledgments	323

14 CMC Materials for Space and Aeronautical Applications 327
François Christin
14.1 Introduction 327
14.2 Carbon/Carbon Composites 328
14.2.1 Manufacturing of Carbon/Carbon Composites 328
14.2.1.1 n-Dimensional Reinforcement 328
14.2.1.2 Three-Dimensional Reinforcement Preforms 329
14.2.1.3 Densification 333
14.2.2 Carbon/Carbon Composites Applications 335
14.2.2.1 Solid Rocket Motors (SRM) Nozzles 335
14.2.2.2 Liquid Rocket Engines (LRE) 337
14.2.2.3 Friction Applications 338
14.3 Ceramic Composites 338
14.3.1 SiC-SiC and Carbon-SiC Composites Manufacture 339
14.3.1.1 Elaboration 340
14.3.2 SiC-SiC and Carbon-SiC Composites Applications 340
14.3.2.1 Aeronautical and Space Applications 340
14.3.2.2 Liquid Rocket Engines Applications 341
14.3.3 A Breakthrough with a New Concept: The Self-Healing Matrix 343
14.3.3.1 Manufacturing of Ceramic Composites 343
14.3.3.2 The Self-Healing Matrix 344
14.3.3.3 Characterization 344
14.3.4 Representative Applications of These New Materials 347
14.3.4.1 Military Aeronautical Applications 347
14.3.4.2 Commercial Aeronautical Applications 349

15 CMC for Nuclear Applications 353
Akira Kohyama
15.1 Introduction 353
15.2 Gas Reactor Technology and Ceramic Materials 354
15.3 Ceramic Fiber Reinforced Ceramic Matrix Composites (CFRC, CMC) 356
15.4 Innovative SiC/SiC by NITE Process 358
15.5 Characteristic Features of SiC/SiC Composites by NITE Process 359
15.6 Effects of Radiation Damage 362
15.6.1 Ion-Irradiation Technology for SiC Materials 363
15.6.2 Micro-Structural Evolution and Swelling 364

15.6.3	Thermal Conductivity 366
15.6.4	Mechanical Property Changes 369
15.7	Mechanical Property Evaluation Methods 371
15.7.1	Impulse Excitation Method for Young's Modulus Determination 372
15.7.2	Bulk Strength Testing Methods for Ceramics 373
15.7.3	Test Methods for Composites 374
15.7.4	Development of Materials Database 378
15.8	New GFR Concepts Utilizing SiC/SiC Composite Materials 379
15.9	Concluding Remarks 381

16 CMCs for Friction Applications 385
Walter Krenkel and Ralph Renz

16.1	Introduction 385
16.2	C/SiC Pads for Advanced Friction Systems 385
16.2.1	Brake Pads for Emergency Brake Systems 388
16.2.2	C/SiC Brake Pads for High-Performance Elevators 388
16.3	Ceramic Brake Disks 391
16.3.1	Material Properties 392
16.3.2	Manufacturing 394
16.3.3	Braking Mechanism 396
16.3.4	Design Aspects 398
16.3.5	Testing 401
16.4	Ceramic Clutches 403

Index 409

Preface

Ceramic Matrix Composites (CMCs) represent a relatively new class of quasiductile ceramic materials. They are characterized by carbon or ceramic fibers embedded in ceramic matrices (oxide or non-oxide) with comparatively low bonding forces between the fibers and the matrix. These weak intefaces, in combination with a porous and/or microcracked matrix, result in composite materials which differ from all other structural materials or composites and show some outstanding properties. Their strain-to-failure is up to one order of magnitude higher than in monolithic ceramics and their low densities result in mass-specific properties which are unsurpassed by any other structural material beyond 1000 °C.

From their research beginnings about 40 years ago, the demands of space technology played the decisive role in the development of CMCs. Hot structures of limited lifetime (e.g. thermal protection systems, nozzles) in aerospace and military applications have been developed in different countries. In recent years, civil and terrestrial requirements became the driving forces and properties and manufacturing processes were consistently improved to transfer CMCs from niche applications to broader markets. Due to their high thermal stability and good corrosion and wear resistance, these composite materials are of increasing interest for long-term applications and damage-tolerant structures in different industrial sectors like ground transportation (e.g. brake and clutch systems), mechanical engineering (e.g. bearings, ballistic protections), and power generation (e.g. burners, heat exchangers).

The goals of further research and development are focused on improvements in the thermal and oxidative stability of the reinforcing fibers and on a considerable reduction of the processing costs. Reasonable costs for series productions are expected by using innovative continuously operated furnaces as they already exist for other structural (monolithic) ceramics. Also, new forming processes for the manufacture of green bodies and new hybrid processes of high reliability are necessary. Beside these fabrication approaches, novel precursors for cheaper ceramic fibers and improvements in the thermomechanical properties of short-fiber reinforced CMCs are key factors to develop CMC materials for wider application.

This textbook provides a comprehensive overview of the current status of research and development on CMCs. It presents data tables, process descriptions, and field reports, giving special emphasis to applications relevant to the respective

Ceramic Matrix Composites. Edited by Walter Krenkel
Copyright © 2008 WILEY-VCH Verlag GmbH & Co. KGaA, Weinheim
ISBN: 978-3-527-31361-7

topics. In this regard, the textbook begins with two chapters on fibers and textile preforms for the reinforcement of ceramic matrix composites, followed by the description of the fiber/matrix interfacial domain of CMCs. In this chapter, data on interfacial characteristics and techniques to measure these characteristics are provided. This is followed by four chapters describing the most important processes used to manufacture non-oxide CMC materials currently. This includes the manufacture of carbon/carbon, the melt infiltration of silicon into carbon/carbon composites, as well as the Polymer Infiltration and Pyrolysis (PIP) and Chemical Vapor Infiltration (CVI) processes. Two chapters on oxide-CMCs with dense and porous matrices, which are promising materials particularly in combustion environments, conclude the processing part of the book.

The following two chapters describe the microstructural modelling and testing of CMCs using different models and methods. These topics are of special interest for designing structural parts and predicting their lifetime, for example by integrating non-destructive testing methods. As all fabrication approaches have certain limitations in terms of size and shape, the following two chapters deal with machining and joining techniques to achieve CMC structures of high integrity. This is followed by chapters providing practical experiences of the application of CMC materials under extreme thermal as well as corrosive conditions. Hot structures in spacecraft and aircraft show the tremendous progress which has been achieved with respect to re-usability and lifetime of CMC structures over the last 20 years. The current stage of development in using SiC/SiC composites as future structural materials in nuclear applications is described in a separate chapter. The most attractive volume market for CMCs currently is the topic of the last chapter. Test results and experiences with high-performance brake and clutch systems equipped with disks and pads of C/SiC composites are presented, demonstrating their superior tribological behaviour in automotive and other applications.

I would like to thank all the authors for their valuable and timely contributions. I am grateful to Roger Naslain, one of the pioneers of ceramic matrix composites, for writing the Foreword of this book. Furthermore, I would like to thank Waltraud Wüst and her team from Wiley-VCH and Petra Jelitschek as well as Angelika Schwarz from my research teams in Bayreuth for their help and cooperation during the publication process.

Bayreuth, Germany   Walter Krenkel

List of Contributors

Rajiv Asthana
University of Wisconsin-Stout
Department of Engineering
and Technology
326 Fryklund Hall
Menomonie, WI 54751
USA

Steffen Beyer
EADS-Astrium Space Transportation
Materials & Processes-Launcher
Propulsion Ariane-Center
81663 München
Germany

François Christin
Snecma Propulsion Solide
Les Cinq Chemins
33187 Le Haillan-Cedex
France

Bernd Clauß
ITCF Denkendorf Körschtalstraße 26
73066 Denkendorf
Germany

Thomas Gries
RWTH Aachen University
Institut für Textiltechnik
Eilfschornsteinstrasse 18
52062 Aachen
Germany

T. Grundmann
RWTH Aachen University
Institut für Textiltechnik
Eilfschornsteinstrasse 18
52062 Aachen
Germany

Jan Marcel Hausherr
University of Bayreuth
Ceramic Materials Engineering
Ludwig Thoma Strasse 36b,
95447 Bayreuth
Germany

Randall S. Hay
Air Force Research Laboratory
Materials and Manufacturing
Directorate
AFRL/RXLN
Wright-Patterson AFB, OH 45433-7817
USA

Bernhard Heidenreich
Deutsches Zentrum für Luft-
und Raumfahrt e.V.
Pfaffenwaldring 38-40
70569 Stuttgart
Germany

Ceramic Matrix Composites. Edited by Walter Krenkel
Copyright © 2008 WILEY-VCH Verlag GmbH & Co. KGaA, Weinheim
ISBN: 978-3-527-31361-7

List of Contributors

Tim Jansen
Universität Dortmund
Institut für Spanende Fertigung (ISF)
Department of Machining Technology
Baroper Straße 301
44227 Dortmund
Germany

George Jefferson
UES Inc., Dayton, OH
4401 Dayton-Xenia Rd.
Dayton, OH 45432
USA

Kristin A. Keller
UES Inc., Dayton, OH
4401 Dayton-Xenia Rd.
Dayton, OH 45432
USA

Ronald J. Kerans
Air Force Research Laboratory
Materials and Manufacturing
Directorate
AFRL/RXLN
Wright-Patterson AFB, OH 45433-7817
USA

Dietmar Koch
University of Bremen
Ceramics – Keramische Werkstoffe
und Bauteile
IW3/Am Biologischen Garten
228359 Bremen
Germany

Akira Kohyama
Kyoto University
Institute of Advanced Energy
Kyoto University
Graduate School of Energy Science
and Technology
Gokasho Uji
Kyoto 611-0011
Japan

Walter Krenkel
University of Bayreuth
Ceramic Materials Engineering
Ludwig-Thoma-Straße 36b
95447 Bayreuth
Germany

Jacques Lamon
Université de Bordeaux-CNRS
Laboratoire des Composites
Thermostructuraux
3 Allée de La Boétie
33600 Pessac
France

Martin Leuchs
MT Aerospace AG
Franz-Josef-Strauss-Str. 5
86153 Augsburg
Germany

Peter Mechnich
German Aerospace Center (DLR)
Institute of Materials Research
Linder Höhe
51147 Köln
Germany

Günter Motz
Universität Bayreuth
Lehrstuhl Keramische Werkstoffe
Ludwig-Thoma-Straße 36 b
95440 Bayreuth
Germany

Ralph Renz
Dr. Jng. h.c. F. Porsche AG
Porschestraße
71287 Weissach
Germany

Stephan Schmidt
EADS-Astrium Space Transportation
Materials & Processes-Launcher
Preopulsion
Ariane-Center
81663 München
Germany

Martin Schmücker
German Aerospace Center (DLR)
Institute of Materials Research
Linder Höhe
51147 Köln
Germany

Mrityunjay Singh
Ohio Aerospace Institute
MS 106-5, Ceramics Branch
NASA Glenn Research Center
Cleveland, OH 44135
USA

Jan Stüve
RWTH Aachen University
Institut für Textiltechnik
Eilfschornsteinstrasse 18
52062 Aachen
Germany

Klaus Weinert
Technische Universität Dortmund
Institut für Spanende Fertigung (ISF)
Department of Machining Technology
Baroper Straße 301
44227 Dortmund
Germany

Roland Weiß
Schunk Kohlenstofftechnik GmbH
Postfach 100951
35339 Gießen
Germany

1
Fibers for Ceramic Matrix Composites

Bernd Clauß

1.1
Introduction

New materials and processing routes provide opportunities for the production of advanced high performance structures for different applications. Ceramic matrix composites (CMCs) are one of these promising materials. By combining different ceramic matrix materials with special suitable fibers, new properties can be created and tailored for interesting technical fields.

This chapter gives an overview on fiber types, which can be used as fibrous components in CMCs [1–5]. The production of these fibers, as well as their structure and properties, will be discussed.

1.2
Fibers as Reinforcement in Ceramics

In CMCs, only fiber components are used that withstand the relatively high temperatures required for the production of ceramics, without significant damage. Other requirements to be met are long-term high-temperature stability, creep resistance, and oxidation stability. The importance of each of these demands depends on the type of application.

Organic, polymeric fiber materials cannot be used in CMCs because of their degradation at temperatures below 500 °C. Also conventional glass fibers, with melting or softening points below 700 °C, cannot be used for this purpose.

Possible candidates for the reinforcement of ceramic materials are polycrystalline or amorphous inorganic fibers or carbon fibers. The term "ceramic fibers" summarizes all non-metallic inorganic fibers (oxide or non-oxide), with the exception of fibers manufactured via solidification of glass melts.

The distinction between ceramic fibers and glass fibers has become more difficult during the last few years, because ceramics produced via new precursor or sol-gel routes can also be amorphous (i.e. "glassy") in structure and the production process can also contain a melt processing step. This means that ceramic fibers can be

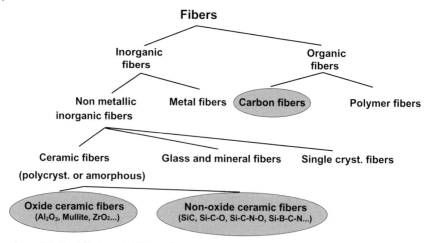

Figure 1.1 Classification of different fiber types.

either polycrystalline, partially crystalline, or amorphous. However, the expression "glass fibers" should only be applied to fibers that are produced via solidification of typical glass melts based on silicate systems. If these melts are produced by using minerals such as basalt, then the fibers should be called "mineral fibers."

Carbon fibers can also be used under certain conditions in CMCs. Although these fibers degrade in an oxidizing atmosphere above 450 °C, they are stable under non-oxidizing conditions up to temperatures of 2800 °C. Carbon fibers have a very good cost-performance ratio, if an environmental protection of the composite allows the use of this fiber type. Therefore, environmental barrier coatings (EBC) is an important field of research in CMCs.

Figure 1.1 shows a general classification of fibers, containing ceramic fibers and carbon fibers.

1.3
Structure and Properties of Fibers

Fibers used in high-performance composites possess superior mechanical properties (and in CMCs also superior thermal properties) compared to "normal" polymeric fibers. This chapter will discuss how this can be achieved with fibers made from different materials, including polymers, glass, ceramics, and carbon.

1.3.1
Fiber Structure

As shown in Figure 1.2, the structure of fibers can be considered from different viewpoints, depending on the "magnification" at which the structure is presented. The example shown is a polymeric fiber.

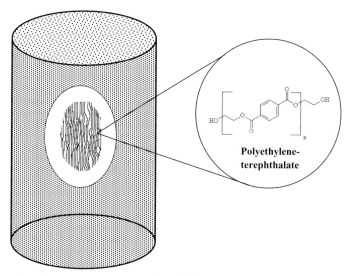

Figure 1.2 Fiber structure (example PET fiber).

At the *molecular level*, the displayed fiber is composed of poly(ethyleneterephthalate) (PET) chain molecules. This "chemical structure" determines the thermal and chemical stability and the theoretical strength of the fiber.

Most fibers also possess *supramolecular structures*, which are determined by molecular orientation and crystalline and/or amorphous regions (e.g. a two-phase crystalline and amorphous structure is indicated in Figure 1.2). These structures are formed during processing and can be significantly influenced by the processing conditions. This "physical structure" has important influence on the thermomechanical properties of fibers.

Finally, the fiber morphology influences the properties at the *macroscopic level*. Important criteria are cross-section, uniformity of the diameter along the fiber, porosity, and structural flaws, as well as surface properties such as roughness and surface energy (which is determined by the chemical structure of the surface). Adhesion to the matrix material and wetting behavior during infiltration steps are strongly influenced by these properties.

1.3.2
Structure Formation

The structure formation in fibers depends not only on the fiber material itself but also on the processing conditions. Process parameters are controlled in such a way that specific supramolecular and macroscopic structures can be obtained.

Important manufacturing processes are meltspinning, dryspinning, and wetspinning, together with modified spinning processes related to these:

- In the melt spinning process, fibers are formed via melts, which are forced through nozzles at high pressure and then solidified by cooling.

- In the dry spinning process, polymer solutions are used, which are also spun through nozzles. In this case, the fiber formation occurs by evaporation of the solvent from the spun solution.

- In the wet spinning process, polymer solutions are also used, but fiber forms by precipitation of the polymer in a liquid precipitation bath.

Important process parameters that influence the structure formation of the fibers are spinning speed, draw ratio, temperature, and other ambient conditions.

If specialty fibers are desired, the spinning process is often followed by after treatments, which will determine the final structure of the fibers.

Such after treatments include cross-linking of melt spun ceramic precursors before pyrolysis, annealing, and sintering of oxide based green fibers in order to achieve ceramic fibers and special surface treatments of carbon fibers.

If ceramic or carbon fibers are to be used in CMCs, the fibers have to be coated in many cases (e.g. with pyrocarbon or boron nitride), which act as interfaces between fiber and matrix.

1.3.3
Structure Parameters and Fiber Properties

The physical properties of fibers are determined essentially by three structure parameters: bond type, crystallinity, and molecular orientation [6].

The energy content of different types of chemical bonds is illustrated in Table 1.1. Covalent and ionic bonds, which can be oriented one-, two-, or three-dimensionally within fibers, have the highest energy content and so determine the mechanical strength and modulus of the fibers. Other bond types are of minor importance in high-performance fibers.

Different fiber types are shown in Table 1.2, which are distinguished by the structural parameters listed. It is obvious that the fibers possess different architectures and the final properties of the fibers (e.g. modulus and strength) are determined by a combination of these structural parameters.

If the bond type and the spatial orientation of the bonds were the main criterion for good mechanical properties, then ceramic fibers and glass fibers with three-dimensional, covalent, or ionic bonds would have far superior properties compared to other fiber types. But since these fibers are isotropic without molecular orientation, they possess lower strength than carbon fibers. This is because carbon fibers have a structure with two-dimensional covalent bonds, showing pronounced crystallinity and high orientation. Based on this advantageous combination of structural parameters, carbon fibers currently display the highest values for strength and modulus. Depending on the processing conditions, carbon fibers can achieve either moduli as high as 600 GPa, or extremely high strength values (above 7000 GPa).

Table 1.1 Chemical bond types and bond energies with examples of materials [6].

Bond type	Energy (kJ mol^{-1})	Examples	
Ionic	800–15 000	NaCl	3D
		ZrO$_2$	3D
		Al$_2$O$_3$	3D
Covalent	200–600 (single bonds)	Aramide	1D
		Graphite (in plane)	2D
		Glass, SiC	3D
Metallic	100–800	Metals	
Hydrogen bonds	20–50	Aramide	
		Aliphat. polyamide	
		Cellulose	
Dipole-dipole	ca. 2	Polyester	
van der Waals	ca. 1	Polyolefines	
		Graphite (between planes)	

Table 1.2 Fiber structures and properties [6].

Fiber type	Polyester polyamide	Aramide fibers from LC phase	Carbon	Ceramic (crystalline)	Ceramic (amorphous) glass
Structure	1D linear 2 phases	1D linear 1 phase	2D layered	3D isotropic	3D isotropic
Bond type	1D covalent, hydrogen bonds (PA), dipole-dipole (PES), van der Waals	1D covalent, hydrogen bonds, van der Waals	2D covalent, van der Waals	3D covalent/ ionic	3D covalent/ ionic
Crystallinity	Medium	Paracyrstalline	Paracrystalline	Polycrystalline	Amorphous
Orientation	Medium	Very high	High	None	None

Polymeric aramide fibers also show high strength values due to their high molecular orientation in the fiber axis, although only one-dimensional covalent bonds and hydrogen bonds are present.

Figures 1.3 and 1.4 show an overview of mechanical properties of different fiber types. Averages are given because, as mentioned before, properties can differ depending on processing conditions. In reality, no perfect fiber structure can be

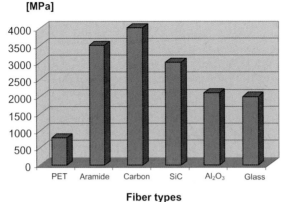

Figure 1.3 Typical tensile strengths (averages) of different fiber types.

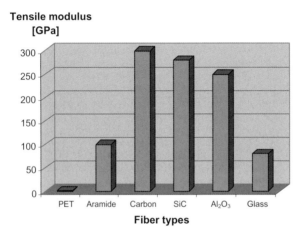

Figure 1.4 Typical tensile moduli (averages) of different fiber types.

obtained during processing, which means that the real properties of fibers are usually far below the theoretical property values calculated for a perfect structure. Therefore, one important goal of a fiber spinning and fiber formation process is to reduce structure imperfections to a minimum by optimized process control. Relatively high prices of high-performance fibers are usually caused by highly sophisticated and complex manufacturing processes rather than by more expensive materials used in the fiber production.

1.4
Inorganic Fibers

1.4.1
Production Processes

The manufacturing processes for inorganic fibers can be divided in two categories: One process is called "indirect," as the fibers or the non-ceramic precursor fibers are not obtained by a spinning process, but by using other fiber materials. The fiber is soaked with a pre-ceramic precursor material, or the precursor material is deposited on the surface. The inorganic fiber is then formed by pyrolysis of the organic template fiber.

In the second route, called "direct process," inorganic precursors (salt solutions, sols, or precursor melts) are directly spun into so-called "green fibers," in some cases by using organic polymer additives.

Another important distinction of production processes is based on the fiber length. There exist production processes for endless fibers as well as for short fibers, with fiber lengths from millimeters to some centimeters. Short fibers are usually produced by fibrillation of spinning dopes using fast rotating discs or by air-blowing techniques.

In addition, a process is conducted in which fabrics made of cellulosic short fibers are saturated with precursors and then pyrolysed and sintered in order to transform the material into a ceramic fiber fabric [7].

Figure 1.5 shows a schematic depiction of the production of ceramic fibers and Figure 1.6 shows a dry spinning line for the production of endless, oxide based green fibers.

1.4.1.1 Indirect Fiber Production

CVD Process In the CVD process, ceramic fibers are formed via gas phase deposition of ceramic materials on carrier fibers. The carrier fiber usually forms the core of the ceramic fiber. Examples of core materials are carbon fibers and tungsten wires.

Relic Process In the so-called "relic process," absorbent organic fiber materials (mostly cellulose-based) are saturated with salt solutions or sols. Afterwards the organic material is burned off and the salt or sol is converted into a ceramic material at high temperature to obtain ceramic fibers.

1.4.1.2 Direct Fiber Production
Direct fibre production processes can be distinguished by the pre-ceramic precursor components used in the spinning dopes:

Spinning Dopes Based on Molecularly Dispersed Precursors In these processes, soluble salts are used in the spinning dopes, which can be converted into ceramics

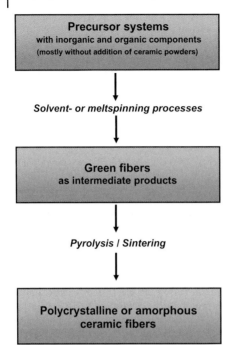

Figure 1.5 Production route of ceramic fibers.

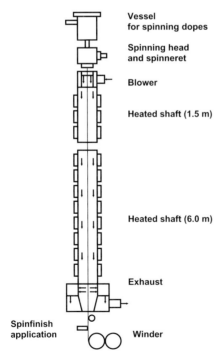

Figure 1.6 Dry spinning facility at ITCF, Denkendorf, for spinning of oxide forming green fibers.

by a calcination step. Although the salts are dissolved in the form of ions, that is, dispersed at the molecular scale, these processes are often incorrectly summarized under "sol-gel" processes.

In addition to the salt, the spinning dope consists of organic polymers such as poly(ethylenoxide), poly(vinylalcohol), or poly(vinylpyrrolidone), in order to achieve the rheological behavior needed for the spinning process. Solvents are usually water or water/alcohol mixtures. In some cases, nano-scaled ceramic particles are added to the dope to control the structure formation during ceramization.

Spinning Dopes Based on Colloidally Dispersed Precursors This process is similar to the one mentioned above, but here colloidal inorganic components are used as pre-ceramic precursors (i.e. particle sols). In this case, the term "sol-gel process" is appropriate. The organic polymers, which have to be added in order to achieve spinnability, are the same as in the above process, with the same solvents also being used.

Spinning Dopes Containing Coarse Ceramic Particles (Ceramic Powders) Sometimes, coarse ceramic particles are added to salt- or sol-based spinning dopes in order to increase the ceramic yield and to reduce the shrinkage during calcinations and sintering. In this case, the process is called the "slurry process."

Spinning Dopes Based on Inorganic Polymers In these so-called "precursor or precursor-polymer processes," the spinning dope consists of either a solution of an inorganic polymer, which can be spun via a dry spinning process or the precursor-polymers are fusible and can be spun using a melt spinning process. Here no addition of organic polymer is necessary since the solution or the melt already possess the visco-elastic rheological behavior needed to be spun into fibers.

The inorganic polymers usually carry organic functional groups such as methyl- or propyl-groups, so that during pyrolysis this organic material also has to be burned off. However, these systems have significantly higher ceramic yields compared to the processes described above. The precursors, which are melt spun, have to be cross-linked (either chemically or by high energy radiation) before pyrolysis, otherwise the material would re-melt and lose its fiber shape if heated above the melting point.

1.4.2
Properties of Commercial Products

1.4.2.1 Comparison of Oxide and Non-oxide Ceramic Fibers
Oxide fibers, currently available commercially, are mostly based on Al_2O_3- or Al_2O_3/SiO_2 ceramics. They possess high values for tensile strength and modulus, and due to their oxidic nature are stable against oxidation at high temperatures.

Unfortunately, even the best polycrystalline oxide fibers are prone to creep under load at 1100 °C. Above this temperature the fibers cannot be used in CMCs for long-time applications. Also oxide fibers tend to form larger grains when kept at high temperatures over long time periods. The larger grains tend to grow at the expense of smaller grains because of diffusion processes at grain boundaries, which can lead to brittle fibers.

Commercially available non-oxide ceramic fibers are based on SiC and Si-C-(N)-O materials, which contain undesirable oxygen to a greater or lesser extent and can also contain Ti, Zr, or Al.

Non-oxide fibers also exhibit high values for tensile strength and modulus (even higher than oxide fibers) and due to their structure, which is amorphous in many cases, they possess lower creep rates at high temperatures compared to the polycrystalline oxide fibers. Disadvantages of these fibers are their susceptibility to oxidation, which leads to fiber degradation in an oxidizing atmosphere over time. The lower the oxygen content of the fiber itself, the better its oxidation resistance.

The production process, which requires an inert atmosphere in most cases, is complex. Especially in the fabrication of low oxygen fibers (<1 wt.-% oxygen), such as the Hi-Nicalon types or Sylramic, sophisticated technologies are needed, which lead to high fiber prices.

The limitations of both oxide and non-oxide fiber types have to be known in order to choose the proper material for the intended application.

Current research activities are focused on the development of oxide fibers with enhanced creep properties und reduced grain growth rates, as well as on the development of non-oxide fibers with improved oxidation stability and lower production costs [8–16].

1.4.2.2 Oxide Ceramic Filament Fibers

Table 1.3 gives an overview of commercially available oxide ceramic filament fibers (i.e. endless fibers). The specifications given are taken from product information of the fiber producers [17–20]. The reported prices are for quantities of over 100 kg of fiber and are usually higher for lower quantities. The "denier" given for the 3 M fibers is a unit of measure for the linear mass density of fibers. It is defined as the mass in grams per 9000 meters. In the International System of Units, the "dtex" is used instead, which is the mass in grams per 10 000 meters. Since the fibers have different densities, the denier does not indicate directly the number of filaments. For Nextel 720, "3000 den" corresponds to about 900 filaments; for Nextel 610, "3000 den" corresponds to about 800 filaments; for Nextel 550 and 440, "2000 den" corresponds to about 700 filaments; and for Nextel 312, "1800 den" also corresponds to about 700 filaments (if fiber diameters of 12 μm are assumed).

Nevertheless, no responsibility can be taken for the correctness of this information.

Figure 1.7 shows the structure of a 10 μm mullite-based oxide fiber developed at ITCF, Denkendorf (Germany), as an example of a new oxidic ceramic fiber [21, 22].

Table 1.3 Overview about commercial oxide ceramic filament fibers.

Producer Fiber	Composition (Wt.-%)	Diameter (µm)	Density (g/cm³)	Tensile strength/ modulus (MPa/GPa)	Production technique/ structure	Approx. price
3M Nextel 720	Al_2O_3: 85 SiO_2: 15	10–12	3.4	2100/260	Sol-Gel/ 59 vol.% α-Al_2O_3 + 41 vol.% Mullite	€790/kg (1500 den) €600/kg (3000 den)
3M Nextel 610	Al_2O_3: >99	10–12	3.9	3100/380	Sol-Gel/ α-Al_2O_3	€790/kg (1500 den) €600/kg (3000 den) €440/kg (10000 den)
3M Nextel 550	Al_2O_3: 73 SiO_2: 27	10–12	3.03	2000/193	Sol-Gel/γ-Al_2O_3 + SiO_2 amorph.	€590/kg (2000 den)
3M Nextel 440	Al_2O_3: 70 SiO_2: 28 B_2O_3: 2	10–12	3.05	2000/190	Sol-Gel/γ-Al_2O_3 + Mullite + SiO_2 amorph.	€500/kg (2000 den)
3M Nextel 312	Al_2O_3: 62.5 SiO_2: 24.5 B_2O_3: 13	10–12	2.7	1700/150	Sol-Gel/Mullite + amorph. or 100% amorph	€260/kg (1800 den)
Sumitomo Altex	Al_2O_3: 85 SiO_2: 15	10/15	3.3	1800/210	Polyaluminoxane/ γ-Al_2O_3	€640–720/kg
Nitivy Nitivy ALF	Al_2O_3: 72 SiO_2: 28	7	2.9	2000/170	Sol-Gel/γ-Al_2O_3	€390/kg (twisted yarn, twists: 10–15)
Mitsui Almax-B	Al_2O_3: 60–80 SiO_2: 40–20	7–10	2.9	Not available	Unknown/δ-Al_2O_3	Price not available

1.4.2.3 Non-oxide Ceramic Filament Fibers

Table 1.4 gives an overview of commercially available (some only in smaller quantities) non-oxide filament fibers [23–26]. Prices for the Nicalon fibers were provided by the North American distributor COI Ceramics Inc. Nevertheless, no responsibility can be taken for the correctness of this information.

Figure 1.8 shows an example of an amorphous non-oxide Si-C-N fiber after temperature treatment at 1500 °C for 12 hours in air. This fiber was produced at ITCF, Denkendorf with a precursor material from the Department of Ceramic Materials Engineering of the University of Bayreuth [27, 28]. Beside the oxidation layer of silicon dioxide at the fiber surface, no further degradation can be observed.

1 Fibers for Ceramic Matrix Composites

(a)
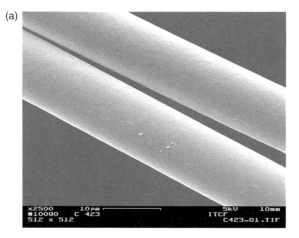

(b)
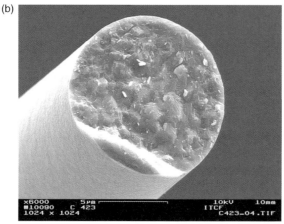

Figure 1.7 Mullite ceramic fibers (produced at ITCF, Denkendorf) as an example for oxide ceramic fibers.

1.5
Carbon Fibers

Carbon belongs to the materials with the highest temperature resistance, if kept in non-oxidizing atmosphere. Under these conditions carbon does not sublimate until temperatures of 3730 °C are reached. The material can be used in technical applications up to 2800 °C.

Well-known modifications of carbon are diamond, graphite, amorphous carbon, and also fullerenes and carbon nanotubes.

In graphite, each atom is bonded trigonally to three other atoms, making a strong two-dimensional network of flat six-membered rings; the bond energies in between the planes are weak (Figure 1.9, left).

Table 1.4 Overview about commercial non-oxide ceramic filament fibers.

Producer Fiber	Composition (Wt.-%)	Diameter (μm)	Density (g/cm^3)	Tensile strength/modulus (MPa/GPa)	Production technique/structure	Approx. price
Nippon Carbon Hi-Nicalon "S"	Si: 68.9 C: 30.9 O: 0.2	12	3.10	2600/420	Polycarbosilane/ β-SiC	€7000/kg >10 kg
Nippon Carbon Hi-Nicalon	Si: 63.7 C: 35.8 O: 0.5	14	2.74	2800/270	Polycarbosilane/ β-SiC + C	€3250/kg >10 kg
Nippon Carbon Nicalon NL-200/201	Si: 56.5 C: 31.2 O: 12.3	14	2.55	3000/220	Polycarbosilane/ β-SiC + SiO$_2$ + C	€1000/kg >10 kg
UBE Industries Tyranno Fiber SA 3	Si: 67.8 C: 31.3 O: 0.3 Al: <2	10/7.5	3.10	2800/380	Polycarbosilane/ β-SiC $_{cryst.}$ + …	€6500/kg >10 kg
UBE Industries Tyranno Fiber ZMI	Si: 56.1 C: 34.2 O: 8.7 Zr: 1.0	11	2.48	3400/200	Polycarbosilane/ β-SiC + …	€1400/kg >10 kg
UBE Industries Tyranno Fiber LoxM	Si: 55.4 C: 32.4 O: 10.2 Ti: 2.0	11	2.48	3300/187	Polycarbosilane/ β-SiC $_{amorph.}$ + …	€1200/kg >10 kg
UBE Industries Tyranno Fiber S	Si: 50.4 C: 29.7 O: 17.9 Ti: 2.0	8.5/11	2.35	3300/170	Polycarbosilane/ β-SiC $_{amorph.}$ + …	€1000/kg >10 kg
COI Ceramics Sylramic-iBN	SiC/BN	10	3.00	3000/400	Precursor-polymer/ SiC/BN and other phases	€10 500/kg >10 kg
COI Ceramics Sylramic	SiC: 96.0 TiB$_2$: 3.0 B$_4$C: 1.0 O: 0.3	10	2.95	2700/310	Precursor-polymer/ SiC and other phases	€8500/kg >10 kg
Specialty Materials SCS-Ultra	SiC on C	140 (with carbon fiber core)	3.0	5865/415	CVD on C-filament/β-SiC on C	€16 400/kg
Specialty Materials SCS-9A	SiC on C	78 (with carbon fiber core)	2.8	3450/307	CVD on C-filament/β-SiC on C	€19 600/kg
Specialty Materials SCS-6	SiC on C	140 (with carbon fiber core)	3.0	3450/380	CVD on C-filament/β-SiC on C	€4850/kg
Tisics Sigma	SiC on W	100/140 (with tungsten wire core)	3.4	4000/400	CVD on W-filament/SiC on W	Price not available

1 Fibers for Ceramic Matrix Composites

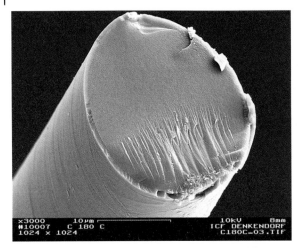

Figure 1.8 Si-C-N-fibers (ITCF, Denkendorf), produced from a polycarbosilazane precursor (University of Bayreuth). The depicted fiber was already treated for 12 h at 1500 °C in air [5].

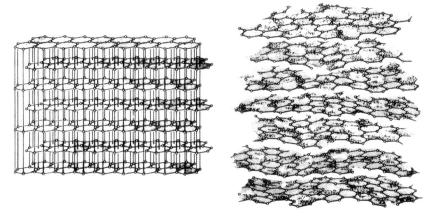

Figure 1.9 Carbon structures. Left: perfect graphite crystal, right: turbostratic structure, which is present in carbon fibers [31].

In carbon fibers, a layered structure is present, but unlike true graphite, the graphitic layers are not neatly stacked, but have a more random arrangement [29, 30]. The layers are also not planar but rather ondulated, so this structure is called "turbostratic" [31] (Figure 1.9 right). It is therefore appropriate to refer to the structure of PAN-based carbon fibers as turbostratic graphite. In fibers derived from mesophase pitch precursors, structures are formed closer to the true graphite structure.

In order to achieve carbon fibers with high tensile strengths and moduli, the carbon planes have to be oriented toward the fiber axis and optimized in their

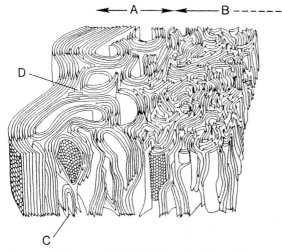

Figure 1.10 Structure of PAN derived carbon fiber with different substructures: A = surface region; B = bulk region; C = "hairpin" defect; D = disclination [32].

structure. This can be done by temperature treatments in an inert atmosphere under tension.

Perpendicular to the fiber axis there is usually no orientation of the carbon planes. This type of structure is present in all carbon fibers, but nevertheless there are differences in terms of degree of orientation, arrangement of the layers in radial direction, interaction in between the planes, and the presence of different structural flaws such as microvoids and other imperfections [32] (Figure 1.10). This leads to fibers with different mechanical properties, from high tensile strength to very high modulus.

The elastic modulus of the fibers depends mainly on the degree of orientation of the planes along the fiber axis, whereas the tensile strength is limited by the number of structural flaws. During temperature treatments up to 1500 °C, the tensile strength tends to be optimized and treatments at even higher temperatures (up to 2800 °C) lead to fibers with high moduli.

For the use of carbon fibers in CMCs, such as C/SiC (carbon fiber reinforced silicon carbide), the fibers have to be protected against oxidation. Otherwise they will deteriorate by oxidative degradation at temperatures above 450 °C. The fibers can be protected against oxidation by the surrounding matrix itself but in most cases the composite has to be protected additionally by a so-called EBC.

1.5.1
Production Processes

1.5.1.1 Carbon Fibers from PAN Precursors

The starting materials for this process are polyacrylonitrile fibers produced especially for carbon fiber manufacturing. The composition of the PAN is different

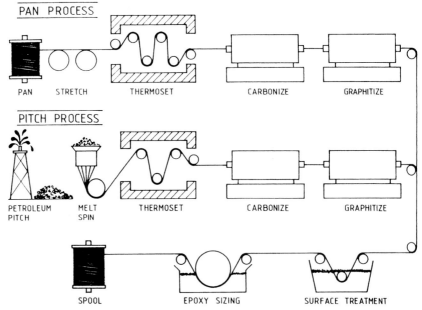

Figure 1.11 Two important production processes for carbon fibers [31].

from the others used for textile applications (different co-monomers are used for the polymerization of acrylonitrile).

The PAN fibers are first treated under tension in a stabilization step in an oxidizing atmosphere at temperatures between 250 and 300 °C. In this process, the linear PAN chain molecules are transformed to a material with cyclic and ladder-like structure elements, which can withstand further treatments at higher temperatures [31] (Figure 1.11, top). In the next step, the fibers are treated at 500 to 1500 °C in an inert atmosphere (nitrogen), also under tension, which leads to carbonization of the fibers. In this process, non-carbon elements are removed as volatile products, yielding carbon fibers with about 50% the mass of the original PAN precursor. During this carbonization, the carbon plains are optimized in their structure (reduction of the number of structural flaws), leading to fibers with high tensile strength.

In additional high temperature steps (graphitization) in a nitrogen or argon atmosphere, the fiber structure can be further improved to produce high moduli. Temperatures up to 2800 °C are applied to arrange the carbon plains and to increase their orientation toward the fiber axis. However, X-ray diffraction patterns show that in this state no true graphitic structure is formed.

Typically fibers, which have been treated only at 1000 °C, exhibit tensile strength values of about 2000 MPa and tensile moduli of about 170 GPa.

After temperature treatments up to 1500 °C, tensile strengths of about 3500 MPa and tensile moduli of about 275 GPa are reached.

At 2500 °C, fibers with reduced tensile strengths of about 2800 MPa are formed (presumably by the evolution of prolate pores during the formation of more compact crystalline structures), but these fibers have very high tensile moduli of 480 GPa.

Even higher values for tensile moduli (up to 600 GPa) can be reached, if the fibers are treated at high temperature under tension.

The PAN process is by far the most important production process for carbon fibers.

1.5.1.2 Carbon Fibers from Pitch Precursors

Pitches derived from coal tar, petroleum residues, or PVC can be used as relatively cheap precursors for carbon fibers.

Pitch is a thermoplastic material and can therefore be extruded directly via melt spinning processes to precursor fibers. These can be subsequently transferred to carbon fibers in processes similar to the one described for the PAN route. This includes a stabilization step between 250 and 300 °C and then carbonization and graphitization at temperatures between 1000 and 2500 °C [31] (Figure 1.11, lower).

In untreated pitches, condensed aromatic structures are present, which are isotropic and randomly distributed. This leads to fibers with low orientation of the carbon plains along the fiber axis, and moderate mechanical properties.

In order to achieve enhanced fiber properties, the pitches are thermally treated at 400 to 450 °C by which liquid crystalline, anisotropic structures are formed, the so-called mesophases.

Precursor fibers, produced from mesophase pitch, exhibit high orientation values of the carbon plains in the direction of the fiber axis and can therefore be transformed to carbon fibers with very good mechanical properties. The yield of carbon from pitches can be above 75 wt.-%.

Tensile strengths values of 3500 MPa and tensile moduli of about 400 GPa can be reached for the carbon fibers based on pitch. Also, this fiber type can be brought to even higher moduli of about 600 GPa by an additional high temperature treatment.

1.5.1.3 Carbon Fibers from Regenerated Cellulose

For this production route, cellulose rayon fibers are used as a carbon source. Rayon fibers are produced by dissolving and spinning of cellulosic materials, such as wood pulp or cotton.

Since cellulose is a polymeric carbohydrate, water evolves during the temperature treatment, leaving a residue of carbon.

The production process is also conducted in steps, including a lower temperature treatment (below 400 °C), subsequent carbonization (up to 1500 °C), and graphitization (at about 2500 °C). The yield of carbon fibers from rayon is low, ranging between 10 and 30 wt.-%.

The mechanical properties of the rayon derived carbon fibers achieved in standard processing (only carbonization) are not as good as for PAN derived fibers.

Typically values of 700 MPa are achieved for the tensile strength and 70 GPa for the modulus.

Unlike in the PAN process, the fibers cannot be held under higher tension during the low temperature treatment. But in the following high temperature treatments, tension can be applied to these fibers, by which good mechanical properties can be reached, that is, tensile strengths up to 2800 MPa and moduli up to 550 GPa. Since the whole process is more complex for the rayon derived fibers, PAN derived types are more commonly used in industry.

1.5.2
Commercial Products

There are a number of companies producing carbon fibers and each has a number of carbon fiber products with different fiber properties and yarn counts. Therefore, no complete market survey of carbon fiber products is given here, but some examples of fiber properties and prices are displayed.

A classification of carbon fibers in terms of their mechanical properties can be conducted by using the IUPAC (International Union of Pure and Applied Chemistry) guidelines (Table 1.5) [33].

Carbon fibers are offered in a wide range of tensile strengths and moduli. Also a wide range of filament numbers in rovings are produced ranging from 1000 (1 K) to 400 000 (400 K) filament bundles. Depending on the fiber properties, the fibers can be priced between 20€/kg up to 1500€/kg.

Table 1.6 shows a survey of carbon fibers from Toray, Toho Tenax, and SGL Carbon [34–36].

Some fiber types are sold in different qualities, that is, a regular type and also a type that is qualified for aerospace applications. Since there are higher quality standards in aerospace applications, the fibers for these applications are 15 to 25% more expensive than those displayed in the table, due to additional quality controls during and after production.

Table 1.5 Classification of carbon fibers.

Classification	Tensile modulus (GPa)	Tensile strength (MPa)	Elongation at break (%)
UHM (ultra high modulus)	>600	–	–
HM (high modulus)	>300	–	<1
IM (intermediate modulus)	275–350	–	>1
LM (low modulus)	<100	low	–
HT (high tensile)	200–300	>3000	1.5–2

Table 1.6 Comparison of carbon fibers from different producers.

Producer Fiber	Diameter (μm)	Density (g/cm^3)	Tensile strength/modulus (MPa/GPa)	Approx. price
Toray Industries T300 (6K)	7	1.76	3530/230	€53/kg
Toray Industries T700 S (12K)	7	1.80	4900/230	€30/kg
Toray Industries T800HB (6K)	5	1.81	5490/294	€250/kg
Toray Industries T 1000G (6K)	5	1.80	7060/294	€240/kg
Toray Industries M60J (6K)	5	1.94	3920/588	€1500/kg
Toho Tenax HTA 5131 (3K)	7	1.77	3950/238	€59/kg
Toho Tenax HTS 5631 (12K)	7	1.77	4300/238	€29/kg
Toho Tenax STS 5631 (24K)	7	1.79	4000/240	€22/kg
Toho Tenax UMS 2731 (24K)	4.8	1.78	4560/395	€95/kg
Toho Tenax UMS 3536 (12K)	4.7	1.81	4500/435	€158/kg
SGL Carbon Sigrafil C (50K)	7	1.80	3800–4000/230	€15–25/kg

Acknowledgments

The support of the companies 3M, Sumitomo Chemical, Mitsui, Nitivy, Nippon Carbon, UBE Industries, COI Ceramics, Specialty Materials, Tisics, Toray Industries, Toho Tenax, and SGL Carbon in terms of providing prices and data of ceramic and carbon fibers is gratefully acknowledged.

References

1 Krenkel, W., Naslain, R. and Schneider, H. (2001) *High Temperature Ceramic Matrix Composites*, Wiley-VCH Verlag GmbH, Weinheim.
2 Lee, S.M. (1993) *Handbook of Composite Reinforcements*, Wiley-VCH Verlag GmbH, Weinheim.
3 Clauß, B. (2000) Keramikfasern – Entwicklungsstand und Ausblick, *Technische Textilien*, **43**, 246–51.
4 Clauß, B. (2001) Fasern und Performtechniken zur Herstellung Keramischer Verbundstoffe, *Keramische Zeitschrift*, **53**, 916–23.
5 Clauß, B. and Schawaller, D. (2006) Modern Aspects of Ceramic Fiber Development, *Advances in Science and Technology*, **50**, 1–8.
6 Blumberg, H. (1984) Die Zukunft der neuen Hochleistungsfasern,

Chemiefasern/Textilindustrie, **34/86** 808 ff.
7 www.zircarzirconia.com (accessed Oct 08, 2007).
8 Belitskus, D. (1993) *Fiber and Whisker Reinforced Ceramics for Structural Applications*, Dekker, New York.
9 Bunsell, A.R. (1988) *Fibre Reinforcements for Composite Materials*, Elsevier Science Publishers B.V., Amsterdam.
10 *Ullmann's Encyclopedia of Industrial Chemistry: Fibers; 5: Synthetic Inorganic*, Vol. A11 (1988) Wiley-VCH Verlag GmbH, Weinheim, pp. 2–37.
11 *Engineered Materials Handbook: Composites, Ceramic Fibers*, Vol. 1 (1987) ASM International, Metals Park, Ohio, pp. 60–5.
12 Cooke, Th.F. (1991) Inorganic fibers: A Literature Review, *Journal of the American Ceramic Society*, **74**, 2959–78.
13 Wedell, J.K. (1990) Continuous Ceramic Fibres, *Journal of the Textile Institute*, **81**, 333–59.
14 Bunsell, A.R. (1991) Ceramic Fibers: Properties, Structures and Temperature Limitations, *Journal of Applied Polymer Science, Symposium*, **47**, 87–98.
15 Wallenberger, F.T. (2000) *Advanced Inorganic Fibers: Processes, Structures, Properties, Applications*, Kluwer Academic Publishers, Dordrecht.
16 Bunsell, A.R. and Berger, M.-H. (1999) *Fine Ceramic Fibers*, Dekker, New York.
17 http://www.mmm.com/ceramics/misc/tech_notebook.html (accessed Oct 08, 2007).
18 http://www.sumitomo-chem.co.jp/english/division/kiso.html (accessed Oct 08, 2007).
19 http://www.nitivy.co.jp/english/nitivy.html (accessed Oct 08, 2007).
20 http://www.mitsui-mmc.co.jp/eindex.html (accessed Oct 08, 2007).
21 Schmücker, M., Schneider, H., Mauer, T. and Clauß, B. (2005) Kinetics of Mullite Grain Growth in Alumo Silicate Fibers, *Journal of the American Ceramic Society*, **88**, 488–90.
22 Schmücker, M., Schneider, H., Mauer, T. and Clauß, B. (2005) Temperature-dependent evolution of grain growth in mullite fibres, *Journal of the European Ceramic Society*, **25**, 3249–56.
23 www.coiceramics.com (accessed Oct 08, 2007).
24 www.ube.de (accessed Oct 08, 2007).
25 http://www.specmaterials.com/silicarbsite.htm (accessed Oct 08, 2007).
26 http://www.tisics.co.uk/fibre.htm (accessed Oct 08, 2007).
27 Schawaller, D. (2001) Untersuchungen zur Herstellung keramischer Fasern im System Si-B-C-N und Si-C-N, PhD Thesis, University of Stuttgart.
28 Schawaller, D. and Clauß, B. (2001) Preparation of Non-oxide Ceramic Fibers in the Systems Si-C-N and Si-B-C-N, in *High Temperature Ceramic Matrix Composites* (eds W. Krenkel, R. Naslain and H. Schneider), Wiley-VCH Verlag GmbH, Weinheim, pp. 56–61.
29 Dresselhaus, M.S., Dresselhaus, G., Sugihara, K., Spain, I.L. and Goldberg, H.A. (1988) *Graphite Fibers and Filaments*, Springer, Berlin.
30 Donnet, J.-B. and Bansal, R.C. (1984) *Carbon Fibers*, Dekker, New York.
31 Buckley, J.D. (1988) Carbon-carbon; An overview, *Ceramic Bulletin*, **67**, 364–8.
32 Bennett, S.C. and Johnson, D.J. (1978) London International Conference on Carbon and Graphite, Society of Chemical Industry, London.
33 http://goldbook.iupac.org/C00831.html (accessed Oct 08, 2007.
34 http://www.torayca.com/index2.html (accessed Oct 08, 2007).
35 http://www.tohotenax.com/tenax/en/products/standard.php (accessed Oct 08, 2007).
36 http://www.sglcarbon.com/sgl_t/fibers/sigra_c.html (accessed Oct 08, 2007).

2
Textile Reinforcement Structures

Thomas Gries, Jan Stüve, and Tim Grundmann

2.1
Introduction

In the previous chapter, ceramic and carbon fiber materials were described. An overview of fibrous materials suitable as reinforcement of Ceramic Matrix Composites (CMCs) was given and specific properties and selection criteria have been described. These fibers and filaments represent the basic structures of all textiles. The connecting element between filaments and textile structure is the yarn. In this chapter, rovings and slightly twisted yarns are described. They represent the most important yarn structures for use as reinforcing elements. The structure inside a yarn can be defined as the microstructure of a textile. On the meso-level, the yarn structure (fiber architecture) is described where the textile structure (yarn architecture) represents the macro-level of a textile (Figure 2.1).

The complexity of the structure increases with the increase in dimension. Inside the composite, the complexity is increased mainly by the matrix material and its formation. However, there is no large increase in dimension due to the near-

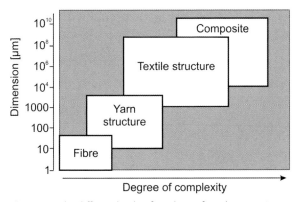

Figure 2.1 The different levels of textile reinforced composites.

Ceramic Matrix Composites. Edited by Walter Krenkel
Copyright © 2008 WILEY-VCH Verlag GmbH & Co. KGaA, Weinheim
ISBN: 978-3-527-31361-7

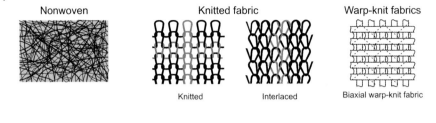
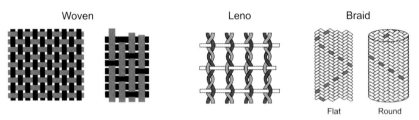

Figure 2.2 Overview of different basic two-dimensional textile structures.

net-shape appearance of the textile structure. This structure can be realized using different textile technologies.

Complex textile reinforcement structures in near-net-shape form (textile preforms) can be produced in one-step or multi-step processes. Textile preforms can be realized using one-step preforming technologies such as three-dimensional braiding, three-dimensional weaving, contour warp knitting, or net-shape weft knitting, if the production process meets the requirements regarding preform geometry and dimension. Usually, sub-preforms can be manufactured that are processed in further preforming steps. In many cases, two-dimensional textile structures are used to create structures of higher complexity by forming and assembling several plies in multistep preforming processes. Figure 2.2 gives an overview of different basic two-dimensional textile structures. In this chapter, they are described in detail, along with three-dimensional textile structures and preforming technologies. The relevance of the textiles and preforming technologies for use in CMC applications will also be discussed.

When using textile structures as reinforcement in composites it is important to realize a yarn position that is as straight as possible, because non-crimped fibers can bear the highest loads and induce the highest stiffness. In several applications, fiber crimp is needed, for example, to achieve high damage tolerance or to realize energy absorption. Therefore, each application requires its own textile structure.

In this chapter, the textile structures are divided into two-dimensional and three-dimensional textiles.

2.1.1
Definition for the Differentiation of Two-Dimensional and Three-Dimensional Textile Structures

In several patents and publications dealing with the topic of "three-dimensional textiles," several different interpretations of textile three-dimensionality can be found. To ease the use of the expressions, two-dimensional textile and three-dimensional textile, a definition has been proposed by Gries *et al.* [1]:

A textile is defined as two-dimensional if it does not extend in more than two directions, neither in yarn architecture nor in textile architecture.

A textile is defined as three-dimensional textile if its yarn architecture and/or its textile architecture extend in three directions, regardless of whether it is made in a one-step-process or a multiple-step-process.

The textile terms used in the definition are defined in Table 2.1.

2.1.2
Yarn Structures

In this Section, selected yarn structures will be introduced and their potential for the processing of brittle high-performance fibers, such as ceramic or carbon fibers in textile processes, will be discussed. In Figure 2.3, an overview of different yarn structures is given.

Twisted and untwisted multifilaments are called rovings. Twisting the roving slightly improves its processability and decreases the risk of filament damage.

Table 2.1 Definition of textile terms [1].

Yarn architecture	Arrangement of yarns on textile level. The yarn architecture is defined as three-dimensional, if it is created by three or more yarn systems or main yarn orientations and no rectangular coordinate system can fit into it, so that one coordination axis is oriented rectangularly to each of the yarn orientations.
Textile architecture	Geometry of the textile. The textile architecture is defined as three-dimensional, if a volume is formed and/or enclosed by the textile structure, regardless of the number of yarn systems and the yarn architecture thus created.
One-step-process	Production of a near-net-shape textile product in a single production step (e.g. three-dimensional-warp knitting, three-dimensional-braiding, etc.).
Multiple-step-process	Production of a near-net-shape textile product in several production steps (e.g. warp knitting and forming, weaving and sewing, etc.).
Near-net-shape	Textile architecture, which dimensions are close to the final products shape. This term is used mostly in the field of composite applications.

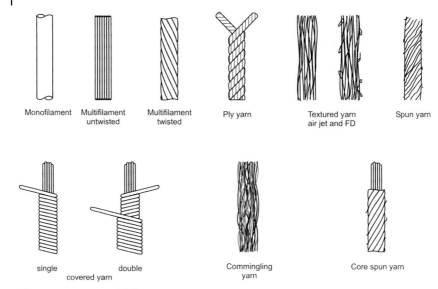

Figure 2.3 Overview of different yarn structures.

Common twist on carbon rovings consists of only 5 to 20 turns per meter. The loss in tensile strength in composite use can therefore be kept low (<5%). Another way to improve the processability of brittle rovings is to protect them with a covering. In covered yarns, this is achieved by winding filaments in one or two layers around the sensitive roving. Instead of using only one yarn material for core and covering, a different kind of material can be chosen for the covering yarn. Such hybrid yarn structures can be manufactured with the help of an air jet, which commingles the filaments of the covering material (commingling yarn).

Another way to realize hybrid yarn structures with a covered core is the core spinning process. During this process, e.g. thermoplastic staple fibers are spun around a core yarn. Here, the surface of the staple yarns is melted, enabling them to stick onto the core yarn. Thickness and density of the covering can be adjusted and protection of the core yarn can be realized with this kind of yarn structure. This protective effect can also be used for joining purposes. ITA has invented a technology for the creation of circular non-crimp fabrics (NCF), whose yarns are joined at the connecting points by melting the covering fibers using ultrasonic welding. With the help of this technology, very brittle glass rovings have been arranged in a textile preform for textile reinforced concrete [2].

2.2
Two-Dimensional Textiles

2.2.1
Nonwovens

According to DIN 61210, nonwovens are defined as flat textile materials either fundamentally or fully consisting of fibers. These fibers can be arranged, or not, in defined directions [3]. Figure 2.4 shows an example of a nonwoven.

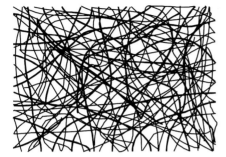
Figure 2.4 Example of a nonwoven structure.

For reinforcement purposes, ceramic or carbon nonwovens are used in CMCs, commonly referred to as fiber blankets.

The manufacturing of nonwovens is subdivided into web formation and bonding. The web formation is further differentiated into four processes:

- hydrodynamic web formation (fiber length 1–20 mm)
- aerodynamic web formation (fiber length 5–40 mm)
- mechanical web formation (fiber length 30–60 mm)
- spunlaid (endless fibers).

In order to create a nonwoven out of the fiber fleece obtained through web formation, it must be solidified. This is either done mechanically (needlefelt, waterjet), chemically, or thermally (calendaring, hot air). Furthermore, nonwovens can also be solidified via warp knitting.

The advantage of fiber blankets for CMCs is the high porosity that eases the infiltration process. Compared to oriented textile structures, the stiffness is very low and so the reinforcing effect is lower. Nonwovens for CMCs are mostly made of cutting converted filament yarns (staple fibers) and solidified by chemical bonding (sometimes also needle felted).

2.2.2
Woven Fabrics

According to DIN 60000, woven fabrics are defined as those made of rectangular crossed yarns of at least two yarn systems, warp and weft, by shed formation [4].

The manner of crossing of weft and warp yarns is called the pattern. The pattern type has a high influence on different fabric properties, such as drapability and shear stiffness. There are three basic patterns, the plain weave, the twill weave, and the atlas weave, and also special pattern types such as the leno weave. Examples of typical two-dimensional woven fabrics are given in Figure 2.5.

There are also special non-crimp woven fabrics available that use thin auxiliary yarns to avoid bending of the reinforcing yarn. The bending, which is caused by the weaving process, is induced only in the auxiliary yarn because its stiffness is

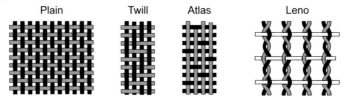

Figure 2.5 Structure of different two-dimensional woven fabrics.

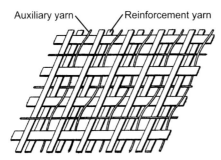

Figure 2.6 Non-crimp woven fabric "Advanced Syncron Weave" [5].

much lower than that of the reinforcing yarn. Figure 2.6 shows a biaxial non-crimp weave with an auxiliary yarn called "Advanced Syncron Weave," from the company ECC GmbH & Co. KG, Heek, Germany.

In comparison to the unidirectional reinforcement structures, the handling and drapability of fabrics are good but the wavelike yarn crimp decreases the mechanical properties of the consolidated textile-reinforced structure. Strength as well as stiffness of fabric-reinforced composites is therefore lower than with unidirectional reinforcements. With the choice of proper weft and warp density as well as suitable patterns, such as Atlas, Advanced Syncron, or leno weave, these disadvantages can be minimized by reducing fiber crimp. Also the spreading of the yarns into a thin tape helps to reduce yarn crimp.

In comparison to NCFs, the handling of woven fabrics is inferior but the drapability is very good, due to the lower shear strength.

The processing of brittle fibers (carbon and ceramic) into woven fabrics is possible but also "yarn-friendly" patterns (e.g. Atlas, Advanced Syncron, or leno weave but not plain weave) and machinery setups have to be carefully chosen. Special attention should be given to the porosity, which is hard to achieve in woven fabrics. The use of lost yarn (e.g. viscose or acrylic yarn) that is removed after processing by heat treatment can help to improve the porosity and thus improve infiltration with the ceramic precursor. This was verified in [6] with the knitting threads of warp-knitted fabrics. For woven fabrics, for example, the auxiliary yarns of the "Advanced Syncron Weave" or the leno yarn of the leno weave can function as lost yarn. Thus, these patterns may have a high potential for the use in CMCs, which has yet to be verified.

2.2.3
Braids

In DIN 60000, braids are described as textile structures of regular appearance with closed selvedges. They consist of at least three yarns (healds) or at least two yarn systems (tubular braids), whose yarns are intertwined and crossed diagonally to the edges of the textile (Figure 2.7) [4].

The angle between the selvedges (i.e. the production direction) and the braiding yarns is called the braid angle (Figure 2.8). This angle can be varied between 20 and 80 degrees. Braids with an angle of 45 degrees look similar to woven fabrics. They do have a comparable yarn architecture regarding yarn crimp, due to intertwining of the braiding yarns. The most common braid pattern is called the Regular Braid. It is comparable to a 1/2-twill woven structure. Other braid patterns are the Diamond Braid (1/1) and the Hercules Braid (1/3) [7]. It is also possible to realize a 2/2-twill-like braided structure by choosing a tandem type of setup of yarn carriers on the machine. In analogy to woven fabrics, the braid pattern influences the drapability of the braids.

In addition to the braiding yarns, axial yarns (0 degrees) can be added inside the braided structure. These yarns are inactive during the production process and are

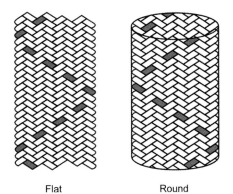

Flat   Round
Figure 2.7 Yarn architecture of biaxial two-dimensional braids.

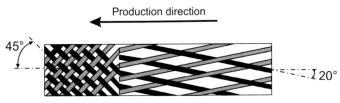

Figure 2.8 Definition of braid angle.

enclosed by the braiding yarns. While the braiding yarns are crimped, the axial yarns remain in a straight position. Braids with axial yarns enclosed are called triaxial braids and those without axial yarns are called biaxial braids. The drapability of triaxial braids is inferior to that of biaxial braids, due to the friction between axial and braiding yarns.

The processing of brittle high-performance yarns such as carbon, glass, or ceramic rovings on braiding machines is common today. Tubular braids of different diameters made of these materials are available on the market. Also special machine components, including yarn carriers and yarn guiding elements for the processing of brittle yarns, have been developed. For the reinforcement of CMC-rods, two-dimensional braids made of carbon or ceramic rovings are of interest.

2.2.4
Knitted Fabrics

According to DIN 60000, knitted fabrics are planar structures, realized by loop formation of one or more yarns or yarn systems. Knitting processes can be differentiated according to the direction of loop formation. The products of a knitting process with loop formation perpendicular to the production direction are called weft-knitted fabrics. Fabrics that are created with loop formation in the production direction are called warp-knitted fabrics [4]. The difference in production process leads to different methods of fabric creation. So one yarn system is sufficient for the creation of a weft-knitted fabric but several yarn systems are needed for the realization of warp-knitted structures.

Both industrial processes provide numerous knitting needles for the fabric creation. But while one yarn passes several needles arranged in a row or a circle during weft knitting, every needle is provided with its own yarn during warp knitting. Knitted structures are the most drapable ones of all two-dimensional textiles. But due to the arrangement of the yarns in loops, these structures do not provide high tensile strength and modulus. Therefore they are not very common as structures for reinforcement, except for a few ballistic applications. Furthermore, the use of brittle high-performance yarns such as ceramic or carbon rovings as knitting yarns is impossible due to high friction and deformation during the knitting processes.

Nevertheless, weft-knits and warp-knits are of high importance in reinforcing composite structures if they are modified. Knitting technology enables the insertion of straight reinforcing yarns in warp and weft direction during production. Mono-, bi-, and multi-axial weft- and warp-knitted fabrics can therefore be realized in economic production processes (Figure 2.9).

Using knitting technology, every warp- and weft yarn is bound with a single knitting loop. Compared to other technologies, the rovings remain undisturbed. As a knitting yarn, an auxiliary yarn made of a thermoplastic, such as Polyethylene or Polyamide, is used. The insertion yarns are not opposed to high load and abrasion during production. Therefore knitted fabrics with weft and warp insertion made of brittle yarns such as glass, carbon, or even ceramic rovings, can be

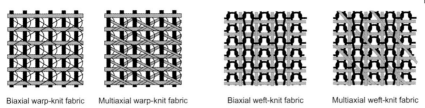

Figure 2.9 Bi- and multiaxial warp- and weft-knitted fabrics.

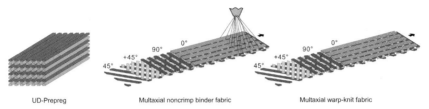

Figure 2.10 Overview of different NCFs.

achieved [6]. A specific characteristic of these structures is that they can have porous structures with spacings between the insertion yarns. This design characteristic has been beneficial when using biaxial warp-knits as structures for reinforcement of concrete. The specific textile design eases the infiltration of the textile reinforcement with the concrete matrix. Because of the comparability to the ceramic matrix processing, these structures show a huge potential for use as reinforcement of CMCs.

A special weft-knitting process with warp and weft insertion, which enables the production of near-net-shape knitted fabrics, has been developed by the Institut für Textil- und Bekleidungstechnik der T.U. Dresden, Dresden, Germany (ITB) [8, 9]. The innovative process is based on conventional flat-weft-knitting technology. Near-net-shape structures, for example, disks with polarorthotrop reinforcing yarns, can be achieved [10].

2.2.5
Non-crimp Fabrics

According to EN 13473-1, non-crimp fabrics (NCF) are defined as textile structures consisting of one or several parallel layers of straight, unidirectional yarns that are not crimped [11].

Multi-axial NCF are two-dimensional textiles consisting of at least two layers of unidirectional reinforcing yarns. These layers are joined by pre-impregnation with resin, by fixation with adhesives, or by using warp-knit seams (Figure 2.10). Adhesives for the joining of NCFs are used in the form of fluid, powder, or a non-woven hot melt. In a multilayer NCF, the layers can have different orientations. Possible orientations are 0, 90, and between 20 and 70 degrees. Furthermore, the yarn count and the kind of material used in each layer can be varied. Present-day industrial production machines can process up to eight layers plus the introduction of two surface fabrics, for example, non-wovens. In NCFs joined by warp-knit seams, the knitting pattern can be varied. The three patterns, chain stitch, tricot lap, and cord lap, influence the drapability of the fabric [12].

Due to the high standard of the machine technology, the processing of brittle high-performance yarns such as carbon and ceramic rovings is enabled with very little fiber damage. Only the joining process using warp knitting can be critical for very brittle yarns because the needles punch directly through the yarns. This can create filament breakage and yarn deflection. Nevertheless, many NCFs with different yarn architecture made of glass, carbon, and aramide rovings, are available today. NCFs made of ceramic rovings are not yet known.

2.3
Three-Dimensional Textiles

Processes for the production of three-dimensional textile structures often enable textile preforms to be achieved in one production step. Many production processes to create three-dimensional textiles have been invented in the last two decades and development is still ongoing. Nevertheless, there are no one-step technologies known for the realization of three-dimensional non-wovens as reinforcement structures for composites. The most relevant technologies for one-step textile preforming are described below.

2.3.1
Three-Dimensional Woven Structures

Different types of three-dimensional woven structures exist. Figure 2.11 gives an overview of some types described in this section.

Three-dimensional woven fabrics are manufactured by the insertion of two picks with double shed opening, being upright to each other. The yarns are fed in at a 0- and 90-degree direction and an additional yarn is positioned orthogonally toward the two others in the z-direction. Three-dimensional fabrics feature quasi-isotropic properties, and a high tensile and compression strength, as well as good bending stability and very good impact behavior [13–16]. The distribution of the reinforce-

Figure 2.11 Structure of different three-dimensional woven fabrics.

ment fibers is uniform in all three dimensions. However, the drapability and elongation behavior of three-dimensional fabrics is poor. Three-dimensional fabrics are applied in textile-reinforced composites with high thermal impacts as well as for structural parts in the areas of car and aircraft construction, and are manufactured by Biteam AB, Gothenburg, Sweden.

Other kinds of three-dimensional fabric structures are the so-called *multilayer woven fabrics*. These are manufactured by the composition of several fabric layers without any spacings in between. The layers are fixed by interlocking or chain warp. The yarns are orientated in a 0- and 90-degree direction and the yarns in the z-direction are variable. Multilayer fabrics can be draped well, featuring good elongation behavior and good tensile, compression, and bending stability. These fabrics are applied in impact-charged multilayer structures made of textile-reinforced plastic (TRP), for example, automobile floor components [17, 18]. 3Tex, Cary NC, USA produces these kinds of fabrics.

Woven spacer fabrics also have a three-dimensional structure. They are manufactured by weaving in upright pile warps and their properties are comparable with those of multilayer fabrics. Only the drapability is somewhat inferior. The distance between the two layers can be adjusted individually and thus be assimilated to the particular application. These textiles show a high resistance against perforation. They are applied in sandwich structures, for example, in TRP lightweight design applications [15, 19].

Figure 2.12 shows an example of a three-dimensional woven spacer fabric produced by 3Tex, Cary NC, USA. Another manufacturer of three-dimensional woven spacer fabrics is Parabeam bv, Helmond, Netherlands.

Tubular woven fabrics are another kind of three-dimensional woven fabric. They are produced by circular or flat shuttle looms. Tubular woven fabrics feature good tensile and compression strength, but low shear and bending stiffness. Tubular woven fabrics are used, for example, as porous vascular grafts in medical technology [21, 22].

Another process to produce three-dimensional textiles is called the *shape weaving process* and was developed by Shape 3 Innovative Textiltechnik GmbH, Wupperthal, Germany [23]. The principle of this technology is the integration of warp and weft yarns of different length during the weaving process. Through these

Figure 2.12 Three-dimensional woven fabric [20].

extra yarn lengths, the textile expands into the third dimension. This opens up the possibility of configuring textile fabrics to a certain shape during the weaving process [24].

There are numerous woven fabric variants of interest for textile-reinforced composite applications. These include multilayer woven fabrics and three-dimensional woven fabrics, as well as flat woven fabrics with three-dimensional design features. Regarding industrial usage, two-dimensional fabrics are the most strongly represented because of good mechanical properties and fast production process, and thus their attractive price.

For CMCs, two-dimensional woven fabrics are already used. R&D work on oxidic CMCs with a porous matrix are accomplished mainly in the United States (University of Santa Barbara CA, General Electric, COI), in Japan (Kagoshima University), and in Germany (DLR, Cologne) [25, 26].

2.3.2
Braids

2.3.2.1 Overbraided Structures

One opportunity to create three-dimensional textiles or textile preforms using braiding technology is overbraiding, a special kind of circular braiding. The creation of the textile is similar to circular braiding but the tubular braid is directly laid down on a net-shaped mandrel. Several layers can be braided onto the mandrel with an individual starting point and endpoint for each layer. Similar to circular braiding, each layer consists of a bi- or tri-axial yarn architecture and braid angles can be varied between 10 and 80 degrees. The orientations of the braiding yarns can be varied in each layer [7]. Also the formation of a layer that only consists of 0-degree yarns is possible. Because the yarn carriers can be equipped with different materials, the production of hybrid structures is possible.

New developments, which are driven by EADS Innovation Works Germany, Ottobrunn, Germany, use the initial division of the two yarn systems of braiding yarns to set up one group with reinforcing yarns and the other with auxiliary yarns. Choosing the suitable auxiliary yarn (e.g. an elastic thermoplast yarn), non-crimp structures can be produced, so-called unidirectional braids (UD-braids). This technique not only leads to minimized filament damage of the reinforcing yarn, but also leads to structures that can concur with UD NCFs. In summary, there is a lot of freedom in the design of the textile preform.

Because the yarns support each other, orientations can be apart from the geodaethic line on the mandrel (Figure 2.13). The braided structure will be stable while it stays on the mandrel. Of course, the braided structure can be removed from the mandrel once a suitable mandrel concept and design has been developed. Depending on the mandrel concept, the braid can also remain on the mandrel as it is transferred to the infiltration step and might become part of the composite or will be removed after curing. For the moving of the mandrel, a multi-axis take-up can be used as well as an industrial robot. Therefore, an automated production of textile preforms of complex geometry can be realized in a one-step process.

Figure 2.13 Overbraiding of net-shaped mandrel.

Figure 2.14 Radial overbraiding machine with robot support.

Recent developments at the August Herzog Maschinenfabrik GmbH & Co. KG, Oldenburg, Germany (Herzog) led to a special design of overbraiding machine. Instead of moving the yarn carriers on the front of the machine body, they are led on the inside of a circular machine body (Figure 2.14). The orientation of the yarn carriers in a radial direction is the reason for the name of this machine type, "radial overbraiding machine." This design leads to reduced filament damage during the braiding process, due to a minimization of yarn movement along yarn guiding elements and reduced yarn-yarn interaction. The result is high production speed and a reliable production process. The processability of brittle high-performance yarns such as glass, carbon, aramide, and ceramic rovings, has been proven by several users of this technology. The most prominent application might be the bumpers of the BMW M6, made of carbon fiber reinforced plastic (CFRP), that is achieved by using overbraided preforms [27]. Other CFRP products using overbraiding technology are crash tubes, rocket engine nozzles, and many more [28, 29].

2.3.2.2 Three-Dimensional Braided Structures

Three-dimensional braided structures do not only expand in the third dimension, but also contain an integral yarn architecture with a three-dimensional yarn course (Figure 2.15). Several three-dimensional braiding technologies do exist, such as the Square-braiding technology, the Four-step braiding technology, and the three-dimensional rotary-braiding technology. Products manufactured using Square-braiding technology are mainly used as seals in machines, often consisting of staple fiber yarns. The Four-step braiding technology enables solid braided profiles of different simple geometry [30]. However, it is limited with regard to geometrical changes and complex cross-sections. Nevertheless, this technology is suitable for the processing of high-performance yarns.

The newest generation of three-dimensional braiding technology is three-dimensional rotary-braiding. Two different types exist: The ITA-Herzog-3D Rotary-braiding technology invented by ITA and Herzog and the 3TEX-3D Rotary-braiding technology invented by 3TEX, Cary NC, USA (3TEX). In both technologies, the transport of the yarn carriers is realized by rotating horngears that turn in counter directions. The ITA-Herzog principle is shown in Figure 2.16. Here the yarn carriers are led inside groves that are cut into the machine table. With the help of switches located between the horngears, it can be decided whether a yarn carrier should be transferred to the next horngear or stay on the initial horngear. Since

Figure 2.15 Model of a three-dimensional braided structure [31].

2.3 Three-Dimensional Textiles | 35

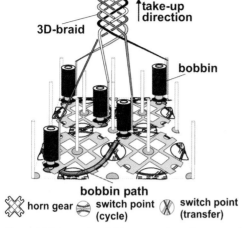

Figure 2.16 Principle of ITA-Herzog three-dimensional Rotary-braiding.

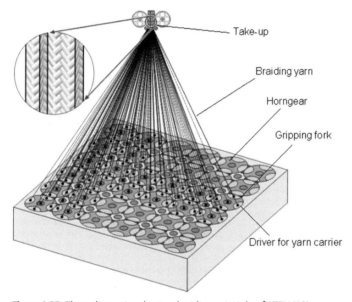

Figure 2.17 Three-dimensional rotary-braiding principle of 3TEX [20].

every switch and horngear can be controlled individually, an individual path of each carrier over the machine table can be given [32]. The machine is computer controlled, therefore complex and variable yarn architectures as well as complex and variable textile architectures can be realized in an automated production process.

The 3TEX-3D Rotary-braiding principle is similar to the ITA-Herzog-3D Rotary-braiding principle. But there are still some main differences. Instead of switches, so-called gripping forks are located between the horngears. The yarn carriers are moved on drivers that fit into the horngears, such that two yarn carriers remain between two horngears at the same time (Figure 2.17). At this point it can be decided whether they should change places or remain on their present horngears [33].

This intertwining step is the main difference between the two technologies. With the ITA-Herzog-3D Rotary-braiding technology, a braiding pattern exists that defines the order in which the yarn carriers pass each other, so avoiding a collision of the carriers. Every technology has its benefits and its disadvantages. The 3TEX technology can use more yarn carriers in the same area than the ITA-Herzog technology. On the other hand, the yarn carriers of the ITA-Herzog technology are larger and have more yarn capacity than the ones of 3TEX. The 3TEX technology might be faster than the ITA-Herzog technology, but it might not offer the same freedom in design of yarn architecture. So the decision for the use of one of the technologies has to be made depending on the desired application.

Three-dimensional rotary-braiding technologies enable the manufacture of textile preforms with integral yarn architecture, continuous changes of cross-section geometry, and dimension in a one-step process. Examples of structures are shown in Figure 2.18, but these can hardly show the large range of geometries and structures that are possible.

Regarding the material, the processing of carbon and glass rovings is state of the art, even if there are still some improvements to be made. The processing of ceramic rovings with these technologies is not yet possible. Investigations of Planck on the interaction of braiding yarns with yarn guiding elements showed that for every yarn material a suitable material for the yarn guiding element does exist, but it has to be discovered [34]. Also an appropriate sizing or a slight twisting of the ceramic rovings could help to strengthen them to withstand the friction that appears in the strong yarn-yarn interaction during three-dimensional braiding and also to enable their processability in three-dimensional braiding processes.

On a machine that is a mixture of three-dimensional braiding and overbraiding technology, preforms made of the ceramic rovings Nextel 720 from 3M Ceramic Textiles and Composites, St. Paul MN, USA have been manufactured. They consist of a tubular braid whose body is an interlocked braided structure. The preform has been successfully processed into a hot gas filter made of CCC [35].

Three-dimensional braids certainly have great potential for use as reinforcement structures in CMC, due to their integral yarn architecture that is often desired by engineers and scientists dealing with CMCs.

Figure 2.18 Examples for three-dimensional rotary braided structures.

2.3.3
Three-Dimensional Knits

2.3.3.1 Multilayer Weft-Knits

In Section 2.2, the weft-knitting process with warp and weft insertion developed by ITB has been described for two-dimensional textile structures. This process also enables the production of textile structures with several layers in one production step. This type of textile structure is called multilayer weft-knit. The layers can consist of weft and warp inserted high-performance yarns that are integrated in a straight position [9]. The layers are connected by the knitting yarns that can be auxiliary yarns. Contoured structures can be realized. Additionally, yarn storage of the insertion yarns can be used to ease the draping of the sub-preforms to the final shape. These structures could be of interest for CMC applications because economic production of contoured sub-preforms can be achieved. Nevertheless, these structures do not provide reinforcing yarns in the thickness direction so their use is limited to applications where no integral yarn architecture is needed.

2.3.3.2 Spacer Warp-Knits

The technology of warp-knitting can be used to combine two different layers of straight reinforcing yarns in such a way that a textile with sufficient dimension stability is realized. Next to multi-axial NCFs, the warp-knitting technology allows the creation of three-dimensional structures. Double needle bar raschel machines fabricate two layers of textiles and a connection via spacer yarns between them. As spacer yarns, auxiliary yarns have to be used because high-performance yarns cannot stand the mechanical loads during the production process.

With this technology, textile preforms with a defined distance can be manufactured in a single process step, by allowing a wide range of design possibilities for the surfaces such as different yarn densities, use of different yarn materials, and many more. The resulting structure can be optimally used as reinforcement for composites in a sandwich design, having the surfaces always kept at the correct position during infiltration. This principle has been successfully tested in textile reinforced concrete applications at RWTH Aachen University [36]. Figure 2.19 shows an example of a warp-knitted spacer fabric. These structures are only suitable as reinforcement for CMCs, if a sandwich design has been chosen or if

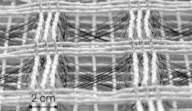

Figure 2.19 Warp knitted spacer fabric.

no reinforcement in thickness direction is needed. In addition to a "simple" sandwich layout of spacer warp-knitted structures also contour warp-knitted spacer fabrics can be realized. In these textile structures, changes in thickness of the spacer fabric can be realized by varying the length of the spacer yarns (e.g. stair-like geometries).

2.4 Preforming

2.4.1 One-Step/Multi-Step Preforming

In today's processing of TRP, two main ways of processing yarn architectures are used: The processing of rovings of textile fabrics pre-impregnated with resin (prepreg-processing) and the processing of dry textile preforms impregnated in a second step, in the shape of the final part (preform-processing). Depending on the requirements and applications, one of these production methods is more suitable than the other. Contrary to prepreg manufacturing, preforming creates a dry textile reinforcement structure already having the characteristics of geometry and fiber orientation required in the final product. This means that the rovings need to be formed and fixed via textile handling and assembly processes to achieve the geometry later required to withstand the load on the component.

For the creation of this inherently stable textile structure, several technologies are available. There are technologies which generate the three-dimensional structure from the yarns in one process step (i.e. three-dimensional braiding, overbraiding, three-dimensional weaving, three-dimensional stitching). These processes are referred to as one-step preforming. However, these technologies are limited to a certain degree of complexity concerning the reinforced textile and thus the geometry of the final product. Examples of typical component geometries, which are created with one-step preforming technologies, are profiles, panels, sandwich structures, and hollow parts.

For the realization of components of high complexity, the yarns are processed in multistep processes. At first, two-dimensional textiles, such as multi-axial NCFs or woven fabrics, and/or three-dimensional subpreforms, such as the above-mentioned three-dimensional braids, are manufactured. These textiles are subsequently processed to complex preforms via handling and assembly processes. A practical example of a complex TRP-component is the so-called rib stiffened panel (Figure 2.20). These panels consist of a flat shell (plane, unidirectional, or multi-axial curved) and of stiffening profiles (e.g. top-hat-, T-, or L-profile). These components are frequently used to function as self-supporting outer skins in lightweight constructions, such as automobiles and aircrafts.

Intermediate textiles for the shell are the above-mentioned flat textiles, which are available as rolled goods. For the ribs, both two-dimensional intermediate textiles as well as sub-preforms can be used.

Figure 2.20 Rib stiffened panel.

In order to process these textiles, the following steps are required:

- cutting
- handling and draping
- joining and conditioning.

These technologies are proven for the processing of carbon rovings. Complex textile preforms made of ceramic rovings have not yet been realized.

2.4.2
Cutting

For automated cutting of textile fabrics made of brittle high-performance yarns, computer controlled cutting machines using different cutting techniques are available. Depending on the material, the fabric type, and the fabric thickness that has to be cut, a suitable cutting device has to be chosen. These range from rotating cutting disks, oscillating knifes, or laser cutters. The cutting disks are supported by ultrasonic movement and role along the path which has to be cut. Oscillating knifes can be moved with high frequency or supported by ultrasonic movement. If necessary, the fabric can be fixed to the cutting table with the help of vacuum applied by covering the fabric with a sheet. Cutting using a water-jet has shown to be unsuitable for cutting of dry textile fabrics. Due to the computer controlled movement of the cutting head, even complex contours can be cut out of the rectangular basis. Due to the abrasive behavior of ceramic fibers, laser cutting is the most suitable technique for the automated cutting of fabrics made of ceramic fibers.

2.4.3
Handling and Draping

The handling and draping of textile fabrics, structures, and sub-preforms is now mainly done manually. This part of the preforming process is probably the most investigated one at present. Methods for gripping and transporting textile structures that have been invented so far include needle gripping, vacuum supported gripping, electrostatic gripping, or ice gripping techniques. But even in TRP-technology, no gripping or handling technology has been found to solve most of the challenges.

2.4.4
Joining Technologies

The most important joining technology for textiles, apart from gluing, is still sewing. Sewing technologies will play an important role in the process chain of future manufacturing of textile reinforced components. Their importance concerning cycle time reduction as well as cost savings can be regarded in line with those of qualified consolidation procedures [37]. The tasks of the sewing techniques within the preforming process range from the insertion of flat reinforcements as well as the fixation of individual components of a preform for the subsequent process steps to the manufacturing of load compatible seam areas. Different sewing technologies are applied in different process steps within the multistep preform production. Sewing technologies are chosen according to the function of the seam, the preform geometry, and the later component loads [38].

The sewing of reinforcement textiles is both a textile and a load-compatible joining process. Thus, a force-flow induced by an upright reinforcement is achieved in the seam area. Local reinforcements can also be applied. Flat textiles are converted to three-dimensional reinforcement structures. Via the application of reinforcement textiles within a component, several special mechanical properties can be realized. Additionally, the application of sewing technologies allows the production of near-net-shape textile structures.

The requirements for a seam in fiber-reinforced components are as follows: Forces and moments need to be transferred through the seam area without losing mechanical properties. High variations in stiffness are not allowed to occur. Different component properties, such as differing thermal expansion coefficients of materials, are not allowed to cause residual stresses. Furthermore, the seam should not cause a mass and dimension variation within the textile-reinforced component. Not all listed requirements for a seam are completely realizable, thus compromises have to be made [39].

Both conventional (e.g. double-step stitch) and one-sided sewing technologies (tufting, blind stitch, ITA sewing technology) are applied [32, 40]. Figure 2.21 shows the seam architecture of relevant sewing technologies. The one-sided sewing technologies have the advantage of good automation and the possibility of sewing in a three-dimensional mould as they are mounted on an industry robot. The infiltration behavior is improved because of the permeability of the seams.

Although seams have a high strength and drapability, there are also disadvantages, for example, the disorientation of reinforcing yarns through the sewing thread. Therefore, alternative joining processes such as adhesive bonding are also applied. Apart from the already mentioned joining processes, the so-called Tailored Fiber Placement (TFP) is used with reinforcement textiles. The reinforcing yarns can be put down purposefully (CAD operated) on a two-dimensional textile. Parallel to the placement, the fibers are stitched onto the textile with a thin

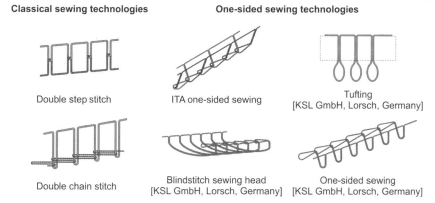

Figure 2.21 Seam architecture of relevant sewing technologies.

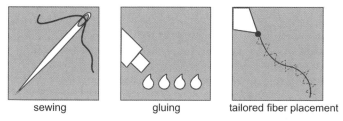

Figure 2.22 Typical joining processes for preforms.

thread. This technology can be used for additional local reinforcements, for example, fittings or holes in the CMC part (Figure 2.22).

2.5
Textile Testing

2.5.1
Tensile Strength

The tensile strength and elongation of textiles are measured by the strip tensile test according to DIN EN ISO 13934-1. A sample of given size from a planar formation is stretched with a constant deformation rate up to the breaking point. The maximum tensile strength F (N), the maximum elongation ε (%), and if necessary the breaking stress and the break elongation are recorded [41].

2.5.2
Bending Stiffness

The bending stiffness of a textile is determined according to DIN 53362. The bending stiffness is defined according to the standard of degree for the resistance a sample opposes the tendency to bend itself when influencing forces (dead weight). The bending stiffness B is accomplished after the Cantilever test and indicated by the unit mN*cm² [42].

2.5.3
Filament Damage

For the reliability of textile processes that process brittle high-performance yarns such as carbon or ceramic rovings, the amount of filament damage is essential. Yarn guides at the machine or the contact of yarns with other yarns have a great influence on filament damage and therefore on the quality of the textile product. Thus, measuring of filament damage becomes increasingly important, while textile preforming processes become increasingly automated. For the measurement of filament damage, there is no standard that exists to date. Filament damage can be measured, for example, with the help of laser optical sensors. Here the yarn is guided over a hook, facilitating the broken filaments to rise, which interrupt a laser beam resulting in a signal. So the number of broken filaments can be counted. At the same time, a pre-adjusted yarn tension can be monitored. Using this method, yarn guides can be compared [43, 44].

2.5.4
Drapability

Two-dimensional fabrics often need to be draped into a final shape. To avoid wrinkling, the measurement of the drapability of a fabric is important, for example, to choose the right knit pattern of multi-axial NCFs. For the measurement of drapability, there are no standards existing as yet. Some effort has been made at ITA, for example, their development of the ITA-drape test rack enables the measurement of shear forces of fabrics. It can be combined with a camera system for the optical detection of wrinkles. But this technique gives limited information about real deformation behavior. In newer methods, tools in the form of domes with different radii are used to perform drapability tests with fabric specimens of different sizes. The appearance of wrinkles is detected by visual inspection and the deformation of the specimen in the different directions is measured. [45]

2.5.5
Quality Management

Different methods for quality management in textile production processes have been classified and described by Wulfhorst [46]. Regarding preforming technologies, some specific criteria have to be taken into account. For near-net-shape manufacturing, the correct position of the flexible textile in the tool, in which conditioning takes place, is of great importance. With a camera system, a two-stage monitoring of the composite position can be accomplished. In the first step, the flat textile is examined overlaid with several gray tone reference pictures for comparison of a sample part for its correct position. For following steps in the process, an automatic correction of the movement coordinates can take place with robotic control. Then an additionally attached light cut sensor examines the topology of the textile for wrinkles, assessing the structure [47].

2.6 Conclusions

Various textile reinforcement structures are available for use in CMCs. Four important criteria have to be considered when choosing a preforming technology:

2.6.1 Processability of Brittle Fibers

All fibers for use in CMCs are more or less brittle. Depending on the textile process, the fibers are exposed to various stresses (tensile stress, bending stress, fiber-fiber friction, fiber-machine friction). Especially bending and shear forces cause damage to brittle fibers. Thus, textile processes that induce a high deflection in the reinforcement fibers are not suitable for CMC preforms. The deflection can be caused either by guiding the fibers through machine elements with a low radius (e.g. knitting needles, shedding, small yarn guides) or by bending forces in the textile (e.g. woven fabrics with a short floating). In summary, textile processes with warp- or weft-insertion (in weaves or knits), fiber-placement, and overbraiding technologies are well suited for processing brittle fibers. The use of brittle yarns as sewing or knitting yarn is not suitable because of the high yarn strain (especially bending strain). Especially in the processing of brittle ceramic fibers, nearly every production process for textile reinforcement structures has to be modified or at least readjusted to a "yarn-friendly" process.

2.6.2 Infiltration of the Textile Structure

As the infiltration of ceramic matrices in textile reinforcement structures is more problematic than the infusion of, for example, epoxy resin, the textiles need to have a defined porosity. The textile processes have different suitability for creating a porous structure. Most suitable are knitted textiles with weft insertion, non-crimp weaves, or NCFs. These textiles can be equipped with auxiliary yarns that are removed by heat treatment (lost yarn). The resulting space improves the porosity for infiltration [6].

It has been shown in [48] that the performance of the chemical vapor infiltrated (CVI) ceramic composites are strongly affected by the yarn architecture. It influences the infiltration and development of the microstructure of the matrix and, therefore, the mechanical properties and the damage mechanisms of the composites.

2.6.3
Mechanical Properties of the Final CMC Structure

The reinforcing yarns are to be placed in a near-net-shape geometry that should be consistent with the strains that occur when the final CMC structure is in use. Different degrees of complexity in the textile structure can be realized by introducing preforming technologies.

Disorientation or damage of the yarns in the reinforcement textile can be caused by mistakes in handling, forming, or assembling processes. This decreases the mechanical properties of the final CMC structure. According to the mechanical requirements of the CMC structure, preforming technologies should be rated by their tendency to yarn disorientation and damage.

2.6.4
Productivity and Production Process Complexity

The introduced preforming technologies strongly differ in productivity and process complexity. According to the requirements of the CMC structure, a compromise of mechanical and geometric properties of the structure and of the productivity of the production process has to be found.

2.7
Summary and Outlook

In this chapter, an overview of requirements and selection criteria for textile reinforcement structures was given. Different textile structures usable for CMC materials were presented. After a short outline about typical joining processes for textile preforms, the most important testing methods for textile structures were described.

- Development trends with reinforced structures go to increasingly complex structures that require further automation of the production but allow a load conforming design of CMC.

- New economical manufacturing processes for infiltratable and near-net-shape preforms with complex geometry become more essential. Great importance lies in economic, automated preform manufacturing, to be able to realize large-scale composite production. By optimizing the production process, uneconomic manufacturing steps can be eliminated and production costs decreased dramatically.

- A further trend is the integration of different functions into composites. An example of this could be the integration of crash elements into structure composites.

- A further point could be continuous health monitoring of composites, in order to recognize delamination at an early stage.

- Future approaches for reinforced composites are further development of simulation and interpretation software for three-dimensional and near-net-shaped structures.

Acknowledgments

We gratefully acknowledge the Deutsche Forschungsgemeinschaft DFG (German Research Foundation) for the financial support of the research project GR 1311/1-3 within the Priority Program 1123.

We also gratefully acknowledge the Arbeitsgemeinschaft industrieller Forschungsvereinigungen, Otto-von-Guericke e.V. (AiF) for the financial support of the research project, AutoPreforms (AiF-No. 14420 N).

References

1 Roye, A., Stüve, J. and Gries, T. (2005) Definition for the differentiation of 2D- and 3D-textiles. Part 1: production in one-stepprocesses. *Technical Textiles*, **48**, H. 4, p. E212-E2142-D and 3-D textiles.

2 Roye, A. and Gries, T. (2003) For industrial production: new 3D-structures for thin walled concrete elements. Techtextil, Techtextil-Symposium, 10*, pp. 1–4, Paper-No. 4.21.

3 N.N. (2003) *Papier, Pappe und Faserstoff 2*, DIN 53120-1 bis DIN 61210, DIN-Taschenbuch 213, Beuth-Verlag GmbH, Berlin und Köln.

4 N.N. (1969) *Textiles, Basic Terms and Definitions*, DIN 60000, 1/1969, Beuth-Verlag GmbH, Berlin und Köln.

5 ECC GmbH & Co. KG, Heek, Germany. http://www.ecc-fabrics.de, latest access on August, 16, 2007.

6 Andersson, C.H., Eng, K., Zäh, W. and Stahl, J.E. (1994) Warp knitted direct oriented structures for pre-shaped composites. Techtextil, Texhtextil-Symposium, 6* Band 3.1, June, pp. 1–7, Paper-No. 3.14.

7 Ko, F.K., Pastore, C.M. and Head, A.A. (1989) *Handbook of Industrial Braiding*, Atkins & Pearce, Covington, KY, USA.

8 Offermann, P., Diestel, O. and Godau, U. (1997) Flat knitting of biaxially reinforced multiple layer fabrics for composites. *Technical Textiles*, **40**, (2), 95, 98–9.

9 Cebulla, H., Diestel, O. and Offermann, P. (2003) *Biaxial reinforced multilayer weft knitted fabrics for composites*. Proceedings of the 6th International AVK-TV Conference, Baden-Baden, paper B10, pp. 1–9.

10 Haller, P., Birk, T., Cebulla, H. and Offermann, P. (2004) Fully fashioned biaxial weft knitted and stitch bonded textile reinforcement for wood connections. *Composites Part B: Engineering*, **37**, (4–5), 278–285.

11 N.N. (2001) Reinforcement–specifications for multi-axial multi-ply fabrics–Part 1: Designation, German version EN 13473-1:2001, Beuth-Verlag GmbH, Berlin und Köln.

12 Hanisch, V., Henkel, F. and Gries, T. (2007) *Performance of textile structures determined by accurate machine settings*. 14th International Techtextil-Symposium, Frankfurt am Main, 11–14 June 2007.

13 Bannister, M., Callus, P., Nicolaidis, A. and Herszberg, I. (1998) The effect of weave architecture on the impact performance of 3-D woven composites. "A new movement toward actual applications," Texcomp 4–4th International Symposium for Textile Composites, Kyoto/J, 12–14 October 1998, Kyoto, Paper No. P2.

14 Bilisik, A. and Mohamed, M. (1994) *Multiaxial 3-D weaving machine and properties of multiaxial 3-D woven carbon/epoxy composites.* "Moving forward with 50 years of leadership in advanced materials," 39th International SAMPE Symposium, Anaheim/USA, 11–14 April 1994, Covina, pp. 868–82.

15 Brandt, J., Drechsler, K. and Arendts, F.-J. (1996) Mechanical performance of composites based on various three-dimensional woven-fibre preforms. *Composites Science and Technology*, **56**, 381–6.

16 Hirokawa, T. (1997) *Development and application of three-dimensional woven fabric.* "From Fibre Science to Apparel Engineering," Proceedings of the 26th Textile Research Symposium at Mt. Fuji, Shizuoka/J, 5–7 August 1997, Hikone, pp. 102–7.

17 Brookstein, D. (1994) Concurrent engineering of 3-D textile preforms for composites. *International Journal of Materials and Product Technology*, **9**, H. 1/2/3, 116–24.

18 Curiskis, J.I., Durie, A., Nicolaidis, A. and Herszberg, I. (1997) *Developments in multiaxial weaving for advanced composites materials.* 11th International Conference on Composite Materials (ICCM-11), Gold Coast/Aus, 14–18 July 1997, Melbourne, Vol. 5, pp. 86–9.

19 Swinkels, K. (1996) *Low-cost production of double wall fiber reinforced plastic tanks by a winding process with a three-dimensional fabric.* 35th Internationale Chemiefasertagung, Dornbirn 25–27 September 1996, Dornbirn.

20 3TEX, Cary, NC, USA. http://www.3tex.com/3weave.clm, http://www.3tex.com/3braid.clm, latest access on August, 16, 2007.

21 Moreland, J. (1994) An overview of textiles in vascular grafts. *International Fiber Journal*, **9** (5), 16–24.

22 Planck, H., Dauner, M. and Renardy, M. (Hrsg.) (1989) *Medical textiles for implantation.* Proceedings of the 3rd International ITV Conference on Biomaterials, Stuttgart, 14–16 June 1989, Denkendorf.

23 Büsgen, W.-A. (1999) Woven 3-dimensional shapes – update and future prospects for new weaving techniques. *Melliand Textilberichte*, **80** (6), 502–5.

24 Büsgen, A., Finsterbusch, K. and Birghan, A. (2006) *Simulation of composite properties reinforced by -D shaped woven fabrics.* 12th International Conference on Composite Materials (ECCM), Biarritz, France, 29 August–1 September 2006.

25 Keith, W.P. and Kedward, K.T. (1997) Shear damage mechanisms in a woven, nicalon-reinforced ceramic-matrix composite. *Journal of the American Ceramic Society*, **80** (6), 357–64.

26 Hirata, Y., Matsuda, M., Takeshima, K., Yamashita, R., Shibuya, M., Schmücker, M. and Schneider, H. (1996) Processing and mechanical properties of laminated composites of mullit/woven fabrics of Si-Ti-C-O fibers. *Journal of the European Ceramic Society*, **16** (2), 315–20.

27 Kümpers, F.-J., Brockmanns, K.-J. and Stüve, J. (2006) 3D-braiding and surround braiding as textile production method for textile preforms. *DWI Reports*, 2006/**130**, ISSN 0942-301X, paper: Kuempers.pdf.

28 Canfield, A.R. (1985) *Braided carbon/carbon nozzle development, AIAA-1985-1096, SAE, ASME, and ASEE.* Joint Propulsion Conference, 21st, Monterey, CA, July 8–10, 1985.

29 Drechsler, K. (2004) Latest developments in stitching and braiding technologies for textile performing, in *SAMPE, International SAMPE Symposium and Exhibition*, Vol. **49**, Society for the Advancement of Material and Process-Engineering, Covina, CA 91724-3748, United States, pp. 2055–67, ISBN 0-938994-96-4.

30 Byun, J.-H. and Chou, T.-W. (1996) Process-microstructure relationships of 2-step and 4-step braided composites. *Composites Science and Technology*, **56** (3), 235–41.

31 Stüve, J., Tolosana, N., Gries, T. (2006) 3D-braided textile preforms – From virtual design to high-performance-braid, in: Society for the Advancement of Material and Process Engineering (Editor): SAMPE Fall Technical Conference: Global Advances in Material and Process Engineering, Dallas 06.–09.11.2006. Covina, CA: SAMPE 2006, Paper: 026.pdf.

32 Wulfhorst, B., Gries, T. and Veit, D. (2006) *Textile Technology*, Hanser, Munich; Hanser Gardner, Cincinnati.
33 Mungalov, D. and Bogdanovich, A. (2004) Complex shape 3-D braided composite preforms: structural shapes for marine and aerospace. *SAMPE Journal*, **40** (3), 7–11.
34 Milwich, M., Dauner, M. and Plank, H. (1995) Optimisation of process conditions and braid structure in braiding reinforcing products using high-performance fibres. *Band- und Flechtindustrie*, **32** (2), 44–55.
35 Lane, J.E. and LeCostaouec, J.-F. (1998) Ceramic composite hot gas filter development. Proceedings of the Advanced Coal-Based Power and Environmental Systems '98 Conference, July 21–23, 1998, Morgantown, WV, USA.
36 Kolkmann, A., Roye, A. and Gries, T. (2006) Combination yarns and technical spacer fabrics – 3D structures for textile reinforced concrete. *DWI Reports*, 2006/**130**, ISSN 0942-301X, Paper: P53_Kolkmann.pdf.
37 Herrmann, A.S., Pabsch, A. and Kleineberg, M. (2000) *Kostengünstige Faserverbundstrukturen – eine Frage neuer Produktionsansätze, Konferenz-Einzelbericht*. 3 Internationale AVK-TV Tagung für verstärkte Kunststoffe und duroplastische Formmassen, 12–13 September 2000, Baden-Baden.
38 Weimer, C., Mitschang, P., and Neitzel, M. (2002) *Continous manufacturing of tailored reinforcements for liquid infusion processes based on stitching technologies*. 6th International Conference on Flow Processes in Composite Materials, 15–16 June 2002, Auckland, NZ.
39 Klopp, K., Anft, T., Pucknat, J. and Gries, T., (2001) Mechanical strength of conventional stitched composite materials. *Technical Textiles*, **44**, E205–7.
40 Gries, T. and Klopp, K. (2007) *Füge- und Oberflächentechnologien für Textilien: Verfahren und Anwendungen*, Berlin, Springer, Heidelberg.
41 N.N. (1999) *Textiles – Tensile Properties of Fabrics – Part 1. Determination of Maximum Force and Elongation at Maximum Force Using the Strip Method*, ISO 13934-1:1999.
42 N.N. (2003) *Testing of Plastics Films and Textile Fabrics (Excluding Nonwovens), Coated or Not Coated Fabrics – Determination of Stiffness in Bending – Method According to Cantilever*, DIN 53362.
43 Knein-Linz, R. and Machatschke, R. (2000) *Verarbeitung von Glasfilamentgarnen im Webprozeß*, Shaker, Aachen.
44 Schneider, M., Clermont, H., Kozik, C., Knein-Linz, R., Müllen, A. and Wulfhorst, B. (1997) Laser filamnet break detector. *Band- und Flechtindustrie*, **34** (4), 100–6.
45 Cherif, C., Kaldenhoff, R. and Wulfhorst, B. (1997) Computer simulation of the drapeability of reinforcement textiles for composites using the finite element method. Composites for the real world: 29th International SAMPE Technical Conference, Orlando, Florida, 28 October–1 November 1997, Covina, Calif., Society for the Advancement of Material and Process Engineering, Vol. 29, 108–21.
46 Wulfhorst, B. (1996) *Qualitätssicherung in der Textilindustrie: Methoden und Strategien*, München, Hanser, Wien.
47 Grundmann, T., Gries, T., Kordi, M.T., Flachskampf, P., Wisner, G. and Kempf, T. (2007) *Automated production of textile preforms for fibre-reinforced-plastic (FRP) parts*. 14th International Techtextil-Symposium, Techtextil, Frankfurt am Main, 11–14 June 2007.
48 Pluvinage, P., Parvizi-Majidi, A. and Chou, T.W. (1992) *Morphological and mechanical characterisation of 2D woven and 3D braided SiC/SiC composites*. American Society for Composites, Technical Conference 7, pp. S. 400–9.

3
Interfaces and Interphases

Jacques Lamon

3.1
Introduction

The fiber/matrix interfacial domain is a decisive constituent of fiber reinforced ceramic matrix composites (CMCs). Depending on the characteristics of the domain, the composite will be either a brittle ceramic or a damage tolerant composite. Thus, several requirements, which may seem to oppose to each other, have to be met:

- Fibers have to be bonded to the matrix, in order to ensure material integrity, and to obtain a continuous medium.
- Fiber failures have to be prevented when the matrix cracks. This is achieved by crack deviation.
- Once deviation of matrix cracks has occurred, the loads still have to be transferred efficiently through the interfaces, so that a certain amount of the applied load is still carried by the matrix.
- Then, in aggressive environments, the fibers should not be exposed to species conveyed by the matrix cracks.

Meeting all of these requirements will lead to high-performance composite materials.

The fiber/matrix interfacial domain may consist of an interface or an interphase. An interface between two phases, or between the fiber and the matrix, can be defined as a surface across which a discontinuity occurs in one or more material properties. An interphase is a thin film of material bonded to the fiber and to the matrix. An interphase also implies the presence of at least two interfaces: one with the matrix and one with the fiber, and more when the interphase consists of a multilayer. The total area of the interface in composites is extremely large. It can be easily shown that it varies inversely with the fiber diameter:

$$I_A = 4V_f \frac{V}{d} \qquad (3.1)$$

Ceramic Matrix Composites. Edited by Walter Krenkel
Copyright © 2008 WILEY-VCH Verlag GmbH & Co. KGaA, Weinheim
ISBN: 978-3-527-31361-7

where V_f is fiber volume fraction, V is the volume of composite, and d is fiber diameter.

Let us assume the fiber volume fraction to be 0.25, fiber diameter to be as small as 10 μm, and the volume of the composite to be $1\,m^3$. Then $I_A = 10^5\,m^2$.

There are two main types of bonding at an interface: mechanical bonding or chemical bonding. Mechanical bonding results from thermally induced residual stresses. For those CMCs manufactured at high temperature, when the matrix radially shrinks more than the fiber on cooling, this will lead to gripping of the fiber by the matrix. Radial gripping can be enhanced when the interface is rough.

An interfacial reaction zone with a certain thickness is involved in chemical bonding. The strength of a mechanical bond is lower than that of a chemical bond.

Interface properties are dictated by the fiber and the matrix that have been selected, since bonding results from chemical reactions during processing or thermal shrinkage during cooling. Therefore the number of routes, which are permitted to meet the above-mentioned requirements for the interfacial domain, is limited by the number of constituents that are compatible. The concept of interphase allows these limitations to be overcome, and the interfacial characteristics to be tailored with respect to composite properties. A certain amount of effort has been directed toward either the optimization of interfacial domain in CVI-SiC/SiC composites, and more recently in C/C composites, or the development of oxide interphases for oxide/oxide composites.

3.2
Role of Interfacial Domain in CMCs

Bonding fibers to a matrix to keep composite material integrity is the first role assigned to interfaces. Any debonding affects a lot of properties, including mechanical properties, thermal properties, and resistance to the environment. Thus, debonding should be avoided in certain situations when interfaces are the weakest element. This is essentially encountered under the following loading conditions:

- Tensile stresses operate perpendicular to fiber axes. Such stresses are either thermally induced stresses due to thermal expansion mismatch and generated during cooling down from the processing temperature [1] or stresses operating on transverse tows in multi-directionally reinforced composites (Figure 3.1) [2].
- Shear stresses generated by shear loads applied parallel to fiber plies [1].

In these situations, very strong interfaces should be preferred. But, when tensile loads are applied parallel to a fiber direction, interfaces are not the weakest element. Thus, when the matrix stiffness is sufficiently large so that a significant part of the load is carried by the matrix, the matrix breaks first. The matrix cracks are perpendicular to the fiber axis (transverse cracks: Figure 3.1). Thus, deviation of

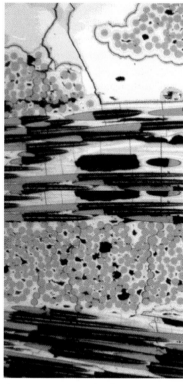

Longitudinal tow

Transverse tow

Figure 3.1 Micrograph of a two-dimensional woven SiC/SiC composite showing debonding in transverse tows perpendicular to loading direction and transverse cracks in the matrix of longitudinal tows oriented in the loading direction (fiber diameter ≈15 μm). The former cracks appeared under low stresses, whereas the latter cracks appear under high stresses.

cracks provides protection to fibers against failure. Crack deviation requires interfaces that can be debonded.

To fully benefit from the load carrying capability of a stiff matrix, the deviation mechanism must be controlled so that load transfer through the interface is not significantly altered by debonding. This can be achieved by tailoring interface properties. When matrix stiffness is small compared to that of fibers, the matrix is subjected to low stresses whereas the fibers carry most of the load (carbon/carbon composites) [2]. If a crack can initiate in the matrix, it is obvious that deviation at interfaces should be sought, in order to avoid failure. But, as discussed below, some combinations such as carbon fibers and a carbon matrix, are initially favorable to debonding. This behavior is essentially controlled by fibers. The influence of fiber/matrix interfaces is less preponderant when compared with the above matrix controlled composites.

3.3
Mechanism of Deviation of Transverse Cracks

The crack deflection mechanism at a weak interface was first proposed by Cook and Gordon [3], who computed the stress field induced by a semi-elliptical crack placed in a single homogeneous material subjected to uni-axial tension. They have shown that the stress state at the crack tip is polyaxial, although a uniaxial tensile stress is applied. In particular, they evidenced that the stress component σ_{rr} parallel to the crack plane (r-axis) reaches a maximum σ_{rr}^{max} in the crack plane at a distance l* from the crack tip. Figure 3.2 shows the stress distribution at the crack tip. Let us consider a fiber/matrix interface perpendicular to the main advancing crack. If the interface tensile strength is less than σ_{rr}^{max}, then the interface can fail in front of the crack tip. From the comparison of maximum stress components σ_{zz}^{max} and σ_{rr}^{max}, respectively with fiber and interface strengths, Cook and Gordon estimated that an interfacial strength of 1/5 or less than that of fiber strength will cause the opening of the interface in front of the crack tip. Deflection of the matrix crack

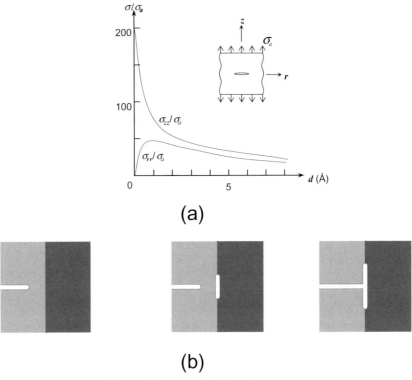

Figure 3.2 (a) stress profile at crack tip in a cell subjected to uniaxial tension; (b) schematic diagram showing crack deviation by debonding at an interface, and coalescence with the advancing crack.

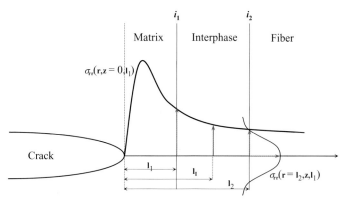

Figure 3.3 Stress state at crack tip (σ_{rr} component) in the presence of an interphase and two interfaces i_1 and i_2 [11].

results from coalescence with the interfacial crack (Figure 3.2). The mechanism has been observed in several combinations of materials [4–10]. Figure 3.3 shows the stress state at the crack tip in the presence of an interphase in a fiber reinforced ceramic [11]. Location of debonding is dictated by interface strengths and interphase transverse strength (perpendicular to the matrix crack direction) respective to the magnitude of σ_{rr}.

3.4
Phenomena Associated to Deviation of Matrix Cracks

Under further loading, local phenomena are allowed by the discontinuities created by interfacial cracks. They induce disturbances in the stress state: overstressing of fibers and a stress lag in the matrix and fibers. A quantitative description of these phenomena is necessary, with a view to better control of composite mechanical behavior. Opening of a transverse matrix crack may cause extension of the interfacial crack (Figure 3.4). It is accompanied by fiber sliding along the debonded interface. The magnitude of this sliding phenomenon is dependent on debond length, surface roughness, and compressive residual stresses in the matrix.

Fibers are locally subjected to higher stresses (Figure 3.5). Depending on the magnitude of fiber/matrix interaction through the interfacial crack, stress on fiber decreases more or less abruptly. Load sharing is optimum outside the debonded portion. Pull out of broken fibers causes further sliding along the debonded interface (Figure 3.4).

The magnitude of these phenomena depends strongly on the size of interface crack. In the presence of long interface cracks, the portion of fibers that carries high stresses is larger, whereas the domain of stressed matrix is smaller and the contribution of matrix to load sharing is reduced, when compared to short interface cracks. This has several implications on the mechanical properties and the

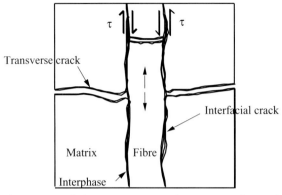

Figure 3.4 Local phenomena associated to matrix cracking and deviation at interfaces.

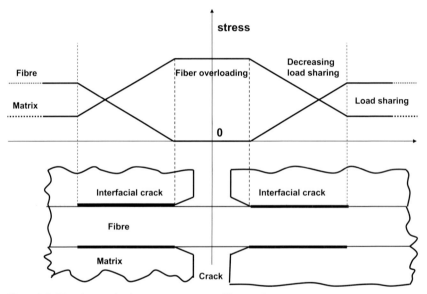

Figure 3.5 Disturbances in the stress state caused by matrix crack deviation at an interface between fiber and matrix.

shape of the tensile stress strain curves. In the presence of long debonded cracks, the composite is weakened as a result of fiber overloading and size effects, whereas energy dissipation associated with sliding is enhanced. In the presence of short debonded cracks, the composite can carry higher loads because of better contribution of matrix to load sharing, and matrix crack density is enhanced as a result of a larger stressed volume of matrix.

Energy absorbing phenomena, such as sliding and multiple matrix cracking, lead to enhanced fracture toughness, whereas contribution of matrix to load

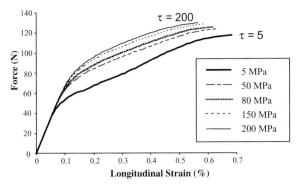

Figure 3.6 Influence of interfacial shear stress on the stress-strain behavior of one-dimensional SiC/SiC composites predictions [12].

sharing leads to enhanced load carrying capacity and ultimate strength. Long interface cracks are detrimental to ultimate strength, whereas short interface cracks are more favorable because they allow both a high toughness and a high ultimate strength to be obtained.

Interface crack size (l_d) varies inversely with interfacial shear stress (τ), according to the following classical equation:

$$l_d = \frac{\sigma \cdot a \cdot r_f}{2V_f(1+a)\tau} \tag{3.2}$$

where σ is the applied stress, $a = \dfrac{E_m V_m}{E_f V_f}$, E_m is the matrix Young's modulus, E_f is the fiber one V_m and V_f are the volume fractions of fiber and matrix, and r_f is the fiber radius.

Figure 3.6 illustrates the influence of interface crack length on tensile stress-strain behavior of SiC/SiC minicomposites [12]. Interface crack length is characterized through τ. Note that the carried force increases with τ, that is, when debonded length decreases.

It is considered that interface weakness varies inversely with debonded length, and that τ is a convenient characteristic of interfacial domain weakness.

3.5
Tailoring Fiber/Matrix Interfaces. Influence on Mechanical Properties and Behavior

The fiber/matrix interfacial domain is a critical part of composites, not only for crack arrest and damage tolerance but also because of transfer of loads from fibers to the matrix (contribution of matrix to load sharing). This latter issue is generally neglected, as most authors essentially focus on crack deviation. The criterion for crack deviation that is recommended is a weak interface. Values of magnitude of weakness are not specified or known. A significant effort has been directed toward the control of interface properties in SiC/SiC composites made via chemical vapor infiltration (CVI) techniques, in order to fully capitalize on the favorable properties of these composites. The control of the interfacial domain involves appropriate interphase and fiber/interphase interface. The concept of composite with a tailored interfacial domain has been established on CVI-SiC matrix composites, using anisotropic interphases made of pyrocarbon (PyC) or a SiC/PyC multilayer, and fibers that had been previously treated to increase the fiber/interphase bond [13–17]. Less interesting results have been obtained with BN [18]. Table 3.1 gives various values of interfacial shear stresses measured on CVI-SiC matrix composites. It should be noted that high interfacial shear stresses in the 100 to 300 MPa range have been measured for the tailored interfacial domains.

The main features that have been observed on SiC/SiC composites may be summarized as follows. In the presence of weak fiber/coating bonds, the matrix cracks generate a single long debond at the surface of fibers (adhesive failure type, Figures 3.7 and 3.8). The associated interface shear stresses are low (Table 3.1). Multiple matrix cracking is limited by the presence of long debonds. The cracks are widely opened (Figure 3.9). The crack spacing distance at saturation as well as the pull-out length tend to be long (>100 µm). Toughening results essentially from sliding friction along the debonds. The fibers carry most of the load, which reduces the composite strength. The corresponding tensile stress-strain curve exhibits a narrow curved domain limited by a stress at matrix saturation, which is distinctive of ultimate strength (Figure 3.10).

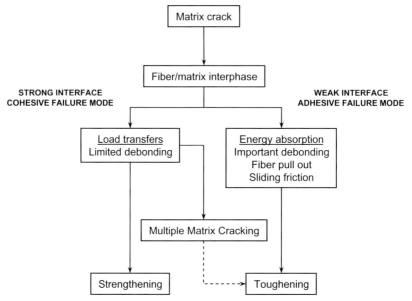

Figure 3.7 Schematic flow diagram summarizing the features associated with interfacial domain strength: weak fiber/matrix interface vs strong fibre coating bond.

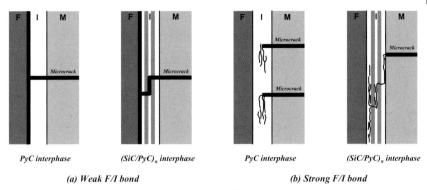

Figure 3.8 Schematic diagram showing matrix interfacial crack patterns when the fiber coating/interface is (a) weak or (b) strong.

In the presence of stronger fiber/coating bonds, the matrix cracks are deflected within the coating (cohesive failure type, Figures 3.7 and 3.8), into short and branched multiple cracks. Short debonds, as well as associated improved load transfers, allow further cracking of the matrix, leading to a higher density of matrix cracks. The crack spacing distance may be as small as 10 to 20 μm. The matrix cracks are slightly opened (Figure 3.9). Sliding friction within the coating, as well as multiple cracking of the matrix, increase energy absorption, leading to toughening. Limited debonding and improved load transfers reduce the load carried by the fibers, leading to strengthening (Figure 3.7). The associated tensile stress-strain curve exhibits a wide curved domain and the stress at matrix cracking saturation is close to ultimate failure (Figure 3.10).

Fracture toughness values (strain energy release rate) increasing from 3 to 8 kJ m^{-2} have been determined on CVI-SiC/SiC composites with weak interfaces or high strength tailored interfacial domains, respectively [15]. A process zone of matrix microcracks forms at the notch tip or at the tip of a pre-existing main macroscopic crack. Microcrack density is commensurate with τ. Experiments have shown that a tailored interfacial domain is also beneficial to lifetime and creep resistance [17, 22, 23]. High strength tailored interphases are based on PyC. Examples of PyC-based interphases are shown in Figures 3.9 and 3.11. Interfacial shear stresses are reported in Table 3.1. Although PyC based interphases are subject to degradation and elimination by oxidation at temperatures above 500 °C, they are a viable concept. It has been shown [17] that SiC/SiC minicomposites, with strong tailored interphases made of a single layer of PyC or a (PyC/SiC) multilayer, exhibited lifetimes that compared favorably with those of minicomposites with BN interphases produced via CVD/CVI. A self-healing matrix is able to protect interphases against oxidation [17]. Lifetimes as long as a few thousand hours have been observed on CVI-SiC/SiBC composites with a self-healing matrix and a PyC interphase.

Table 3.1 Available interfacial shear stresses (MPa) measured using various methods.

	τ (MPa)	Specimens	Methods	Ref
C/PyC/SiC	13–30	Minicomposites	Equation 3.4	[19]
Nicalon/BN/SiC	40–100	Microcomposites	Push in/push out	[18]
	40–140	2D composites	Push out	
Nicalon/BN/SiC	15–35	Minicomposites	Hysteresis loops	[20]
Hi-Nicalon/BN/SiC	5–25			
Sylramic/BN/SiC	70			
Nicalon/PyC/SiC	12	2D composites	Equation 3.4	[21]
Untreated fibers	8		Equation 3.5	
	0.7–4		Hysteresis loops	
	14–16		Push out	
	3		Hysteresis loops	
	4–20		Model of tensile behavior	
	21–115		Hysteresis loops	
	40–80	Minicomposites	Model	
Hi-Nicalon/PyC/SiC	50–12		Hysteresis loops	
Untreated fibers	34–81	Minicomposites	Equation 3.4	
	29–71		Equation 3.5	
	100		Model	
Nicalon/(PyC/SiC)$_4$/SiC	9	2D composite	Hysteresis loops	
Untreated fibers	12–28		Push out	
Nicalon/(PyC/SiC)$_{10}$/SiC	35–100	Minicomposite	Hysteresis loops	
Untreated fibers	25–89		Equation 3.4	
	22–78		Equation 3.5	
	100		Model	
Nicalon/PyC/SiC	203	2D composite	Equation 3.5	
Treated fibers	140		Equation 3.4	
	190–370		Hysteresis loops	
	100–270		Push out	
Nicalon/(PyC/SiC)$_2$/SiC	150	2D composite	Hysteresis loops	
Treated fibers	133		Push out	
Nicalon/(PyC/SiC)$_4$/SiC	90	2D composite	Hysteresis loops	
Treated fibers	90		Push out	
Nicalon/PyC/SiC	46–127	Minicomposites	Hysteresis loops	[17]
Treated fibers	54–160		Equation 3.4	
	50–140		Equation 3.5	
	150		Model	
Nicalon/(PyC/SiC)$_{10}$/SiC	113–216	Minicomposites	Hysteresis loops	[17]
Treated fibers	61–117		Equation 3.4	
	54–103		Equation 3.5	
	350		Model	

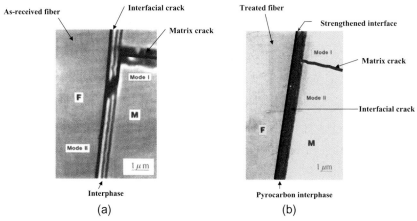

Figure 3.9 Micrographs showing crack deviation in: (a) a SiC/SiC composite with a weak fiber/interphase interface; (b) in a SiC/SiC composite with a strengthened fiber/interphase interface [15].

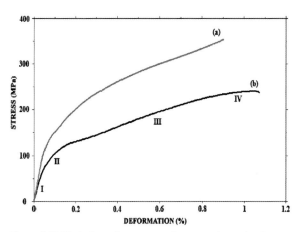

Figure 3.10 Typical tensile stress-strain curves determined on two-dimensional woven SiC/SiC composites with a PyC interphase with: (a) a tailored interfacial domain (treated fibers); (b) a weak fiber/coating interface (as-received fibers). I: elastic deformation; II: matrix multiple cracking; III: fibre deformation; IV: fibre failures.

3.6
Various Concepts of Weak Interfaces/Interphases

Various types of interphases have been proposed. Unlike the high strength tailored interphases, research was driven only by the criterion of crack deflection and, secondarily by resistance to oxidation.

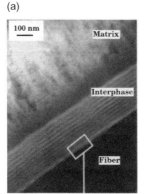
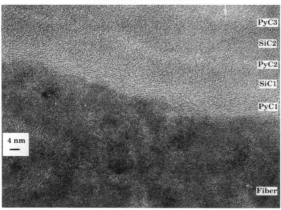

Figure 3.11 TEM micrographs of a multilayered interphase: (a) interfacial zone; (b) HRTEM micrograph of the interfacial bond between the fiber and the interphase [17].

NASA developed an *in-situ* grown BN interphase on Sylramic SiC fibers. A thermal treatment in a controlled nitrogen environment allows mobile boron sintering aids in the Sylramic fiber to diffuse out of the fiber and to form a thin *in-situ* grown BN layer on the fiber surface [20]. This Sylramic-iBN SiC fiber reinforces SiC/SiC composites with a CVI-SiC matrix with porosity filled via infiltration approaches, such as melt infiltration of silicon metal. These composites have been reported to exhibit a high oxidation resistance when compared to those with a CVD/CVI produced BN coating [24]. An interfacial shear strength ≈70 MPa has been reported [20] (Table 3.1).

Several *oxide coatings* have been developed for oxide fibers reinforced with oxide matrix composites: $LaPO_4$ (monazite), $CaWO_4$ (Scheelite), and $NdPO_4$. Of these, monazite has been the most widely studied. Indentation testing and push-out testing have shown that these coatings provide the weak interface function in a dense oxide matrix [25]. Furthermore, an interesting feature was evidenced during push-out of oxide fibers coated with monazite. The monazite coating exhibited a permanent deformation. Contribution of this plasticity to composite mechanical properties has not been investigated. The use of monazite coatings in alumina composites was found to increase the composite life, as compared to a coatingless control composite, by a factor of several hundreds. Results of tensile tests or four-point bending tests suggest that oxide coatings are viable interphase candidates. But the strain to failure of these composites is modest (<0.2%), indicating a limited damage tolerance capability [26].

The porosity localized in fiber **porous coatings** is expected to provide crack deflection. It is postulated that cracks will generally meander to connect pores and hence arrive at fibers in pores, imposing less stress concentration on the fibers than in the dense matrix system. Various oxide-based porous coatings have been examined, including zircon [27], zirconia [28], and rare earth aluminates [29]. It is considered that the porous coating approach does not seen viable for small diameter filaments, because the corresponding scale of porosity required cannot be expected to be stable [26].

The term "fugitive" indicates that the coating material, such as molybdenum or carbon, can be removed after composite fabrication, for instance, through oxidation forming a gap at the fiber/matrix interface. **Fugitive coatings** have been demonstrated in several composite systems [30], such as sapphire-reinforced YAG and Nextel 720/CAS composites. Two-layer coatings of fugitive carbon + zirconia have also been shown to provide a weak interface in a sapphire/alumina composite [31]. Molybdenum does not provide significantly improved properties over fugitive carbon coatings [26]. For fugitive coatings there is an obvious sacrifice in strength, stiffness, and thermal conductivity in directions perpendicular to the fiber axes. Furthermore, since the interfacial domain is extremely weak, the contribution of matrix to load carrying is small. This may eliminate fugitive-interface composites from consideration for many applications.

Research has been conducted on other interfacial coating concepts. They will be briefly mentioned: easy-cleaving coatings including phyllosilicates [32], hexaluminates [33] and layered perovskites [34], ductile coatings [35], segregation weakened interfaces, and reactive coatings [36].

3.7
Interfacial Properties

Among the properties that characterize the behavior of interfacial domains, the following are important for interface design or prediction of composite mechanical behavior: the interfacial strength (tensile or shear strength), the coefficient of friction, the interfacial shear stress, and the interfacial debond energy. The interfacial tensile strength is defined as the resistance to debonding under a tensile stress perpendicular to fiber axis. The interfacial shear stress is the stress required to initiate an interfacial crack. The interfacial shear strength is the stress required for fiber sliding in the debonded interface. The tensile properties characterize initiation of debonding, whereas the shear properties refer to load transfers. The tests used for the determination of these properties are aimed at producing tensile or shear loading conditions onto the interfacial domain. Surprisingly, authors focused essentially on shear properties, although they were interested in crack deviation and were not too concerned about the load transfers through interfacial cracks.

Data on interfacial tensile strength are generally not available in the literature. A few techniques have been proposed to generate stresses perpendicular to interfaces:

- Transverse bend tests: a three-point bend test with fibers aligned perpendicular to the specimen length [37].

- Transverse tensile tests: a tensile test with a fiber direction aligned perpendicular to the specimen length [38].

- Brazilian tests: diametral compression of disks generates tensile stresses, which operate perpendicular to fiber direction when plies are parallel to loading direction.

- Flexural tests on bilayered specimens: this technique allows determination of interfacial tensile properties of model material combinations [39].

These tests give an estimate of the interfacial tensile strength. Improvement of models is required. By contrast, significant effort has been devoted to measurement of shear properties. Most of the test techniques were developed first for polymer matrix composites. They consist in pushing or pulling one or several fibers out of the matrix. Interfacial properties are derived from the force-displacement curve and have been extended to CMCs. Many methods involving the pressing of an indenter on a fiber cross-section have been devised. The push-in and the push-out tests have been the subject of abundant literature [40–43]. Figure 3.12 shows typical single push-out curves measured on SiC/SiC composites with either a weak or a tailored interfacial domain [44]. Alternative methods are based on the tensile behavior of a model composite, such as a micro-composite (a unidirectional composite reinforced by a single fiber) and a mini-composite (a unidirectional composite reinforced by a single multifilament tow) [12, 45, 46]. Interfacial characteristics are determined from features induced by the presence of interfacial cracks: hysteresis loops during unloading-reloading cycles, matrix cracking, or nonlinear deformations.

The interfacial shear stress is related to the width of hysteresis loops ($\delta\Delta$) measured during unloading-reloading cycles, by the following [46]:

$$\tau = \frac{b_2 N(1-a_1 V_f)^2 R_f}{2 V_f^2 E_m} \left(\frac{\sigma_p^2}{\delta\Delta}\right)\left(\frac{\sigma}{\sigma_p}\right)\left(\frac{1-\sigma}{\sigma_p}\right) \qquad (3.3)$$

with

$$a_1 = \frac{E_f}{E_c}$$

$$b_2 = \frac{(1+v)E_m(E_f+(1+2v)E_c)}{E_f[(1+v)E_f+(1-v)E_c]}$$

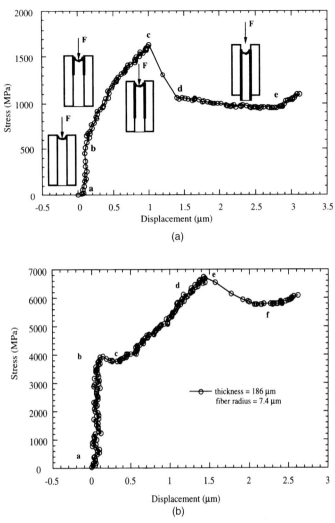

Figure 3.12 Single-fiber push-out curves measured on SiC/C/SiC composite samples reinforced with as-received fibers: (a) [ab: elastic deformation of fiber; bc: stable fiber debonding; cd: unstable debonding; de: fiber sliding] or treated fibers; (b) [ab: elastic deformation of fiber; bc: fiber debonding; cd: fiber loading; de: stable debonding; ef: unstable debonding; f: fiber sliding] [44].

where σ is the applied stress in the unloading-reloading sequence that corresponds to $\sigma\Delta$, σ_p the initial stress level at unloading, E_c the Young's modulus of the minicomposite, R_f the fiber radius, the Poisson's ratio ($\nu = \nu_m = \nu_f$), E_m the Young's modulus of the matrix, E_f that of fiber, and V_f the fiber volume fraction. τ is derived from the $\delta\Delta - \sigma$ data measured during the last unloading-reloading sequence before ultimate failure of the mini-composites. The number of matrix cracks in gauge length, N, is determined from SEM inspection of the mini-composites after failure.

The interfacial shear stress is also related to the spacing distance of the matrix cracks at saturation (l_s) and to the corresponding stress σ_s by the following equations [47, 48]:

$$\tau = \frac{\sigma_s R_f}{2 V_f l_s \left(1 + \frac{E_f V_f}{E_m V_m}\right)} \tag{3.4}$$

$$\tau = \frac{\sigma_s R_f V_m}{2 V_f l_s} \tag{3.5}$$

The interfacial shear stress is extracted from the stress-strain curve using a model of nonlinear deformations that has been detailed and validated in previous works [12, 16]. The model involves constituent properties and flaw strength parameters for the description of the multiple matrix cracking process. τ is estimated from the comparison of predicted stress-strain curves with the experimental ones.

Interfacial shear stresses available in the literature are given in Table 3.1. The data exhibit a certain scatter and dependence on the measurement method. Push-in and push-out tests provide values for a single fiber loaded over a short distance (less than a few hundreds of microns). On the contrary, the tensile tests on mini-composites lead to values that characterize a much larger volume of material. Anyway, an unambiguous trend can be identified from the interfacial shear stresses summarized in Table 3.1: τ is less than 100 MPa for SiC/SiC composites reinforced by as-received SiC fibers, and it can be as high as 350 MPa for tailored interfaces obtained with treated SiC fibers.

3.8
Interface Control

Since the interfacial domain determines composite damage tolerance and mechanical properties, composite design should integrate the selection of interfacial domain that will allow crack deflection and further load carrying capability of the matrix. Design of composites is generally based on an empirical approach involving both processing and testing. The interface characteristics result from the interaction between the matrix and the fiber during processing. Thus, data on interface resistance are generally not available.

There are some general guidelines associated with choosing an interface that will allow crack deflection. One of the most often cited requirements is that the ratio of fracture energies of the interface and the fiber be less than ≈ 0.25 [49]. But this criterion cannot be used for composite design, since data on fracture energy

3.8 Interface Control

of the interface are not available. Debond energy can be determined using the previously mentioned testing methods. But the material has to be made first. It is worth mentioning that data on debond energy are obtained with a certain uncertainty and a significant scatter [46].

According to the stress criterion for a Gordon's crack (debonding), a deviation potential can be defined as [50].

$$\sigma_i^c \leq \sigma_2^c \frac{\sigma_{rr}^{max}}{\sigma_{zz}^{max}} = \sigma_{ic}^+ \quad (3.6)$$

where σ_{rr}^{max} is the maximum of σ_{rr} in the interphase or at the interface (Figure 3.3), and σ_{zz}^{max} is the maximum of σ_{zz} in the fiber (Figure 3.2), σ_2^c is the tensile strength of fiber, and σ_i^c is the interface tensile strength.

σ_{ic}^+ depends on properties of bonded materials: strength of fiber and elastic constants of both materials through σ_{rr}^{max} and σ_{zz}^{max}. σ_{ic}^+ represents the maximum permissible strength value for debonding. Thus, materials that maximize σ_{ic}^+ should be selected.

Figure 3.13 shows the master curve from which we can determine σ_{ic}^+, knowing fiber strength and constituent properties. This master curve represents the

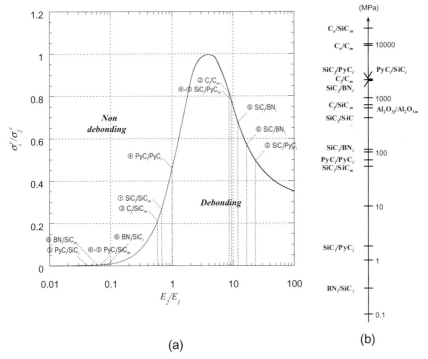

Figure 3.13 Master curve (a) and scale of debonding potentials (b) for various couples of materials: the cracked material (material 1) is cited second. Index: m refers to matrix; f to fiber; I to interphase; n to nanotube.

dependence of $\dfrac{\sigma_{rr}^{max}}{\sigma_{zz}^{max}}$ on elastic constants. A deviation potential chart is also plotted in Figure 3.13 for various combinations of materials. Four groups of deviation potentials σ_{ic}^+ can be identified in Figure 3.13:

- Very high deviation potentials ($\sigma_{ic}^+ > 10\,000$ MPa) obtained with C nanotubes. It is worth pointing out that efficiency of nanotubes requires that each matrix crack impinges on a nanotube and that the nanotube be longer than the interface crack. Deviation is highly probable.
- High deviation potentials ($\sigma_{ic}^+ > 1000$ MPa) obtained with PyC or BN coatings on SiC or C fibers: crack deviation can be expected at the coating/fiber interface. High interface strengths are permissible.
- Intermediate deviation potentials ($\sigma_{ic}^+ \approx 100$ MPa). Deviation is less easy, as low interface/interphase strength or low interphase transverse strength are required. PyC$_i$/PyC$_i$ refers to a deviation within pyrocarbon interphase.
- Low deviation potentials ($\sigma_{ic}^+ < 2$ MPa): deviation requires very weak interfaces. Deviation is not probable. If deviation occurs, the composite will exhibit very poor properties.

3.9
Conclusions

Interphases/interfaces are critical constituents of fiber reinforced CMCs. They are assigned several functions, which appear opposite from a qualitative point of view. These functions depend on several factors, including loading conditions, environment, and fiber orientations. The importance of the interfacial domain depends on the respective properties of fibers and matrix. It is less significant when matrix stiffness is small compared to fibers. It is crucial when the matrix is able to carry a substantial part of applied loads (stiff matrix).

Quantitative criteria for interfacial domain characteristics are required in order to develop composites with excellent properties. But most researchers look only for weak interfaces. This criterion is not sufficient as such. Significant effort has been devoted to developing a tailored interfacial domain in SiC/SiC composites. Interfacial domain characteristics can be tailored via an anisotropic interphase (i.e. a carbon coating or a multilayer) and a strengthened interphase/fiber bond. Debonding is easier with certain combinations of materials. For this purpose, the contrast in elastic moduli and the strength of fiber must be large. Thus, the behavior of the interfacial domain can be controlled by a sound selection of constituents.

In those composites with a matrix stiffer than fibers, the contribution of the matrix to load sharing must be considered also. This phenomenon requires short interfacial cracks, which implies either strengthened interfaces, tough coating, or both. It was obtained in SiC/SiC composites when using an anisotropic pyrocarbon interphase and a strengthened fiber/interphase interface. Similar effect was not observed with a BN interphase.

This pyrocarbon based tailored interphase demonstrated to be a viable concept.

Various weak interface concepts have been proposed in the literature. Much emphasis was put on processing. Interface characteristics are not well documented.

Owing to the importance of the interfacial domain in composite performances, relations between interfaces characteristics and composite mechanical behavior must be established. For this purpose, appropriate characteristics must be measured, so that interfacial domain selection can be made possible, and then introduced into the composite design process.

References

1 Siron, O. and Lamon, J. (1998) *Acta Materialia*, **46** (18), 6631–43.
2 Lamon, J. (2001) *Composites Science and Technology*, **61**, 2259–72.
3 Cook, J. and Gordon, J.E., (1964) *Proceedings of the Royal Society*, **78A**, 508–20.
4 Theocaris, P.S. and Milios, J. (1983) *Journal of Reinforced Plastics and Composites*, **2**, 18–28.
5 Lee, W., Howard, S.J. and Clegg, W.J. (1996) *Acta Materialia*, **44** (10), 3905–22.
6 Clegg, W.J., Blanks, K.S., Davis, J.B. and Lanckmans, F. (1997) *Key Engineering Materials*, **132–136**, 1866–9.
7 Zhang, J. and Lewandowski, J.J. (1997) *Journal of Materials Science*, **32**, 3851–6.
8 Warrier, S.G., Majumdar, B.S. and Miracle, D.B. (1997) *Acta Materialia*, **45** (12), 4969–80.
9 Kagawa, Y. and Goto, K. (1998) *Materials Science and Engineering*, **A250**, 285–90.
10 Majumdar, B.S., Gundel, D.B., Dutton, R.E., Warrier, S.G. and Pagano, N.J. (1998) *Journal of the American Ceramic Society*, **81** (6), 1600–10.
11 Pompidou, S. and Lamon, J. (2007) *Composites Science and Technology*, **67**, 2052–60.
12 Lissart, N. and Lamon, J., (1997) *Acta Metallurgica*, **45**, 1025.
13 Naslain, R. (1993) *Composites Interfaces*, **1** (3), 253–86.
14 Jouin, J.M., Cotteret, J. and Christin, F. (1993) *Proceedings of 2nd European Colloquium "Designing Ceramic Interfaces"* (ed. S.D. Peteves), Office for Official Publications of the European Communities, Luxembourg, pp. 191–202.
15 Droillard, C. and Lamon, J. (1996) *Journal of the American Ceramic Society*, **79**, 849–58.
16 Bertrand, S., Forio, P., Pailler, R. and Lamon, J. (1999) *Journal of the American Ceramic Society*, **82** (9), 2645–475.
17 Bertrand, S., Pailler, R. and Lamon, J. (2001) *Journal of the American Ceramic Society*, **84** (4), 787–94.
18 Rebillat, F., Lamon, J. and Guette, A. (2000) *Acta Materialia*, **48**, 4609–18.
19 Dupel, P., Bobet, J.L., Pailler, R.and Lamon, J. (1995) *Journal de Physique III France*, **5**, 937–51.
20 DiCarlo, J.A., Yun, H-M., Morscher, G.N. and Bhatt, R.T. (2005) *Handbook of Ceramic Composites* (ed. N. Bansal), Kluwer Academic Publishers, pp. 77–98.
21 Lamon, J. (2005) *Handbook of Ceramic Composites* (ed. N. Bansal), Kluwer Academic Publishers, pp. 55–76.
22 Pasquier, S., Lamon, J. and Naslain, R. (1998) *Composites Part A*, **29A**, 1157–64.
23 Rugg, K.L., Tressler, R.E. and Lamon, J. (1999) *Journal of the European Ceramic Society*, **9**, 2297–303.
24 Yun, H.M., Gyekenyesi, J.Z., Chen, Y.L., Wheeler, D.R. and DiCarlo, J.A. (2001)

Journal of the American Ceramic Society, **22** (3), 521–31.

25 Morgan, P.E.D. and Marshall, D.B. (1995) *Journal of the American Ceramic Society*, **78** (6), 1553–63.

26 Keller, K.A., Jefferson, G. and Kerans, R.J. (2005) *Handbook of Ceramic Composites* (ed. N. Bansal), Kluwer Academic Publishers, pp. 377–421.

27 Boakye, E., Hay, R.S., Petry, M.D. ,and Parthasarathy, T.A. (1999) *Ceramic Engineering and Science Proceedings A*, **20** (3), 165–72.

28 Holmquist, M., Lundberg, R., Sudre, O., Razzell, A.G., Molliex, L., Benoit, J. and Alderborn, J. (2000) *Journal of the European Ceramic Society*, **20**, 599–606.

29 Cinibulk, M.K., Parthasarathy, T.A., Keller, K.A. and Mah, T. (2000) *Ceramic Engineering and Science Proceedings*, **21** (4), 219–28.

30 Keller, K.A., Mah, T., Cooke, C. and Parthasarathy, T.A. (2000) *Journal of the American Ceramic Society*, **83** (2), 329–36.

31 Sudre, O., Razzell, A.G., Molliex, L. and Holmquist, M. (1998) *Ceramic Engineering and Science Proceedings*, **19** (4), 273–80.

32 Demazeau, G. (1995) *Materials Technology*, **10**, 43–58.

33 Cinibulk, M.K. and Hay, R.S. (1996) *Journal of the American Ceramic Society*, **79** (5), 1233–46.

34 Fair, G., Shenkunas, M., Petuskey, W.T. and Sambasivan, S. (1999) *Journal of the European Ceramic Society*, **19**, 2437–47.

35 Wendorff, J., Janssen, R. and Claussen, N. (1998) *Journal of the American Ceramic Society*, **81** (10), 2738–40.

36 Hay, R.S. (1995) *Acta Metallurgica*, **43** (9), 3333–48.

37 Chawla, K.K. (1993) *Ceramic Matrix Composites*, Chapman & Hall, London.

38 Rollin, M., Lamon, J. and Pailler, R. (2006) Proceedings 12th European Conference on Composite Materials, Biarritz, France, 29 August-1st September 2006, CD ROM, edited by J. Lamon and A. Torres Marques.

39 Pompidou, S. and Lamon, J. (2007) *Ceramic Engineering and Science Proceedings*, **27** (2), 207–16.

40 Wereszczak, A.A., Ferber, M.K. and Lowden, R.A. (1993) *Ceramic Engineering and Science Proceedings*, **14** (7–8), 156–67.

41 Hsueh, C.H. (1993) *Journal of the American Ceramic Society*, **76** (12), 3041–50.

42 Kerans, R.J. and Parthasarathy, T.A. (1991) *Journal of the American Ceramic Society*, **74** (7), 1585–96.

43 Marshall, D.B. (1992) *Acta Metallurgica et Materialia*, **40** (7), 1585–96.

44 Rebillat, F., Lamon, J., Naslain, R., Lara-Curzio, E., Ferber, M.K. and Besmann, T.M. (1998) *Journal of the American Ceramic Society*, **81** (4), 965–78.

45 Guillaumat, L. and Lamon, J. (1996) *International Journal of Fracture*, **82**, 297–316.

46 Lamon, J., Rebillat, F. and Evans, A. (1995) *Journal of the American Ceramic Society*, **78** (2), 401–5.

47 Marshall, D.B., Cox, B.N. and Evans, A.G. (1985) *Acta Metallurgica*, **33** (11) 2013.

48 Aveston, J., Cooper, G.A. and Kelly, A. (1971) Conference Proceedings of the National Physical Laboratory, Vol. **15**. IPC Science and Technology Press Ltd., Surrey, UK.

49 Evans, A.G. and Marshall, D.B. (1989) *Acta Metallurgica*, **37** (10), 2567–83.

50 Lamon, J. and Pompidou, S. (2006) *Trans Tech Publications*, **50**, 37–45.

4
Carbon/Carbons and Their Industrial Applications
Roland Weiß

4.1
Introduction

In the late 1960s, a lack of suitable high temperature materials led to the development of carbon/carbon (C/C) materials [1, 2]. Due to the high costs of these materials, first developments in the United States and Europe were limited to military applications such as rocket nozzles and re-entry parts for missiles [3–8].

Carbons, graphites, and their modifications are typical high temperature materials. Their intrinsic high thermal stability (>3000 °C) and low density (<2.2 g cm^{-3}) make carbon-based materials one of the most promising candidates for high temperature applications.

The requirement for higher mechanical properties at high temperatures for US Air Force and NASA research programs resulted in the reinforcement of carbon, hence the development of C/Cs. The C/Cs used in rocket nozzles [3–8] and nose tips for re-entry components were reinforced with carbon fabrics made from low-modulus rayon precursor fibers. Therefore, these C/Cs were of low performance quality compared to the C/C grades now available.

In this chapter, the influence of fibers, reinforcement patterns, matrix systems, and manufacturing parameters will be discussed.

Since the early days of C/Cs, the manufacturing technique for industrial applications has been mainly based on the reinforcement of the polymers and their conversion to C/C via heat treatments. The first matrix systems were selected for their high carbon yields, which is also valid for all the manufacturing techniques up to the present day.

4.2
Manufacturing of C/Cs

The manufacturing technique finally used depends on the geometry, the size, and the number of parts, together with the mechanical and thermal requirements of the components. Figure 4.1 shows the parameters that can be modified for the

Ceramic Matrix Composites. Edited by Walter Krenkel
Copyright © 2008 WILEY-VCH Verlag GmbH & Co. KGaA, Weinheim
ISBN: 978-3-527-31361-7

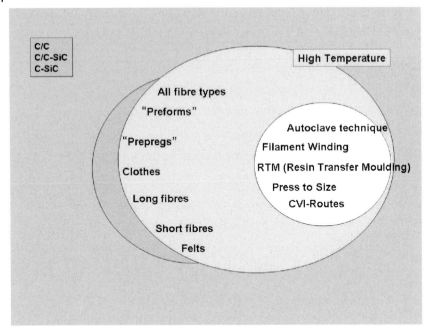

Figure 4.1 Parameters for C/C and CMC production.

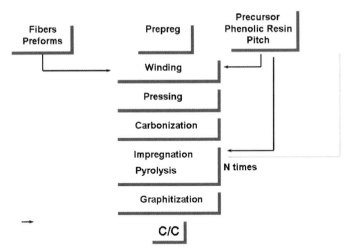

Figure 4.2 Schematic manufacturing cycle of C/C [9].

production of C/C or C/SiC materials. Figure 4.2 shows a typical manufacturing cycle for C/Cs.

The carbon fibers and their orientation in the composites dominate the mechanical properties. The matrix is responsible for the load transfer in a composite and determines the physical and chemical properties of the C/C material. Prepregs (pre-impregnated carbon fibers) are mainly used as a commercially available semi-product in composite manufacturing for press molding or autoclave procedures.

4.2.1
Carbon Fiber Reinforcements

High performance C/C materials are tailored to the mechanical requirements of the final composite structure. Three different types of carbon fibers are commercially available, such as high tensile (HT), high modulus (HM), and intermediate modulus (IM) fibers. These fibers are based on pitch or PAN (Polyacrylonitride), whereas rayon based C-fibers are only used in a few specified applications, such as for military or space structures.

Table 4.1 gives an overview of some of the commercially available C-fibers. The broad variety of the mechanical properties of C-fibers allows composites with optimized strength or with maximum stiffness. The best utilization of these mechanical properties can be realized in UD (unidirectional) C/C composites (Table 4.2). The mechanical properties of the structural components can be varied

Table 4.1 Properties of commercially available C-fibers.

Manufacturer	Product name	Precursor	Filament count	Density (g/cm³)	Tensile strength (MPa)	Tensile modulus (GPa)	Strain to failure (%)
Amoco (USA)	Thornel 75	Rayon	10K	1.9	2520	517	1.5
	T300	PAN	1, 3, b, 15K	1.75	3310	228	1.4
	P55	Pitch	1, 2, 4K	2.0	1730	379	0.5
	P75	Pitch	0.5, 1, 2K	2.0	2070	517	0.4
	P100	Pitch	0.5, 1, 2K	2.15	2240	724	0.31
HEXEL (USA)	AS-4	PAN	6, 12K	1.78	4000	235	1.6
	IM-6	PAN	6, 12K	1.74	4880	296	1.73
	IM-7	PAN	12K	1.77	5300	276	1.81
	UHMS	PAN	3, 6, 12K	1.87	3447	441	0.81
Toray (Japan)	T300	PAN	1, 3, 6, 12K	1.76	3530	230	1.5
	T800H	PAN	6, 12K	1.81	5490	294	1.9
	T1000G	PAN	12K	1.80	6370	294	2.1
	T1000	PAN	12K	1.82	7060	294	2.4
	M46J	PAN	6, 12K	1.84	4210	436	1.0
	M40	PAN	1, 3, 6, 12K	1.81	2740	392	0.6
	M55J	PAN	6K	1.93	3920	540	0.7
	M60J	PAN	3, 6K	1.94	3920	588	0.7
	T700	PAN	6, 12K	1.82	4800	230	2.1
Toho Rayon (Japan)	Besfight HTA	PAN	3, 6, 12, 24K	1.77	3800	235	1.6
	Besfight IM 60	PAN	12K, 24K	1.80	5790	285	2.0
Nippon Graphite Fiber Corp. (Japan)	CN60	Pitch	3K, 6K	2.12	3430	620	0.6
	CN90	Pitch	6K	2.19	3430	860	0.4
	XN15	Pitch	3K	1.85	2400	155	1.5

Table 4.2 Mechanical properties of UD-C/C composites [10].

	Flexural strength (MPa)	Flexural modulus (GPa)	Strain to failure (%)	Tensile strength (MPa)	Tensile modulus (GPa)
C/C with HT-C-fibers	1200	220	0.55	1100	250
C/C with HM-C-fibers	600	480	0.15	700	480

over a broad range and be adapted to the requirements gained by design criteria from finite element calculations. Therefore, load carrying structures are modeled via finite element analysis to minimize the structural weight and to guarantee a high service lifetime. The C-fibers can be aligned according to the mechanical load requirements.

However, for cost effective manufacturing cycles, carbon fabrics are used, which are now commercially available. These include all fiber types, as well as the desired fiber orientations (Figure 4.3).

Although the transverse strength properties are low and thereby a limiting factor in the load capability of C/Cs, three-dimensional-reinforcement architectures or two-and-a-half-dimensional reinforcements are only used in a few critical applications in space or military structures. The first automated three-dimensional-fiber preforms were manufactured by Brochier in France. The technology was transferred to the US defense company, Textron Specialty Materials, in 1983/1986 [11].

Nowadays, three-dimensional and two-and-a-half-dimensional fiber preforms are manufactured by many companies worldwide, such as FMI, Brochier, Aerospatiale, EADS, SNECMA, Carbone Industries, and others. However, these preforms are often restricted to internal use and are not often commercially available.

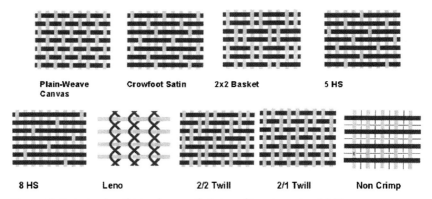

Figure 4.3 Typical carbon fabrics for manufacturing of two-dimensional C/Cs.

Furthermore, the preform costs increase with the increasing number of desired fiber reinforcement directions in multiple-directional reinforced preforms. Therefore, industrially used C/Cs are mostly based on a two-dimensional reinforcement, which itself is based on carbon fabrics or non-crimped lay-ups (Figure 4.3).

4.2.2
Matrix Systems

The carbon matrix in C/C composites has to transfer the mechanical loads to the reinforcing fibers, even if the mechanical strength of the matrix can be ignored. However, the literature shows that the matrix is multifunctional and as important as the reinforcement itself. The matrix contributes to the physical, chemical, and mechanical properties [12–15]. In particular, the failure behavior strongly depends on the microstructure of the matrix system [12].

The most important criteria for the selection of the matrix (carbon) precursors are:

1. High carbon yield
2. Minimal shrinkage during pyrolysis
3. Easiest processing for all types of manufacturing routes, i.e. RTM (resin transfer molding), prepreg manufacturing, filament winding
4. Low solvent content
5. High degree of pre-polymerization with lowest viscosity
6. Availability from multiple sources
7. Low in costs
8. High pot and storage lifetimes.

A high carbon yield is required so as to obtain a completely densified C/C, without the need for additional re-impregnation/re-carbonization cycles (Figure 4.2). Fully densified principally applies to C/Cs with an open remaining porosity below 10%. This goal can be reached by using matrix precursors with carbon yields beyond 84%. Higher carbon yields are always combined with a reduced shrinkage during pyrolysis and thereby with minimized damage of the C/Cs [12]. Solvents as well as a low degree of pre-polymerization reduce the carbon yield. Easy processing is important because the manufacturing of the green bodies (CFRP = carbon fiber reinforced polymers) has to be performed as precisely as possible. All damage occurring during the production of the semi-product CFRP are detectable later in the final C/C component.

Three types of matrix systems are used for the industrial manufacturing processes for C/Cs: thermosetting resins, thermoplastic precursors, and gas phase derived carbons or ceramics (Figure 4.2).

4.2.2.1 Thermosetting Resins as Matrix Precursors

Phenolics, phenolic furfuryl alcohols, furnaces, polyimides, polyphenylenes, and polyarylacetylenes have been used as thermosetting resins [1, 11, 16]. At the beginning of C/C development in 1960 in the United States, phenolics and their modifications were already available. The resins were cheap due to their application in glass reinforcement, ease of process, and a reasonable char yield. Nowadays, phenolic resins are available with carbon yields up to 74% (resols).

Additionally, other types of precursor systems such as isocyanates are now used, due to their high carbon yield and better pyrolysis behavior, as these resins tend to be thermally stable up to 400 °C. Therefore, pyrolysis starts later in comparison to standard thermosets and leaves the possibility of manufacturing a more perfect green body with a lower rate damage and reduced shrinkage.

In the late 1970s, these improvements of thermal stability and increased carbon yield resulted in the development of special types of polyimides with a carbon yield above 84%. Also, another thermosetting resin called polyarylacetylene was used by the Aerospace Corporation [11] and in research activities in Europe [17]. Both thermosets enabled a C/C within a single pyrolysis cycle, a time and cost saving criteria. However, the high price of the polymers (up to some hundreds of $/kg) was inappropriate and even more strikingly, was the poor handling qualities of the polymers. Initially, all thermosetting resins form a glassy carbon after pyrolysis, which cannot be fully graphitized. The influence of the microstructure of the carbon will be discussed later (Section 4.2.4).

4.2.2.2 Thermoplastics as Matrix Precursors

Low cost precursors are isotropic pitches. All pitches belong to the second type of matrix precursors, the thermoplastics. However, high carbon yields are only feasible in cases of high pressure carbonization (Table 4.3), with the disadvantage of higher carbonization costs and often a limitation in size and geometry. Pressureless carbonization of isotropic pitches results in carbon yields comparable to low cost phenolics.

The development of mesophase pitches and their industrialization by Ashland, Mitsubishi, and Osaka has offered a new class of matrix systems. Their polyaromatic structure, with a high carbon/hydrogen ratio, is responsible for their superior carbon yield (Table 4.3). However, the higher the carbon yield, the higher will

Table 4.3 Polymer precursors, their carbon yield, and processability.

Matrix-precursors	Carbon yield (%)	Processibility
Phenolics		
Novolacs	50–55	Good
Resoles	50–55	Good
Furanes	70–75	Medium
Polyarylacetylene	85–90	Medium
Pitches		
MPP + MPP-cokes	85–95	Bad
Isotropic pitch (with high pressure carbonization)	>85	Medium
Isotropic pitch (carbonization without pressure)	50–55	Good

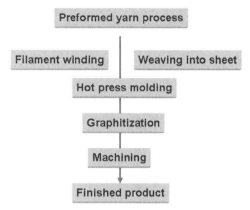

Figure 4.4 Yarn prepreg process of Accross [19–21].

be the viscosity and consequently the melting temperature. Therefore, in their early development, mesophase derived pitches were only used as matrices in C/Cs for hot pressing techniques, a further limitation for low cost industrial applications. A second disadvantage was the thermoplastic behavior. Cross-linking of these pitches was only feasible by oxidation or treatments with sulfur. Therefore, manufacturing of prepregs with pure mesophase pitches was not feasible. Nowadays, such MPP are already commercially available as a pre-oxidized type and thereby suitable for prepregging via powder technology to be used in a one-step pyrolysis cycle in C/C manufacturing (Figure 4.2).

Industrial applications of such a one-step cycle have been put into reality by Schunk [18] and Accross [19, 20]. Schunk formed the bulk mesophase during the molding step up to 400 °C and added binders such as sulfur or binder pitches during impregnation to the pitches responsible for bulk mesophase. A standard method is the blending of MPP with other carbon precursors (Figure 4.4). In principle, the carbon fiber bundle is coated with a blend of pitch and coke. A thermoplastic coating (sleeve) surrounds the carbon fiber bundle to encase the powder blend. The carbon fiber bundle can be directly used to make all textile performs (Figure 4.3) and then processed to C/C according to the manufacturing schemes shown in Figures 4.2 and 4.4. The patents of Schunk and Accross are acknowledged throughout the C/C community.

4.2.2.3 Gas Phase Derived Carbon Matrices

C/Cs can also be manufactured via CVI (Chemical Vapor Infiltration) techniques (Figure 4.2).

The CVI-process is based on gas phase pyrolysis of gaseous species deposited on the surface (pore walls) of the fibrous preform. The process itself is controlled by the deposition parameters and the in-pore diffusion rates. The microstructure of the deposits is mainly determined by the surface of the substrates and the deposition parameters. The CVI process for C/C and other CMC-composites is extensively described in [11, 16]. The CVI parameters, as in pore diffusion and deposition kinetics, are described in [22]. The homogeneity of pore fillings with a minimum of closed porosity is the key factor for fast CVI densification procedures. The most common CVI process is the isothermal impregnation method.

There are four basic CVI processes:

- Isothermal CVI
- Thermal gradient CVI
- Pressure gradient CVI
- Rapid CVI.

All four methods have advantages and disadvantages. All methods are used, modified with one another, or combined with polymer impregnation techniques, depending on the requirements of the composite part. The four basic methods are described in the following:

Isothermal CVI The fiber preforms are heated in a furnace up to the CVI temperature, which can be in the range between 800 and 1200 °C. The furnace temperature and pressure (typically below 100 mbar) are kept constant throughout the complete deposition process. Typical resident times for CVI impregnation of C/C brakes for aircraft are 120 to 200 hours. C/C brake manufacturers such as Dunlop have a furnace capacity of about 10 tons of C/C brakes per process cycle and furnace. The used precursor gas is methane, which can be used in the Dunlop process simultaneously for CVI impregnation and furnace heating.

The Pyrocarbon (PyC) often closes the open porosity on the outer surface before the final densification is carried out. Therefore, the C/C parts have to be removed from the furnace as an intermediate step and machined to open the closed pores. The parts have to be reheated before the CVI process continues up to the required final densification. The number of intermediate machining steps depends on the thickness of the C/C component to be densified via CVI. The machining step can be performed up to three times in cases of C/C components with a thickness of 40 mm.

The isothermal CVI process has an excellent reproducibility. The resulting PyC matrix possesses a high density, high modulus, and good graphitizing ability. The process is time-consuming but the cost is competitive, as proven by the suppliers of C/C brake disks such as Dunlop, SEP, Hitco, Bendix, Allied Signal, Carbone Lorraine, and others.

Thermal Gradient CVI The thermal gradient method is normally performed as a cold wall CVI procedure. The fibrous preform is placed on a graphite tool (mandrel), which is inductively heated (Figure 4.5). The highest temperature in the fibrous preform is at the interface graphite mandrel/fibrous preform and temperature decreases toward the cold wall. The deposition rate increases with increasing temperature, so densification starts at the interface. The deposited PyC increases the thermal conductivity of the fibrous preform and the highest temperature region moves toward the cold wall. The deposition front follows the thermal front.

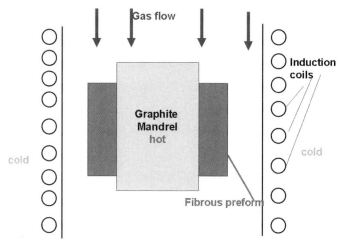

Figure 4.5 Principle of thermal gradient CVI.

Therefore, only large individual parts are densified via thermal gradient methods, such as rocket nozzles [23].

Pressure Gradient CVI The pressure gradient CVI method is a modification of the isothermal process. The fiber preform is sealed in a gas-tight tool and heated in the furnace to the deposition temperature. The deposition gas, usually methane for C/C, is forced to flow through the fiber preform. The pressure drop in the flow direction, caused by the porous fiber preform, increases with increasing deposition. In contrast to isothermal CVI processes, the deposition rate is independent of diffusion effects due to the forced flow and increases with decreasing pressure. The main drawback of this method is the limitation to single item processing. Nevertheless, this method and a combination of thermal and pressure gradient CVI was licensed by MAN in Germany, from Oak Ridge National Laboratory, TN, USA.

Rapid CVI Methods Two different methods of rapid CVI densification for C/Cs have been developed and can be applied industrially for full densifications of C-fiber preforms.

The first method was developed by CEA in France, the so-called film boiling method [24]. The porous C-fiber preform acts as a carbon susceptor and is fully immersed in a hydrocarbon liquid, for example, cyclohexane or toluene. The carbon yield is higher with toluene. The liquid boils during inductive heating and the vapor penetrates into the porous structure. The thermal decomposition of these hydrocarbons vapors forms a carbon deposit on the inner surface of the porous structure and densifies the preform via CVI. The vapor, which is not decomposed, is cooled, and can be used again [25]. A complete densification of C/C brakes with a two-and-a-half-dimensional C-fiber preform can be completed

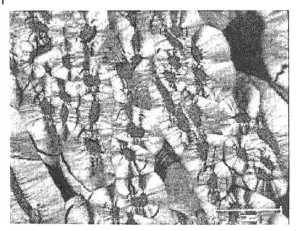

Figure 4.6 Microstructure of the carbon deposited (optical microscopy with polarized light) [26].

within 10 hours [26]. The carbon matrix is highly orientated and can easily be graphitized (Figure 4.6) [26].

The CVI reactor is very simple and made of glass, because the C-fiber preform is cooled over the complete densification cycle by the surrounding liquid precursor. A second method was developed by Huettinger et al. [27, 28]. The porous C-fiber preform, often carbon felt, is placed in an isothermally heated furnace with an outer graphite tool, which only leaves a small gap for the forced gas flow of the precursor gas between graphite tool and porous C-fiber preform. The relation of inner and outer surface volume dominates the deposition chemistry and kinetics [27–30].

The rapid CVI method is controlled by kinetics, allowing the gaseous species to diffuse into the porous preform. However, reactions inside the preform structure can occur by deposition, even before diffusion back to the gas flow has come to an end. A complete overview of C/C manufacturing by CVI was given by J. Diefendorf, a pioneer of CVD and CVI techniques [31].

As explained here, the selected matrix precursors have a dominant influence on the manufacturing process. CVI methods can be used, if C/C parts have to be manufactured near-net shape, because no shrinkage occurs during PyC deposition in the porous structure of fiber preform. In the case of matrix precursors, which have to be carbonized, such as pitches or thermosetting polymers, the shrinkage can be reduced by increased carbon yields of the precursor.

Industrial manufacturing processes are selected for their cost effectiveness. Therefore, polymer precursors are widely used. Polymer precursors can be applied for all technologies known from fiber reinforced polymers (Figure 4.1). These composite technologies are available at all C/C manufacturers. The special knowledge of the manufacturers centers mainly on the precursors used, how the different manufacturing techniques are combined, and the parameters for the final heat treatment (HTT) of the C/C.

4.2.3
Redensification/Recarbonization Cycles

The mechanical properties of C/Cs increase with an increasing degree of densification (Table 4.4). Therefore, it is important to control the pore filling behavior during re-impregnation and re-carbonization, and to avoid the formation of high amounts of closed pores. The pore filling mechanisms have been well described in the literature (Figure 4.7) [32]. In the case of CVI densification, the inner walls of the pores are directly coated with carbon. The advantage of CVI impregnations is their ability to fill up large pores. The disadvantages are the small pore-mouth diameters, the closing of which results in closed porosity (Figure 4.7).

In the case of liquid re-impregnations, the pore filling depends on different parameters:

- impregnation pressure
- viscosity of the polymer
- carbonization pressure
- type of polymer

Low impregnation pressures, in combination with low viscosities of the polymer, fill larger pores easier than smaller ones. However, during the curing process, or carbonization, the polymers flow out from the pores. In the case of pitch impregnation, this can be prevented by applying high pressures (up to 100 bars) during carbonization, whereas for impregnations with thermosets, the curing of the polymers is preformed under pressure (up to 40 bars).

Table 4.4 Influence of densification on the mechanical properties of two-dimensional C/Cs.

Grade	Density (g cm^{-3})	Porosity (%)	Flexural strength (MPa)	Tensile strength (MPa)
CF222	1.55	<8	240	200
CF222/2	1.4	20	140	160
CF226	1.5	<8	150	180
CF226/2	1.4	20	120	120

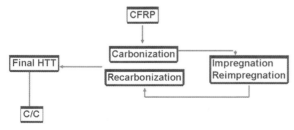

Figure 4.7 Carbon pore-filling mechanisms [32].

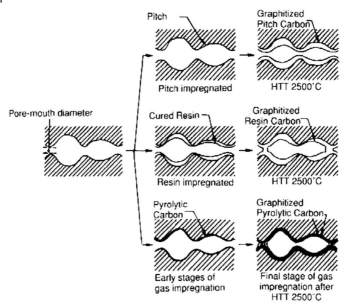

Figure 4.8 Recarbonization/reimpregnation cycles.

The disadvantage is the encapsulation of the condensates (in the case of thermosets) or pyrolysis gases (in the case of pitch), which can result in the damage of the composite itself. In principle, re-impregnations with pitches are better for composites with large pores, because the carbon shrinks during pyrolysis onto the surface of the inner wall pore of the (Figure 4.7). Carbonized and graphitized resins (thermosets) shrink off from the inner wall surface, therefore influencing graphitization behavior during the HTT (Figure 4.8).

4.2.4
Final Heat Treatment (HTT)

The final heat treatment (HTT) determines the chemical and physical properties of C/Cs. The influence of the HTT on the microstructure is well described in the literature [33–35]. The HTT is performed from 1700 to more than 3000 °C. Increasing temperatures result in increasing graphitization of the fibers and the matrix, which can be investigated by TEM [33]. Pitch or CVI derived matrix carbon shows better graphitization than carbons derived from phenolic resin. Although phenolic resins form a glass-like carbon, they can be graphitized by *in-situ*-mechanical stress during HTT (Figure 4.9a, b) [36]. The degree of graphitization determines electrical and thermal resistivity. Increasing graphitization improves the electrical and thermal conductivity, which is important for heating elements or first-wall materials in fusion reactors. However, the failure behavior is also affected. High degrees of graphitized matrix carbon result in a pure shear failure of the matrix, with the

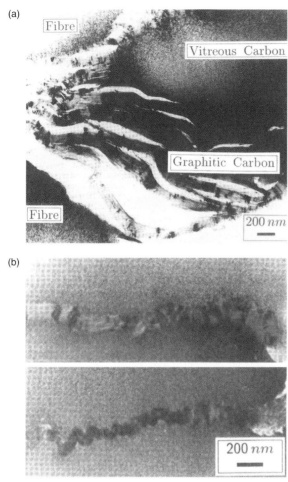

Figure 4.9 (a) Transverse cross-section of UD C/C. The fibers are surrounded by stress-oriented graphite, which split into smaller crystallites easily. Thus cracks are deflected around fibers and vitreous carbon. (b) Two examples of closely packed fibers with equidistant surfaces.

loss of ductility of the C/C. However, the creep resistance increases with the increasing final heat treatment temperature.

In the case of C/Cs with thermosets as matrix precursors, the final HTT dominates the mechanical properties as well as the fracture behavior of the C/C composites [35, 37–39]. The failure behavior can be modified and tailored from pure brittle failure [37], with an HTT of 1000 °C to a pure shear failure (Figure 4.10) [37].

At intermediate temperatures, between 1800 and 2600 °C, a mixed mode of shear failure of the matrix and delamination of the fiber/matrix interface [35] is observed,

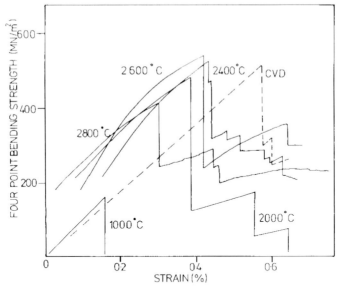

Figure 4.10 Fracture behavior of a carbon-carbon composites made from a furan precursor, with respect to HTT [37].

resulting in high energy of fracture [38, 39], the fracture energy increasing from 0.3 kJ m^{-2} (HTT 1000 °C) to 4 kJ m^{-2} (HTT 2700 °C [38]. However, the maximum of fracture energy depends on the type of reinforcing C-fibers as well as the polymer precursor [39]. Therefore, the best ductility (pseudo-ductility) of C/Cs can be tailored to more convenient temperatures, to reduce the manufacturing costs. The optimization of the mechanical and physical properties depends on the final applications.

4.3
Industrial Applications of C/Cs

As mentioned above, the application of C/Cs started with rocket nozzles and reentry parts for military or space projects, in about 1960. The first industrial products were C/C brakes for civil aircraft developed by Super Temp Division of B.F. Goodrich and were licensed by Dunlop. Today, C/C brake disks are used for most civil aircraft and manufactured worldwide by companies such as Dunlop, SEP, Hitco, Bendix, Carbone, and others. The matrix is PyC (CVI) and the materials are finally heat treated to achieve an optimum ductile failure behavior. The market volume is estimated with some 50% of the total C/C used worldwide.

The brake disks have to be protected against oxidation. This can be achieved by design and construction of the braking system itself, that is, by avoiding as much oxygen contact as possible with the C/C surface or by minimizing the contact with air during braking. In such an optimized design the carbon must be highly graphitized up to approximately 2500 °C to obtain the best oxidation resistance together with a reasonable ductility of the material. Oxidation protection of C/C has to be applied for all materials used in an oxidative atmosphere above 420 °C. The principles of oxidation protection are described in the following.

4.3.1
Oxidation Protection of C/Cs

All oxidation protections for C/Cs basically include three components to protect against oxidation attack over long-term applications (Figure 4.11). These three protection systems are an inner bulk protection, an outer multilayer CVD-coating, and finally a glass sealing layer serving as a surface coating.

The first patent on oxidation protection of carbons was given to the National Carbon Company in 1934 [40]. Already by then, first oxidation protection systems included a bulk protection by formation of SiC and a sealing layer of B_2O_3. CVD coatings were unknown at this early stage of oxidation protection of carbons. However, the principles of protecting carbon materials are still valid today.

4.3.1.1 Bulk Protection Systems for C/Cs

The oxidation protection systems depend on the oxidation conditions, such as temperature, atmospheres, architecture of reinforcement, matrix systems, and required lifetimes. C/Cs, applied in oxygen containing atmospheres below 650 °C, can be protected by pure bulk treatments. C/Cs are oxidized according to the following chemical reactions:

(1) $C_{(s)} + O_{2(g)} \rightarrow CO_{2(g)}$

(2) $2C_{(s)} + O_{2(g)} \rightarrow 2CO_{(g)}$

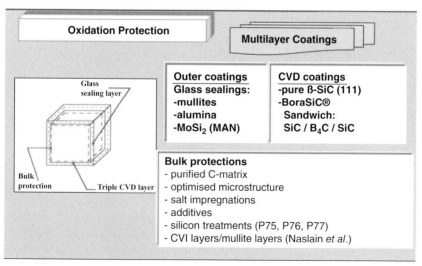

Figure 4.11 Multilayer coated C/C.

At higher temperatures, the second chemical reaction is preferred (Bouduard's reaction), whereas at lower temperatures, CO_2 is formed. The oxidation attack starts on the atoms at the edge of the carbon material, which means crystalline defects in the carbon structure. Therefore, the carbon matrix is more sensitive against oxidation than the reinforcing C-fibers, which are described in detail in [41]. Oxidation rates at low temperatures can be reduced by increased crystallinity and purification of the carbon matrix by higher final HTT (Figure 4.12) [16, 41].

Furthermore, it can be concluded from the Arrhenius plots (Figure 4.12) that the oxidation mechanism depends on temperature. At lower oxidation temperatures (<800 °C), the oxidation rate is controlled by the chemical oxidation attack (kinetics), whereas at higher temperatures (low 1/T-values), the burn-off-rate is limited by diffusion effects, which was also confirmed by oxidation studies on graphite materials [42]. The active sites in the carbon matrix can be reduced by increased final HTT or blocked by contaminations applied as salt impregnations.

Industrially available are those grades possessing phosphoric salt impregnations, which are applied as standard C/C materials in manufacturing hollow glasses at applied temperatures of approximately 700 °C without any further oxidation protection. The mechanical properties of these grades are given in Table 4.5. The influence of the salt impregnation can be shown by comparison of the oxidation behavior of CF264 (Figure 4.13, without salt) and CF264Q (Figure 4.14, with salt). Salt impregnation inhibits the oxidation rate measured as weight loss via thermo-gravimetrical analysis. Furthermore, it can be concluded from these results, that salt impregnations decrease the influence of higher oxygen flow rates and thereby oxidation attack. CF264Q is used in hollow glass manufacturing up

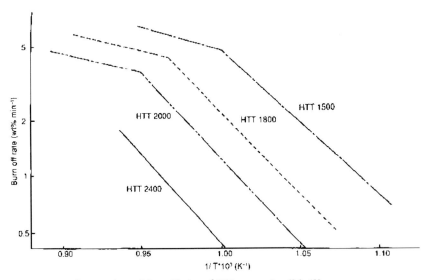

Figure 4.12 Arrhenius plots of the oxidation of C/C composites [16, 41].

Table 4.5 Properties of C/C grades with bulk protection based on impregnations with phosphoric salts.

Grade	Reinforcement	X_F (%)	γ (g cm^{-3})	P (%)	Flexural strength (MPa)	ILSS (MPa)	HTT (°C)
CF260Q	C/C	55	1.40	8	80	7	2100
CF264Q	C/C	55	1.45	8	80	8	2400

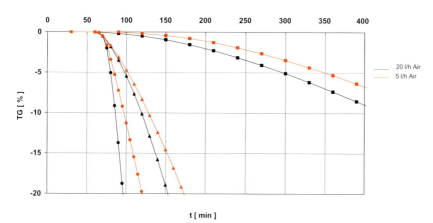

Figure 4.13 Oxidation behavior of CFC 264 without salt impregnation.

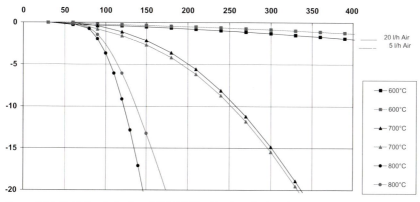

Figure 4.14 Oxidation behavior of CFC 264Q with phosphoric salt impregnation.

to 2000 hours, if the contact times and contact areas are sufficiently low. The oxidation rate depends on temperature, time, and the surface area attacked by oxygen.

As already mentioned and shown in Figure 4.11, bulk treatments are often performed with silicon to form SiC, which possesses superior oxidation protection behavior compared to salt impregnations. Silicon treatments of the bulk material are well known [10, 43–46] and different methods are industrially available (Figure 4.15). The advantages and disadvantages of the different methods are given in Figure 4.15. Silicon treatments can be performed with Si vapor, by melt impregnation under vacuum and pressure (P76-process), by capillary impregnations via liquid phase reaction (P77, Figure 4.15) called the LSI-process (liquid silicon impregnation), by pack cementation, and by a combination of cementation and liquid phase impregnation (P75 process; Figure 4.15). The advantages and disadvantages of the different methods are given in Figure 4.15.

The silicon treatment via P75 is the only procedure that can be used for fully densified C/Cs without any dimensional change of the treated C/C [10]. The degree of matrix conversion from carbon to SiC depends on the reactivity of the carbon, open porosity, and density. A broad range of C/C over C/C-SiC up to C/SiC-materials can be realized according to the final requirements. The higher the amount of formed SiC is, the better the results for oxidation resistance. However, increasing degrees of conversion are always combined with increasing brittleness of the composites, including fiber damage by chemical reactions between C-fibers and silicon.

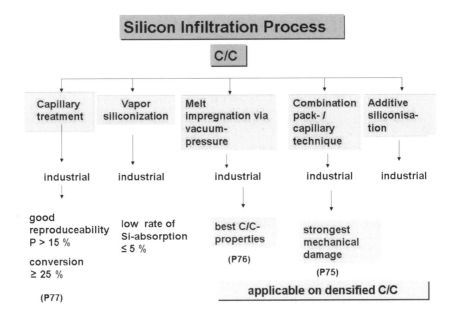

Figure 4.15 Bulk protection methods via silicon treatments.

Silicon treatments of C/C are used not only for bulk protection but also for improvement of the wear behavior for tribological applications. The best bulk protection is obtained by multilayer CVI coatings on the inner surface of bulk materials. Such methods have been described and used [47–49], in particular for C/SiC and SiC/SiC components applied in space or aircraft structures. The combination of nano-scaled multilayers with PyC, SiC, BN, or B_4C, directly applied on the internal surface of fiber preforms, improves their oxidation resistance as well as the ductile failure behavior of the high temperature composites. Boron containing intermediate layers are able to form boron oxide (B_2O_3), even at low oxidation temperatures, resulting in self healing behavior of the matrix during oxygen attack. However, the same oxidation protection of the bulk material can be obtained by using additives with boron (Figure 4.11).

A further method for bulk protection is PIRAC (Powder Immersion Reaction Assisted Coating), an *in-situ* CVI-method developed by Prof. E. Gutmanas, Technion, Haifa. An overview on bulk protection of C/C with chromium powder is given in the thesis [50]. PIRAC coatings with titanium can also be used as bulk protection for C/Cs (Figure 4.16). The formed TiC filled the intra-laminar cracks in the composites and thereby prevented strong oxidation attack by closing possible diffusion paths for the oxygen. However, for long-term oxidation resistance outer CVD coatings must also be applied. The PIRAC-method is a simple and cost-efficient CVI technique. The samples or components, which have to be protected, are packed in a powder mixture of metal and powder of the same metal halogenide, or a pure halogenide such as jodium. The package is vacuum-tight by, for example, metal foil and heated up in a conventional furnace to the CVI reaction temperature, preferably below 1100 °C. The furnace itself can be used under inert gas. Enclosed in the metal foil under low pressure, a metal sub-

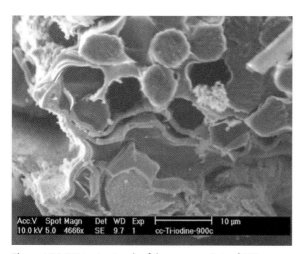

Figure 4.16 SEM mircograph of the cross-section of C/C PIRAC treated showing the inter-fiber cracks filled by titanium carbide.

Figure 4.17 SEM micrograph showing the oxidation scale on the surface of coating obtained by oxidizing coated SiC at 1000 °C for 2 h in static dry air [51].

halogenide is formed that diffuses as a CVI reaction inside the C/C, the halogenide acting as a catalyst. The bulk protection itself shows a gradient layer and can be modified from composition (Figure 4.17) [51].

A SiC-protected C/C was treated with a mixture of Cr and J_2 and oxidized after the PIRAC treatment for 2 hours at 1000 °C. As can be seen, the PIRAC treatment can also be applied for outer coatings.

4.3.1.2 Outer Multilayer Coatings

CVD-SiC coatings are normally used as outer multilayer coatings of C/Cs (Figures 4.2 and 4.17). The mismatch in CTE of the bulk C/C and the CVD-SiC coating results in thermally-induced crack formations during cooling down from CVD-coating temperature to room temperature (Figures 4.11 and 4.18). An increase in the CVD-SiC layer thickness results in a decrease of the crack distance, which means increased crack densities due to the induced higher thermal stresses [52]. The formed cracks are diffusion paths for oxygen in the case of oxidation. However, the thermally induced cracks can be almost completely closed by heating the samples to the CVD coating temperatures.

Therefore, the oxidation rate of multilayer coated C/Cs depends not only on physical parameters such as reaction kinetics and diffusion rates, but also on crack opening and crack closing effects, which are influenced by the coating parameters and the temperature. It is well known from oxidation kinetics, that the maximum oxidation rate for multilayer coated C/C is observed at about 800 °C [51]. This maximum is caused by the ratio of internal surface area available for oxidation (by crack opening) and highest oxidation kinetics.

However, for long-term applications, such cracks are a limiting lifetime factor. Therefore, self-healing effects have to be used to close these cracks. The self-healing capacity can be tailored according to the requirements, from the viewpoint of oxidation temperatures and atmospheres. Although the thermodynamic and kinetic stability of the different oxidation protections will not be discussed in this

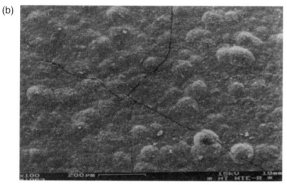

Figure 4.18 (a) Equidistant crack-formation on a CVD-SiC-coated C/C perpendicular to the reinforcement [52]. (b) Irregular crack formation on the surface of a CVD-SiC coated C/C.

chapter, some general points are necessary to understand the principles of oxidation protection.

Silicon carbide (SiC) is applied as the standard CVD-SiC coating (Figure 4.11). SiC forms a silica layer (SiO_2) in oxidizing atmospheres, with a parabolic growth of the layer thickness. SiC is selected as the most favored oxidation protection, because the SiO_2 layers possess the lowest diffusion rate of oxygen. Therefore, the formation of silica during oxidation is an additional self-healing layer (glass sealing layer). The disadvantage of such a layer is the fact that a completely covered SiO_2 layer can be obtained at temperatures above 1000 °C, whereas at lower oxidation temperatures (700–1000 °C) only isles of SiO_2 are formed on the attacked carbon surface accessible by the thermally induced cracks in the multilayer coating. Glass formers are needed for lower temperatures, such as with boron derived additives. Therefore, the bulk can be additionally protected, with borides as additives. Another

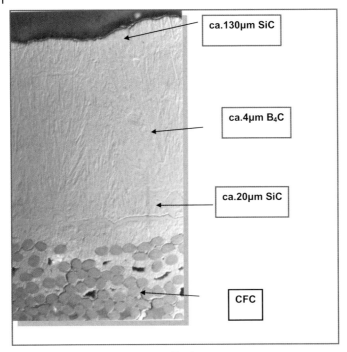

Figure 4.19 BoraSiC coating with self healing capacity [9, 53].

possibility is to include a boron oxide former, such as B_4C, in the multilayer coating (Figure 4.19) [9, 53].

Even at low temperatures, boron carbide is able to form a boron oxide, which can close the cracks in the CVD multilayer (Figure 4.20) [53]. The cracks on the top surface of the multiplayer BoraSiC coating are healed (Figure 4.20). The crack is closed under formation of B_2O_3, boron silicate, and SiO_2, as proved by EDX. In principle, borides can be used up to 1600°C as self-healing glasses. However, the boron oxide is neither stable under water vapor flow, nor thermally stable at temperatures above 1200 to 1250°C, due to evaporation of boron silicates. Therefore, additional outer glass sealing layers are necessary, with superior high temperature behavior.

4.3.1.3 Outer Glass Sealing Layers

As mentioned above, multilayer CVD-SiC coatings (Figure 4.11) form *in situ*, during oxidation, a dense silica layer at temperatures above 1000°C. The diffusion rate of oxygen through the SiO_2 layer depends on the layer thickness, which increases with increasing oxidation time, and the oxidation temperature. The diffusivity of oxygen in silica is extremely low compared to its diffusivity in B_2O_3. The diffusion rate in SiO_2 of oxygen is seven orders of magnitude lower [55, 56], resulting in a superior oxidation protection.

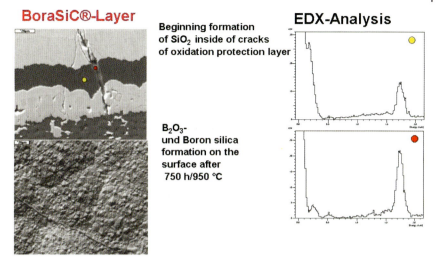

Figure 4.20 Self healing capacity of BoraSiC multilayer coatings [54].

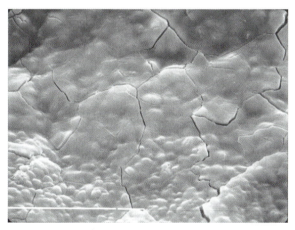

Figure 4.21 SEM of the protected C/C sample after 1000 h at 1530 K in 80 l h^{-1} O$_2$; sealing thickness 3 to 4 μm (Magnification: 10 mm = 5 μm).

Figure 4.21 shows a multilayer coated C/C with an *in-situ* formed silica layer. The cracks visible at a.T. are closed during oxidation. The sample did not show any weight loss after 1000 hours at 1530 K [44]. Nevertheless, glass sealings based on silica have their limitations. The first one is the compatibility with the CVD-SiC layer underneath. The thermal shock resistance of these glass layers is limited by their thickness. Spallations can be observed in cases of layer thicknesses exceeding 8 μm. Furthermore, crystallization occurs, reducing the lifetime of the glass sealing. The viscosity of SiO$_2$ is one order of magnitude or more higher than B$_2$O$_3$

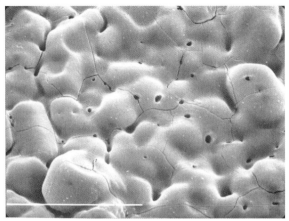

Figure 4.22 Perforation of SiO_2 glass sealings on multiplayer CVD-SiC coated C/C after 1000 h at 1530 K in air.

and thereby the sealing efficiency is extremely reduced. A non-stoichiometric CVD-SiC coating with an excessive amount of Si shows a formation of pores in the SiO_2 glass sealing, due to the following chemical reaction (Figure 4.22):

$$Si_{(s)} + SiO_{2(s)} \rightarrow SiO_{(g)}$$

Silica glass sealings can be replaced by glass sealings with $MoSi_2$ and their modifications, which have a lower viscosity and so a better sealing effect. At temperatures above 1800 °C, the SiO_2 glass sealings are thermally unstable. Therefore, different high temperature refractory oxides have to be applied, such as ZrO_2, Y_2O_3, or HJO_2. However, the oxygen diffusivity is some orders of magnitude higher than in SiO_2 glass sealings. A further disadvantage of silica is the insufficient corrosion resistance against water vapor with high flow rates, as required for application in gas turbines. Therefore, environmental barrier coatings (EBC) are under development to overcome these problems. For most of the industrial applications oxidation protection is not required, because C/Cs are normally applied in inert gas or under vacuum.

4.3.2
Industrial Applications of C/Cs

As already mentioned, the driving forces for industrial applications of C/Cs have been the developments for space and military applications, followed by C/Cs for military and civil aircraft brakes (e.g. 1971 for Concorde, 1970 for F-15 fighters). Subsequently, the use of C/C brakes has been exploited in Formula 1 racing cars by Hitco (1980), SEP (1981), A.P. Racing, Le Carbon Lorraine, and others. These C/C brake disks are sensitive to oxidation and can only be used for a single race. The Formula 1 disks are reinforced with chopped fibers, carbon felts, or carbon cloth with a CVI derived PyC matrix. The same is valid for C/C clutches in racing cars.

However, these C/Cs could not be used for automotive applications due to their limited lifetime. Therefore, in 1979, developments were started in Europe to replace asbestos brake pads with metal impregnated C/Cs to improve their wear and oxidation resistance [57–59]. Although these developments were successful from a technical viewpoint, they were never realized due to the high costs of these materials in mass applications. These materials were already optimized from the viewpoint of improved wear resistance and thermal conductivity by silicon treatments and combinations with copper impregnations. However, industrial application took an additional 20 years. The restrictions to use in luxury cars and requirements of improved safety and lifetime concepts enabled the first CMC brake disks to be used in Porsche cars at the beginning of 2000.

Today, CMC brake disks for different luxury cars are commercially available from S.G.L. Carbon Brakes for Porsche and Audi, and from Brembo for Ferrari and Mercedes.

Approximately 50 000 to 70 000 CMC brake disks were manufactured in 2006. The broad application on cars with lower costs is still limited due to the strong competition with brake disks made of steel. The manufacturing costs of steel disks are below 10/disk, whereas CMC disks with a weight up to 5 kg/disk are above 500/disk. Although CMC disks are superior in performance, lifetime, friction behavior, and safety aspects, their application appears to be feasible only in luxury cars within the next few years.

The manufacturing technologies are in excellent progress (Figure 4.23) [10]. The different manufacturing steps are already, or rather, can be automated to a certain

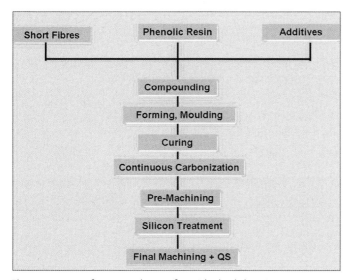

Figure 4.23 Manufacturing scheme of CMC brake disks [10].

degree. Compounding is executed batch-wise in a mixing device or continuously in an extruder. Forming and molding can be made semi-continuously with an automated molding device. The requested molding times depend on the process and can be reduced to times below 15 min/disk. Near-net-shape techniques minimize the machining costs, in particular for the final machining steps. Carbonization can be performed continuously, whereas continuous silicon treatments are still under development. However, the price target given by metallic disks can never be reached, due to the costs of raw material and the processing costs.

The complex shape caused by internal vented disks can be achieved by different manufacturing techniques (Figure 4.24) [10]. The complete machining of the cooling channels before or after the silicon treatment step is too expensive for a production process and was only used for the development phase so as to be as flexible as possible. Industrially usable techniques for making cooling channels are given in Figure 4.24.

The most flexible method is the joining of three parts with a cover, a bottom plate, and a middle sector, which contains all cooling channels. The middle part can be water-jet cut after the molding step. This technique allows the use of simple molding tools, and due to the reduced thickness of a single part with approximately 13 to 14 mm, extremely short molding cycles. The disadvantage is the waste of the water-jet cut pieces, resulting in higher material costs. The molding process, as well as the water-jet cutting step, can be fully automated. Water-jet cutting can be avoided by net-shape pressing of the middle section, which requires an additional pressing tool. In the case of minimized material waste, two pieces can be net-shape molded and then mechanically jointed. Chemical joining (welding) takes place during the silicon treatment step (Figure 4.23).

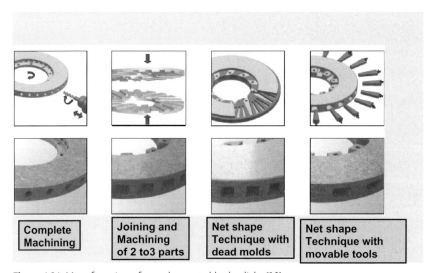

Figure 4.24 Manufacturing of complex vented brake disks [10].

Another possibility is net-shape molding of a single part by using dead moulds or movable tools (Figure 4.24). The disadvantages are that movable tools are more complex and have to be cleaned after each molding cycle, or in the case of dead moulds, be replaced. Furthermore, thicker green bodies require longer molding and curing cycles, a disadvantage from the viewpoint of cost-effectiveness.

As shown by Figure 4.24, the tribological surface can be adapted according to the requirements by using a surface coating, which is combined with increased costs. The excellent friction behavior of C/C-SiC, the silicon treated C/C, is used for emergency braking systems such as high-speed elevators, high-speed turning, milling machines, and amusement rides. However, the silicon treatment contributes not only to an increased friction coefficient. C/C-SiC grades can also be tailored for a superior sliding behavior due to the extreme low wear rates and the possibility to combine hard carbide with a soft graphitic matrix. A typical example for low friction and low wear applications are C/C-SiC sliding elements in the high-speed train, Transrapid, running in Shanghai (Figure 4.25) [60].

The requirements are wear rates below $0.1\,mm\,km^{-1}$ sliding length, with a friction factor below 0.3 against an iron track at a maximum speed of $500\,km\,h^{-1}$. The C/C-SiC sliding elements possess high mechanical strength, low thermal conductivity, and must be biocompatible and environmentally compatible due to the wear losses. The good sliding behavior in combination with good mechanical properties was realized by a newly developed silicon treatment (P75, see Figure 4.15), a combination of pack cementation and capillary impregnation. The excellent sliding

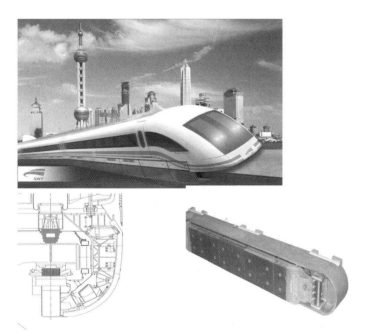

Figure 4.25 Sliding elements for the Transrapid [60].

Figure 4.26 Bearing and sliding ring made from C/C-SiC for petrochemical pumps.

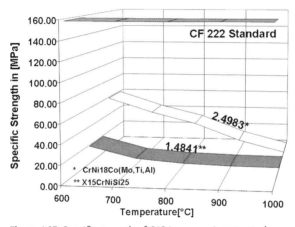

Figure 4.27 Specific strength of C/C in comparison to steel.

behavior can also be used in petrochemical pumps as bearings and sliding rings (Figure 4.26). The advantage of these parts for pumps is their superior failure behavior (pseudo-ductile) in comparison to monolithic ceramic elements. The industrial application of C/Cs is based on the specific properties of these lightweight materials. C/Cs possess the best specific mechanical high temperature properties of all materials (Figure 4.27).

The strength and stiffness of C/Cs increase with increasing temperature. The improvement in strength and stiffness from a. T to 2000 °C is about 25 %. Therefore, mechanical finite element analyses based on room temperature properties include an additional safety factor for high temperature applications. C/Cs are predestined for all high temperature applications under inert gas or vacuum atmospheres, such as furnace materials, hot pressing dies, handling tools for hollow glass manufacturing, semi-conductor industry, and photovoltaic applications.

4.3.2.1 C/Cs for High Temperature Furnaces

High performance vacuum furnaces are fully equipped with C/C materials. Thermal insulation based on hard felts, CVI, or resin impregnated soft felts, are used as standards in most furnaces (Figure 4.28). The hard felt boards are protected at the edges with U- and L-shaped profiles (carbon cloth reinforced) with a standard length of 1, 1.5, or 2 m. The profiles are used for protection against handling damages, against gas erosion effects during gas quenching, and for assembling the thermal insulation with C/C screws and nuts. Screws, bolts, and nuts are used for fixation and joining of mechanical loaded bars and pillars within the furnace. A broad database on the mechanical properties of C/C joining elements exists for tensile as well as for shear load introductions (Figure 4.29) [61].

The furnace chamber itself is often equipped with a C/C chamber (Figure 4.30). The C/C furnace chamber shown in the figure is $2000 \times 1000 \times 1000 \, mm^3$, with a total weight of 30 kg. The advantage of a furnace chamber is the improved homogeneity with respect to the temperature distribution. The maximum temperature deviation within a furnace chamber can be limited to about 10 °C over the full range of required temperatures. However, the used heater system has to be adapted to avoid heat sink effects. A further advantage of a C/C chamber compared to a graphite one is the required lower thickness, resulting in a reduced dead mass and better heating and cooling efficiencies. The disadvantage of the stiffness

Figure 4.28 Interior of an industrial furnace.

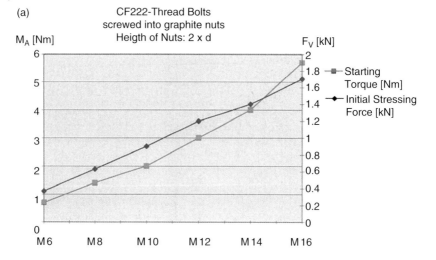
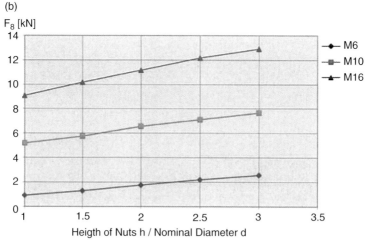

Figure 4.29 Mechanical properties of C/C screws [61]. (a) Starting Torque and initial stressing force of C/C bolts in dependence of the nominal diameter. (b) Breaking loads of C/C-bolts as a function of the ratio height of nut to nominal diameter.

reduction due to the reduced wall thickness can be overcome by stiffening the whole component with profiles (Figure 4.30).

Heating elements for vacuum and inert gas furnaces can be tailored from electrical properties, by using C/C materials as resistivity heaters. Their specific electrical resistivity is dominated by the resistivity of the fibers and their thermal post

Figure 4.30 C/C chamber for vacuum and inert gas furnaces for temperatures up to 2400 °C.

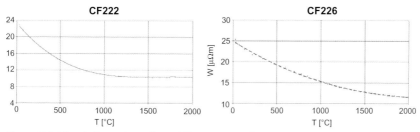

Figure 4.31 Electrical resistivity of two different C/C grades as a function of application temperature.

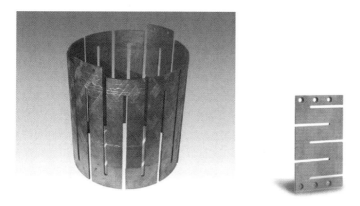

Figure 4.32 Meander-shaped tubular and flat heating elements.

treatments. Figure 4.31 shows the electrical resistivity of two different C/C grades, with a two-dimensional reinforcement depending on the application temperature. Furthermore, the resistivity can be tailored by the design of the heating elements (Figure 4.32). The total resistivity of the heating element can be calculated according to the following equation:

$$R_{total} = \frac{p \times L}{A}$$

where P is the specific electrical resistivity, L is the length of the heating element, and A is the cross-section. The length and cross-section can be optimized by the design of the heating elements within the given limitations of the geometry of the furnace facilities. The specific electrical resistivity depends on fiber orientation, fiber volume fraction, fiber type, and the final heat treatment of C/C. Thereby, a broad variety of total resistivity can be made according to the requirements of the customer. However, the higher the final application temperature and therefore what the final heat treatment has to be, the lower the specific resistivity. As mentioned above, an important target in high performance vacuum and inert gas furnaces is their homogeneity of the temperature distribution. Therefore, in inert gas furnaces, a forced ventilation is also required, which can be achieved by ventilators made of C/C applicable up to extreme temperatures above 1500 °C (Figure 4.33) [62].

The superior mechanical behavior, such as HT strength in fiber direction, is used in chains made of carbon for sinter furnaces (Figure 4.34). The whole chain is made of C/C. The maximum load can be overtaken by the UD reinforced chain links with a specific tensile strength beyond 800 MPa. The maximum load is limited by the C/C axes and their cross-sections. The chain links, the axes, and

Figure 4.33 C/C ventilators for forced inert gas conversion.

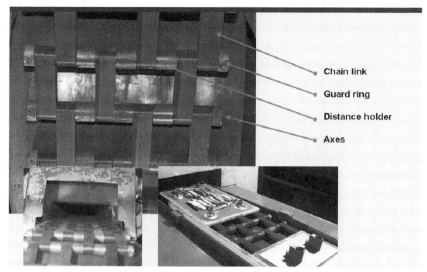

Figure 4.34 Chain made of C/C with a total length of 38 m applied in a sinter furnace.

the distance holders are filament winded and near-net-shape manufactured. The parts have only to be machined according to the required length or width, respectively. C/C chains can be applied for vertical as well as for horizontal transportation in a furnace. The main disadvantage is their sensitivity against wear loss and different gas atmospheres.

As already mentioned, C/C materials are attacked by oxygen at temperatures beyond 420 °C. In sinter furnaces, such oxidations are suppressed by using inert gases, a mixture of nitrogen (N_2) and hydrogen (H_2) as endo- or exo-gases. In the case of sintering, iron alloys, or thermal treatments of steel, a tailored carburization of the metal alloy is required, which often forms water as an intermediate product. Furthermore, open and closed furnaces sometimes contain water within the applied insulation materials, due to the uptake at ambient temperature, if the furnace is cooled down. This low amount of water cannot always be completely removed during the heating phase in the temperature range below 420 °C. Thereby, two typical chemical attacks can occur and affect the C/C materials inside the furnace:

(1) $C + H_2O \leftrightarrows CO + H_2$ $\Delta H = -131.38\,kJ/Md$ at T ≈ 800–1000°C

(2) $C + 2H_2 \leftrightarrows CH_4$ $\Delta H = -74.86\,kJ/Md$.

The most severe reaction is the formation of methane (reaction 2), because oxygen or water attacks can be controlled by *in-situ*-measurements of the water concentration in the furnace and an improved preheating to remove the remaining humidity from the thermal insulation materials. Table 4.6 gives an overview of the

Table 4.6 Overview of maximum application temperatures for thermal treatments of iron alloys in furnaces with C/C materials.

Atmosphere	Maximum application temp. (°C)	Applicability of C/C
N_2	1080 (1100)	+++
H_2	1200	+++/○
Exo-gas	1050	+++
Endo-gas	1050	+++
Vacuum	1050	○
O_2	<420	+++

+++, applicable without restraint; ○, applicable with limitations.

maximum application temperature of C/C materials for thermal treatments of iron alloys with respect to chemical resistance, including uncontrolled carburization effects of the metallic alloys.

In the case of hydrogen, the methanization reduces the lifetime of the C/C parts. This can be suppressed by using a gas mixture of hydrogen and nitrogen with low hydrogen content. An alternative method is the treatment of the C/Cs with Si, to form a C/C-SiC material, which reduces the attack of hydrogen by some orders of magnitude. The disadvantage is the formation of silicides with compounds within the metallic alloy, which has to be prevented. Therefore, the silicon treatment is limited to selected alloys and to a maximum temperature of approximately 1100 °C for thermal treatment of metals.

The limitations in vacuum are mainly caused by carburization effects of the metallic alloy due to the increased diffusion activity of the carbon under vacuum. This behavior can be improved and shifted to higher temperatures by using an improved crystallinity of the matrix, which can be achieved by CVI/CVD coatings with highly orientated PyC. Such coatings (impregnations) enable maximum application temperatures up to 1200 °C.

4.3.2.2 Application for Thermal Treatments of Metals

Load carrying C/C systems are industrially used as fixtures for thermal treatments of metallic parts such as vacuum brazing of steel or stainless steel components, for annealing of high-quality steels for gas turbine compounds, low pressure carburizing, gas quenching of automotive parts, and case hardening of steel parts with subsequent oil quenching (Figure 4.35). The industrial motivation to use C/C fixtures instead of metallic ones is based on the following advantages:

- no distortion of C/C-trays, resulting in handling by roboting (increased efficiency, reduced manpower costs);
- no brittleness, even after some years of application;
- increased lifetime of the fixtures;

Introduction of water-jet cut C/C plates leads to:

- doubling capacity
- increased heat efficiency
- shorter cycle times
- longer lifetime

Figure 4.35 C/C–fixture with waterjet cut C/C plates for case hardening of rods with subsequent oil quenching.

| Conventional cast metal tray Distortion due to thermal cycling | Carbon / Carbon tray design in service for 8 years without any distortion |

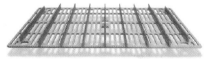

gross weight: app. 30 kg / distortion after 9 month in service

Dimensions: 900 x 600 x 4 / gross weight: 2 kg / load: 20 kg

Figure 4.36 Comparison of C/C and cast metal trays.

- improved energy efficiency by reduced weight compared to steel fixtures;
- increased number of trays by reduced thickness of the C/C fixtures;
- increased number of rods by optimized design.

All these advantages contribute to an improved efficiency and reduced costs in thermal treatments of metallic components. The return of investment for the more costly C/C fixtures in comparison to metallic ones is below 2 years. Therefore, this is one of the fast growing markets for industrial C/C-applications.

A comparison of a cast metal tray and a C/C tray is given in Figure 4.36. As can be seen, even after 8 years in service, no distortion can be detected in the C/C tray. The weight saving is 28 kg per tray (Figure 4.36) and 336 kg per batch (Figure 4.35).

This illustrates the advantage of reduced dead mass per batch and the enormous potential of energy saving over a service period of 8 years, with up to 6 to 7 batches per day. With increasing energy costs, the use of C/C fixtures will be even more attractive and superior from the viewpoint of lifecycle and manufacturing costs.

The trays shown in Figure 4.36 are used for the same metallic parts as for case hardening. The reduced height of the C/C trays enabled additional stacks in comparison to the metallic ones (Figure 4.35) Furthermore, the number of connecting rods per tier could be increased by the thinner distance holder made from C/C. The overall efficiency improvement resulted in doubling of connection rods per batch. This demonstrates the economical and ecological superiority of C/C fixtures compared to metallic ones. The C/C fixtures can be tailored to the required load capacity (Figure 4.37). The C/C trays are designed according to customer requirements and can be used in their existing facilities and furnaces.

A new development for vacuum brazing of turbine components (Figure 4.38) is the Unigrid system of Schunk. This lightweight grid (~2 kg) is based on a textile preform made via tailored fiber placement (TFP), without any joinings manufactured near-net shape. The manufacturing process can be performed via RTM or RI-methods (Resin Infusion). The direct contact of carbon and gas turbines has to be avoided to prevent uncontrolled carburization. Therefore, CarboGard protective tiles based on alumina are available for the UniGrid system. The high flexibility of the UniGrids is shown in Figure 4.39.

The load capacity can be brought into line with these requirements by using graphite supports connecting the trays. The graphite supports can be fixed wherever necessary (i.e. in the middle, at the corners). The load capacity increases with reduction of the free length between two supporting points. The system can be assembled by the final end user and therefore be modified for a variety of different applications. The use of C/C fixtures for thermal treatments of metals will be

Figure 4.37 C/C fixtures for heavy loads, for example, brazing of heat exchangers for cars.

Vacuum brazing of turbine components

Figure 4.38 Schunk UniGrid-system with CarboGard protective tiles.

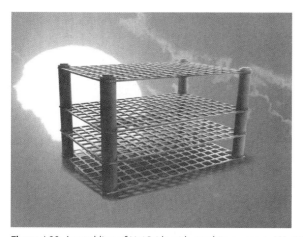

Figure 4.39 Assembling of UniGrids with graphite supports [63, 64].

widespread in the future, due to the high cost efficiency for automotive suppliers. The second fast growing market is the field of photovoltaic industry.

4.3.2.3 Application of C/C in the Solar Energy Market

The growth of the global solar energy market is shown in Figure 4.40 [60, 65]. This market is mainly politically driven, with a worldwide lead by Germany. C/C structural components are necessary for the high temperature processes in manufacturing photovoltaic cells such as re-crystallization and CVD-coatings, and for making polysilicon as one of the most important raw materials. The shortage of polysilicon on the market resulted in tremendous investments worldwide of some billions of US dollars. The planned increase in the worldwide capacity for polysilicon is given in Table 4.7. The announced capacity covers up to 2012, the demand of

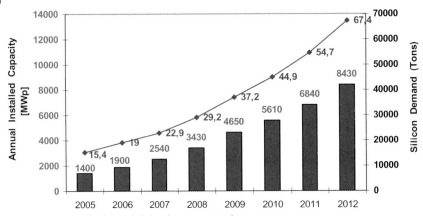

Figure 4.40 Growth of the Global Solar Energy Market (Annual Installed Capacity). Source: EPIA-Report 2006.

Table 4.7 Planned capacity of polysilicon.

Capacity	2007	2010
Hemlock	7700	36 000
Wacker	6500	21 500
REC	5300	12 500
Tokuyama	5200	5 400
MEMC	3800	8 000
Mitsubishi	2850	3 500
Sumitomo	800	1 600
Total China	130	6 000
Elkem	–	10 000
Others	–	4 500

Source: Flynn & Bradford Prometheus Institute "Publications from semiconductor companies."Announced capacity in 2010: 109 000 tons. Demand: PV, 45 000 tons.

semiconductor industry as well as photovoltaic applications. The need for C/Cs is planned to increase from 2007 to 2012, by up to more than 600%. Therefore, most of the investments of European C/C manufacturers are concentrated on these applications. Typical requirements are:

- high purity levels
- CVI/CVD coated C/Cs, to avoid carbon particle release
- coatings for improved corrosion resistance against SiO-attacks
- ductile
- cost efficient
- low methanization rates.

These requirements can only be fulfilled by C/Cs with inner and outer protection systems. Typical structural components are heaters, thermal insulations, heat

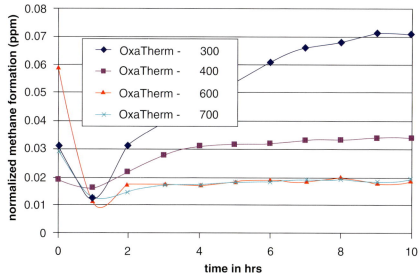

Figure 4.41 Methanization rate of thermal felt insulations normalized in comparison to standard C/C hard felt boards.

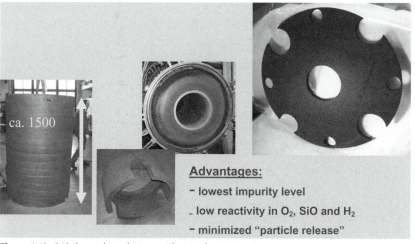

Figure 4.42 C/C thermal insulations with mixed matrix systems.

shields, crucibles, handling tools, and carrier systems for solar cells. Most of the C/Cs for these applications are based on mixed matrix systems, including CVI and resin densifications [60]. Hard felt boards as thermal insulations possess have a limited lifetime of some months in hydrogen or SiO environments, whereas thermal insulations with mixed matrix systems survive more than two years under the same conditions. The methanization, as well as the oxidation rate of such high performance C/C insulations, is superior to standard grades (Table 4.7, Figure 4.41). The methanization rate can be reduced to 2% in comparison to standard hard felt boards. Some typical thermal insulations are shown in Figure 4.42.

Crucibles for pulling of single crystals are made of C/Cs with mixed matrix systems as well as heating elements, to reduce their oxidation by SiO attacks (Figure 4.43). Comparable chemical requirements are valid for etching processes of wafers by using felt susceptors (Figure 4.44) [66]. Felt susceptors are necessary for oxidative etching as well as for doping processes of wafers. Therefore, the susceptor must be highly purified, oxidation resistant, and should not be contaminated by the doping species. The gas flow characteristics have to be homogeneous in order to guarantee excellent homogeneity of the treated wafer surface.

For future applications in fixation and fusion reactors, C/Cs are required with a high thermal conductivity combined with a low erosion and sublimation rate. Therefore, new development activities are undertaken to dope C/C with carbides and to maximize their thermal conductivity. Although the potential for such modified C/Cs is high, an industrial application will not be feasible within the next 5 to 10 years.

Figure 4.43 Crucibles made of C/C with mixed matrix systems for pulling of single crystals.

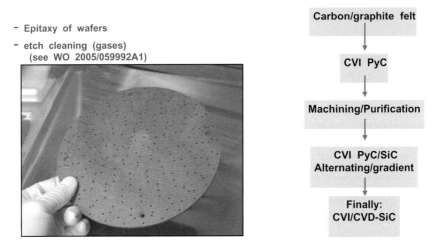

Figure 4.44 C/C felt susceptor with mixed matrix systems.

References

1. Thomas, C.R. and Walker, E.J. (1978) Proceedings of the 5th London International Carbon and Graphite Conference, Vol. 1, p. 520.
2. Thomas, C.R. (1993) *Essentials of C/C Composites*, Elsevier Science Ltd., pp. 1–35.
3. Fitzer, E. and Burger, A. (1971) International Conference on Carbon Fibres, their Composites and Applications, London, paper no. 36.
4. McAllister, L.E. and Taverna, A.R. (1976) Proceedings of the International Conference on Composite Materials, Vol. 1, Met. Soc. of AIME, New York, p. 307.
5. Lamieq, P.J. (1977) Proceedings of the AIAA/SAE 13th Propulsion Conference, paper no. 77-882, Orlando.
6. Fitzer, E., Geigl, K.H. and Hüttner, W. (1978) Proceedings of the 5th London International Carbon and Graphite Conference, Vol. 1, p. 493.
7. Girard, H. (1978) Proceedings of the 5th London International Carbon and Graphite Conference, Vol. 1, Soc. Chem. Ind., London, p. 483.
8. Thomas, C.R. and Walker, E.J. (1978) Proceedings of the 5th London International Carbon and Graphite Conference, Vol. 1, Soc. Chem. Ind., London, p. 520.
9. Weiss, R. (2001) Carbon fibre reinforced CMCs: manufacture, properties, oxidation protections, in *High Temperature Ceramic Matrix Composites (HTCMC)* (eds W. Krenkel, R. Naslain and H. Schneider), DGM-Wiley-VCH, pp. 440–56.
10. Weiss, R. (2006) *CMC and C/C-SiC-Fabrication*, CIMTEC, Sicily.
11. Schmidt, D.L. (1996) Wright Laboratory WL-TR-96-4107.
12. Peters, P.W.M., Lüdenbach, G., Pleger, R. and Weiss, R. (1994) Influence of matrix and interface on the mechanical properties of unidirectional carbon/carbon composites, *Journal of the European Ceramic Society*, **13**, 561–9.
13. Takano, S., Kinjo, T., Uruno, T., Tlomak, T. and Ju, C.P. (1991) Investigation of process-structure-performance relationship of unidirectional reinforced carbon-carbon composites. *Ceramic Engineering and Science Proceedings*, **12** (9–10), 1914–30.
14. Zaldivar, R.J., Rellick, G.S. and Yang, J.M. (1991) Studies of fiber strength utilization in C/C composites. In Extended Abstracts. 22nd Carbon Conference, pp. 400–1.
15. Peters, P.W.M., Schmauch, J., Weiß, R. and Baser, M.G. (1992) *The strength of carbon fibre bundles, loose and embedded in a polymer and carbon matrix*. In Proceedings of ECCM V, Bordeaux, 7–10 April 1992, pp. 157–63.
16. Savage, G. (1993) *C/C Composites*, Chapmann & Hall.
17. Hüttner, W. (1980) Parameterstudie zur Herstellung von kohlenstofffaserverstärkten Kohlenstoffverbundkörpern nach dem Flüssigimprägnierverfahren, Doctorate thesis, University of Karlsruhe.
18. DE 2714364 (1977) Verfahren zur Herstellung von Kohlenstofffaserverstärkten Kohlenstoffkörpern.
19. Preformed yarn, useful in forming composites articles and process of producing same, JP 1986-182483.
20. Process for preparing a flexible composite material, JP 1985-276440.
21. Brochures of Accross (2007) http://www.acrosscfc.com/aboutcfc.htlm (accessed 2007).
22. Diefendorf, R.J. and Sohda, Y. (1983) Extended Abstracts Program 17th Biennial Conference on Carbon, **17**, 31.
23. Kotlensky, W.V. (1973) *Chem. Phys. Carbon*, **9**, 173.
24. Process and apparatus for thermolytically dissociating water, CEA, US Patent No. 4696809.
25. Rovillain, D., Trinquecoste, M., Bruneton, E., Derré, A., David, P. and Delhaes, P. (2001) Film boining chemical vapor infiltration – An experimental study on C/C composite materials. *Carbon*, **39**, 1355–65.
26. Davin, P.C., Blein, J., Gachet, C. and Robin-Brosse, C. (2007) Carbon and Ceramic Matrix Composites Materials by

Film Boiling Process, 16th Intern. Conf. On Composite Materials, **16**, 736–7.

27 Becker, A. and Hüttinger, K.J. (1996) Chemistry and kinetics of CVD of PyC II, III. *Carbon*, **36**, 177–211.

28 Benzinger, W. and Hüttinger, K.J. (1999) Chemistry and kinetics of chemical vapor infiltration of pyrocarbon – Infiltration of carbon fiber felt, *Carbon*, **37** (6), 941–6.

29 Doug, G.L. and Hüttinger, K.J. (2002) Consideration of different mechanisms leading to pyrolytic carbon of different textures. *Carbon*, **40**, 2515–28.

30 Hu, Z.J. and Hüttinger, K.J. (2002) Mechanisms of carbon deposition – a kinetic approach, Carbon, **40** (4), 624–8.

31 Diefendorf, J. (2006) *C/C composites produced by chemical vapor deposition*, Cocoa Beach Conference.

32 Jenkins, G.M. and Kawamura, K. (1976) *Polymeric Carbons, Carbon Fibre, Glass and Char*, Cambridge University Press.

33 Pleger, R. Vom monolithischen Kohlenstoff zum Verbundwerkstoff: Analyse der Strukturvariation und Gefügeentwicklung mittels hochauflösender Durchstrahlungselektronenmikroskopie, Fortschrittsberichte VDI: Reihe 5, Grund- und Werkstoffe, Kunststoffe.

34 Wanner, A. (1991) Structure and properties of CMCs. Doctorate Thesis, University of Stuttgart.

35 Weiss, R. (1990) *Plenary lecture*. Tsukuba Carbon Conference, Japan.

36 Peters, P.W.M., Lüdenbach, G., Pleger, R. and Weiss, R. (1994) Influence of Matrix and Interface on the Mechanical Properties of Unidirectional Carbon/Carbon Composites, *Journal of the European Ceramic Society*, **13**, 561–9.

37 Adams, D.F. (1974) Elstoplastic Crack Propagation in a Transversely Loaded Unidirectional Composite, *Journal of Composite Materials*, **8**, 38–54.

38 Zhao, J.X., Bradt, R.C. and Walker, P.L., Jr, (1981) Ext. Ast. 15th Biennial Conf. on Carbon, Am. Chem. Soc, Washington, DC, p. 274.

39 Eckert, K. (1992) Einfluss unterschiedlicher Endglühbehandlungstemperaturen auf die Duktilität von HT-, IM- und HM-faserverstärkten C/C-Werkstoffen. Diploma Thesis.

40 Johnson, H.V. (1934) US Patent 1.948.382.

41 Yasuda, E., Kimura, S. and Shilsusa, Y. (1980) Oxidation behaviour of Carbon Fiber/Glassy Carbon Composites, *Transactions of the Japanese Society of Composites Materials*, **6** (1), 14–23.

42 Walker, P.L., Rusinko, F. and Austin, L.G. (1959) *Advances in Catalysis*, **11**, 164.

43 Chown, J., Dencon, R.F., Singer, N. and White, A.E.S. (1963) Refractory coatings on graphite, in *Special Ceramics* (ed. P. Popper), Academic Press, p. 81.

44 Huettner, W., Weiss, R., Dietrich, G. and Meistring, R. (1990) *Space applications of advanced structural materials*. Proceedings of the ESA Symposium, ESTEC, Noordwijk (NL), 21–23 March 1990, pp. 91–5.

45 Weiss, R. (1994) Oxidation behaviour of carbon/carbon composites, *High temperature processes*, **3**, 351–6.

46 Huettner, W., Weiss, R. and Scheibel, T. (1986) *Oxidation resistance of C/C-C/C at 1530 K, Abstract 19*. Biennal Conference on Carbon 25–30 June1986, Penastate University, USA.

47 Cavalier, J.C., Berdoyes, I. and Bouillon, E. (2006) Composites in Aerospace Industry. *Advances in Science and Technology*, **50**, 153–62.

48 Naslain, R.R. (2006) Processing of non-oxide ceramic composites. *Advances in Science and Technology*, **50**, 64–74.

49 Naslain, R.R., Pailler, R., Bourrat, X., Bertrand, S., Heurtevent, F., Dupel, P. and Lamouroux, F. (2001) Synthesis of highly tailored ceramic matrix composites by pressure-pulsed CVI, *Solid State Ionics*, **141/142**, 541–8.

50 Eliezer, R. (2003) Coating of graphite and C/C composites via reaction with Cr-powder. MSc Thesis, Technion, Haifa.

51 Weiss, R. and Lauer, A. (2004) Final Report, on the Nationally Funded Project: Gradierte CVD- und PIRAC-Multibeschichtungen auf C/C als Korrosions – und Oxidationsschutz durch innovative Hochtemperaturprozesse, FKZ 03N5039B.

52 Weiss, R. (2001) Final Report: RESTAND, EC Contract No: SMT4-CT97-2200.
53 Weiss, R. (2007) CMC and C/C-SiC fabrication, paper presented on Internet. Workshop on Nanostructured HT/High Performance Materials, 10 September 2007, Sardar Patel Univ., Vidyanagar, Indien.
54 Mao, C. (2004) Hithex Report 36.
55 Sucor, E.W. (1963) *JACS*, **46**, 14.
56 Grigorev, A.L. and Polishchuk, D.J. (1973) *Fiz. Aerodisp. Sist.*, **8**, 87.
57 Kehr, D. (1983) Entwicklung von asbestfreien Reibbelägen auf Kohlenstoffbasis in Form von C/C und Kohlenstoff-Metallcarbidverbundwerkstoffen. BMBF-Projekt 01VM039, from 08/1979 to 07/1983; BMBF.No.FB-T85-014.
58 Fitzer, E., Fritz, W., Gkogkidis, A. and Moergenthaler, K.D. (1986) Proceedings, Carbon 86 Conference, Baden-Baden.
59 Brite-Euram Project (1990) MA1E-0068, Entwicklung von CFC-Werkstoffen mitpartiell reduziertem Gehalt an C-Fasern für Anwendungen im PKW-Bremssystem.
60 Weiss, R. (2007) HTCMC6. 4–7 September 2007, New Delhi,
61 Weiss, R. (1992) Development of oxidation resistant C/C, Final Report, BMBF Reference No. 03M1019 BO.
62 Weiss, R. (2007) DGM-Fortbildungsseminar, Universität Bayreuth,
63 Verfahren zur Herstellung eines Faserverbund-Bauteils sowie Vorrichtung zur Herstellung eines solchen, EP 1106334 (2001).
64 US-Patent 7,175,787 (2007).
65 European Photovoltaic Industry Association (EPIA) (2006) EPIA-Report.
66 WO 2005/059992A1 (2005) Felt susceptors.

5
Melt Infiltration Process
Bernhard Heidenreich

5.1
Introduction

For industrial applications, CMC materials based on SiC or SiSiC matrices and C- or SiC fibers are most common. Four different process routes are used for the manufacture of these so-called C/SiC, C/C-SiC, or SiC/SiC materials: Chemical Vapor Infiltration (CVI), Polymer Infiltration and Pyrolysis (PIP) also called Liqind Polymer Infiltration (LPI), sintering by hot isostatic pressing, which are described in detail by other authors in this book, and Melt Infiltration (MI), also called Liquid Silicon Infiltration (LSI).

CVI- and LPI-CMC typically offer high mechanical strength and strain capability. However, the main drawbacks of these manufacturing methods are long process times, due to the low growth rates of the SiC, especially for CVI, and multiple densification cycles needed for LPI. Additionally, due to high manufacturing costs, CVI and LPI derived CMC are generally limited to aerospace applications.

To overcome these disadvantages, new processes have been developed since the mid-1980s, based on the infiltration of molten metals in porous, fiber reinforced preforms, offering a fast build-up of the ceramic matrix. The MI process of SiC matrix based CMC is derived from the industrial manufacturing technology of SiSiC materials, developed in the 1960s. Thereby, porous preforms of SiC particulates, held together by organic binders, are infiltrated with molten Si, resulting in dense SiSiC composites. Using C-fiber fabrics or C-felts, even thin-walled SiSiC structures could be achieved by fully converting the carbon fibers to SiC [1]. However, it has also been observed that high modulus C-fibers could withstand the siliconization process, leading to non-brittle fracture behavior [2, 3].

To capitalize on the high tensile strength and fracture elongations of the C- and SiC fibers, which are significantly higher than those of SiC bulk ceramics, further development focused on the protection of the fibers from attack by the molten silicon as well as on weak embedding of the fibers into the brittle SiC matrix. Therefore, tailored C/C preforms have been developed by several companies such as Sigri Great Lakes Group (SGL Group), Schunk Kohlenstofftechnik (SKT), and

---
Ceramic Matrix Composites. Edited by Walter Krenkel
Copyright © 2008 WILEY-VCH Verlag GmbH & Co. KGaA, Weinheim
ISBN: 978-3-527-31361-7

Dornier, now EADS Astrium, in the 1980s and 1990s. At DLR, C/C-SiC materials, and structural parts based on the low-cost LSI process, have been in the focus since 1988. Originally developed for thermal protection systems (TPS) of reusable spacecraft [4], new application fields beyond aerospace could be opened for these CMC materials.

The development of MI-infiltrated SiC/SiC started more than 20 years ago [5] and was driven by high temperatures as well as long-term applications in stationary gas turbines and aero-engines. In 1993, General Electric (GE) presented the so-called Toughened Silcomp material, based on coated SiC fibers [6]. In several US development programs, for example, EPM (Enabling Propulsion Materials; 1994–1999), UEET (Ultra Efficient Engine Technology Programme; 1999–2005), NGLT (Next Generation Launch Technology), MI SiC/SiC materials were further developed [7] and first exemplary components, including combustion chamber liners, turbine shrouds, and even turbine blades, were field tested under real conditions, resulting in various commercially available MI SiC/SiC materials.

This chapter focuses on C- and SiC fiber reinforced MI SiSiC materials offering high strength and fracture toughness. Relatively brittle C/SiC and C/C-SiC materials, characterized by the transformation of most of the C-fibers to SiC, for example, CESIC materials (ECM) as well as MI SiSiC materials based on wooden preforms, so-called biomorphic SiSiC, are not described. Due to their early stage of development and limited data available, CMC materials based on ZrC and CuTiC matrices, presented by Ultramet and DLR, respectively, as well as C/SiC materials recently introduced by M Cubed, are not considered here. Although the direct metal oxidation (DIMOX) process [8] is based on the MI of metals into a porous fiber preform, it is not described in this chapter because, to the knowledge of the author, these materials are not used in commercial applications up to the present day.

5.2
Processing

The manufacture of CMC materials and structural parts via MI generally can be subdivided into four main process steps:

1. Coating of fiber tows to obtain fiber protection and weak fiber matrix interface;
2. Manufacture of a near-net-shape fiber reinforced green body;
3. Build-up of a porous, fiber reinforced preform with preliminary C and/or SiC;
4. Build-up of SiSiC matrix by infiltration of molten Si.

Various MI processes have been developed for the manufacture of so-called C/SiC, C/C-SiC, and SiC/SiC materials, which mainly differ in the way the fibers are protected and embedded in the SiC matrix, as well as in the way the green body and the porous, fiber reinforced preform are realized (Figure 5.1).

For MI CMC materials based on SiC matrices, generally all types of C fibers can be used, leading to material variations with significantly different properties. Due

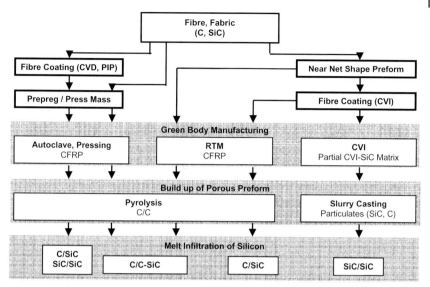

Figure 5.1 Schematic overview of different methods for producing MI-CMC.

to the high temperatures well above 1425 °C during MI processing, as well as the high service temperatures, the range of SiC fibers is limited to high temperature stable variants, that is, fiber qualities with low oxygen and free carbon content, for example, Hi-Nicalon, Hi-Nicalon Type S (Nippon Carbon. Co. Ltd), as well as on nearly stoichiometric SiC fibers such as Tyranno SA (UBE Industries Ltd.) and Sylramic (ATK COI Ceramics, Inc.). For building up the SiC matrix, preliminary C matrices based on polymer precursors, for example, phenolic resins, C and SiC particulates or CVI-derived C as well as Si granulate, for example, highly pure CVD-Si or low-cost Si qualities based on mineralogical manufacturing processes, are used.

5.2.1
Build-up of Fiber Protection and Fiber/Matrix Interface

Due to the high reactivity of molten Si, direct contact to the C or SiC fibers usually has to be avoided. Additionally, a weak embedding of the brittle fibers in the brittle matrix is mandatory to obtain characteristic CMC properties, such as high strength, fracture toughness, and thermal shock resistance. To ensure both, the fiber protection and the weak fiber matrix interface, three different methods are used for industrial production (Figure 5.2):

1. Polymer impregnation of fibers and subsequent curing/pyrolysis (PIP);
2. CVI of fibers and fiber preforms;
3. *In-situ* fiber coating.

Fiber protection via PIP is widely used for fabric as well as short fiber reinforced materials, for example, leading to so-called Sigrasic materials from SGL [9]. For

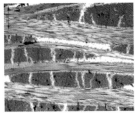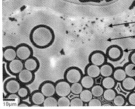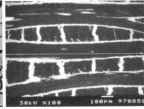

Figure 5.2 Microstructure of MI CMC based on different methods for fiber protection and weak fiber/matrix embedding. Left: C/SiC (SGL) based on PIP fiber coating showing dense C/C bundles (dark gray) embedded in SiSiC matrix (gray: SiC, light gray: Si), Middle: SiC/SiC (GE) based on CVI fiber coating (black) with single fibers embedded in SiC matrix. Right: C/C-SiC based on *in-situ* fiber protection, characterized by dense C/C bundles (black), embedded in SiC matrix (gray).

the manufacture of short fiber reinforced C/SiC brake disks, endless fiber bundles or tows are impregnated with phenolic resin, which is then cured and pyrolysed, embedding the fiber filaments in a dense carbon matrix ($\phi_{C,matrix} \approx 50$ mass-%). Subsequently, the coated fibers are cut to different lengths, resulting in a C/C-like raw material.

Using CVI for fiber coating, a thin interphase layer is deposited on each fiber filament, resulting in C/SiC and SiC/SiC materials with the filaments mainly embedded individually in the SiC matrix. Whereas fibers and fabrics can be coated continuously, dry fiber preforms based on fabric stacks or three-dimensional preforms are coated in batch processes. For C-fibers, usually C coatings with a layer thickness of ~0.1 μm are applied. SiC fibers are typically coated with C or BN (~0.1–1 μm). Due to their significant higher oxidation stability, Si-doped BN coatings on Hi Nicalon fibers, as well as on *in-situ* coated Sylramic-iBN fibers [7], are preferred for long-term use at high temperatures in an oxidizing atmosphere. To avoid degradation of BN fiber coatings, caused by contact with molten Si, an additional overcoating of SiC or Si_3N_4 is applied. Typical thicknesses for SiC overcoatings are 0.5 to 5 μm [10]. CVI fiber coating generally is used for SiC/SiC materials by GE (Hypercomp) and NASA, as well as for C/SiC and C/C-SiC materials developed by EADS Astrium (Sictex, Sictex Si) [11].

Time consuming and costly fiber coatings are not necessary if particularly suitable precursors, which offer a strong fiber matrix bonding, are used for the manufacture of the CFRP perform [12], leading to a segmentation of each fiber bundle into dense C/C bundles during pyrolysis. During the subsequent MI, only the fibers on the outer surface of the C/C bundles are contacted to Si and transformed to SiC, whereas the fibers inside the C/C bundle are well protected. This cost-efficient method is the basis of the LSI process, which has been developed at DLR and which has already been transferred to FCT Ingenieurkeramik GmbH for the serial production of friction materials. Similar processes are used by SKT and Brembo Ceramic Brake Systems SpA.

5.2.2
Manufacture of Fiber Reinforced Green Bodies

Two different methods are used mainly for the manufacture of C- and SiC-fiber reinforced green bodies in near-net-shape geometry. In the first method, polymer matrix-based green bodies, such as carbon fiber reinforced plastics (CFRP), are manufactured, whereas in the second method, CVI is used, mainly for the build-up of green bodies of SiC fiber reinforced materials.

For the manufacture of polymer-based preforms, well-known techniques, such as Resin Transfer Moulding (RTM), autoclave techniques, and warm pressing are used. In the following, these processes are described in detail for the manufacture of C/C-SiC materials at DLR.

The RTM process is based on dry fiber preforms built up by stacking of cut fabric sheets and one-dimensional plies or even on three-dimensional preforms. The preforms are put into rigid, closed moulds and infiltrated with a polymer precursor ($p_{max.} \approx 0.1\,MPa$). After curing the polymer ($T_{max.} = 210\,°C$, $p_{max.} = 2.2\,MPa$) the CFRP green body, characterized by high fiber content of about 60 vol.-% and low porosity (e' <2–5%), can be taken out of the mould.

Autoclave and warm pressing techniques are used for the manufacture of CFRP as well as for SiC fiber reinforced green bodies. Thereby, pre-impregnated fiber plies are manufactured by wet drum filament winding (one-dimensional prepreg) or by resin impregnation of fabrics. The resulting prepregs are cut into sheets and laminated into open moulds or on cores, and are subsequently compressed and cured in an autoclave ($p_{max.} = 0.8\,MPa$, $T_{max.} \approx 180–250\,°C$) or by warm pressing. For short fiber reinforced C/SiC or C/C-SiC materials, carbon fibers are mixed together with liquid or powdery polymer precursors. Subsequently, the press masses are put in a mould, compressed ($p \leq 5\,MPa$), and cured ($T_{max.} = 250\,°C$), leading to near-net-shape CFRP green bodies with a fiber content of about 50 vol.-%. Typical process times are 2 hours for warm pressing, 6 hours for autoclave, and up to 30 hours for RTM.

Using CVI for the manufacture of SiC fiber reinforced green bodies, dry fiber preforms are heated up ($T_{max.} \leq 1100\,°C$) and infiltrated by gaseous precursors, usually a mixture of methyltrichlorosilane (process gas) and hydrogen (catalyst). Thereby a small amount of SiC matrix, only as much as necessary to stabilize the fiber preform, is deposited on the previously coated fibers, leading to rigid green bodies with an open porosity of about 34 vol.-%.

The composition and properties of the final CMC material are mainly determined by the green body. To meet the specific requirements of different applications, tailored green bodies have been developed, leading to a wide range of different MI-CMC materials.

For components used at temperatures below about 1200 °C, for example, brake disks and pads, Si-rich materials can be favorable, due to the high thermal conductivity of silicon. Therefore, green bodies with low fiber contents, leading to high open porosities and broad micro-channels, are used.

For high temperature applications and long-term use, where the melting temperature of silicon is reached or even exceeded, the presence of free Si has to be reduced to a minimum. This can be obtained, for example, by high fiber contents and polymer precursors with high carbon yield, leading to a low open porosity and narrow micro-channels after pyrolysis, resulting in Si contents below 5 vol.-% for C-fiber reinforced SiC materials. Starting from SiC fiber preforms with lower fiber contents and therefore higher porosity, low Si contents can be obtained by using precursors with SiC and C-particulates added, as well as by two phase precursors, that is, different precursors with high and low C and/or SiC yield, or pore forming agents, building up a favorable microstructure after pyrolysis, ideal for the forming of a stoichiometric SiC matrix [10, 13, 14].

5.2.3
Build-up of a Porous, Fiber Reinforced Preform

In the third process step, a porous preform is generated, providing a C source for the subsequent build-up of the SiSiC matrix by chemical reaction (Reactive MI) or by bonding of SiC particulates (Non-Reactive MI) with the infiltrated molten Si. For reactive MI, high emphasis is put on tailoring the porous preform to ensure an entire and homogeneous infiltration of the part as well as the build-up of a SiC matrix to be as stoichiometric as possible. To obtain this, the polymer-based green bodies are pyrolysed in an inert gas atmosphere at $T > 900\,°C$ and the polymer matrix is converted to an amorphous C matrix, leading to porous C/C, or SiC/SiC preforms, whereas the CVI derived green bodies are infiltrated with a C and/or SiC-containing slurry.

In the case of fabric-based CFRP green bodies, used for the LSI process at DLR, the volume contraction of the polymer matrix ($\approx$50–60 vol.-%) is not corresponding to the low volume change of the green body, which is limited by the fiber reinforcement. For example, two-dimensional fabric stacks lead to a contraction of the wall thickness of about 8 to 10% and of only 0.5% in the in-plane directions. Due to the resulting tension stresses in the matrix, micro-cracks are formed, and a characteristic, translaminar channel system of interconnected open porosity (e' ~20–30%) [15] is obtained in the C/C preform, which can be infiltrated easily by molten metals. Thereby, the original CFRP fiber bundles are segmented into several C/C bundles, consisting of carbon fiber filaments, embedded in a dense carbon matrix (Figure 5.3). Overall process duration of the pyrolysis varies between 190 hours for large and thick walled preforms based on laminated fabrics, down to less than 40 hours for small-sized, cut fiber reinforced parts with random fiber orientation.

SiC fiber reinforced green bodies derived by CVI typically consist of SiC fibers, coated with a C or BN interphase and a SiC overcoating. To provide a carbon source, necessary for the build-up of the final SiC matrix via MI, the residual porosity of about 34 vol.-% is filled by slurry casting. The slurry is based on water and SiC particulates for Non-Reactive MI or a combination of SiC and C-particulates, for Reactive MI and is infiltrated by applying gas pressures of about

Figure 5.3 SEM figures of the cross-section before (left) and after (middle) pyrolysis of the CFRP preform, showing dense C/C bundles (right).

Table 5.1 Physical properties of molten silicon.

Physical Properties	Unit	Value
Density ρ	g cm^{-3}	2.33–2.34 (20 °C)
Melting temperature	°C	1414–1420
Surface stress σ	N m^{-1}	0.72–0.75 (1550 °C in Vacuum)
Wetting angle ϑ	°	30–41 (compared to SiC in Vacuum)
		0–22 (compared to C in Vacuum)
Dynamic viscosity η	10^{-4} Pas	5.10–7.65 (1440 °C)
		4.59–6.38 (1560 °C)

6 MPa [13]. Using a defined ratio of SiC and C particulates with well adapted grain sizes, SiC/SiC materials with low contents of free Si ($\phi_{Si} \geq 5$ vol.-%) and low open porosities (e′ < 6%) can be achieved after drying the preform and subsequent MI [16]. For high strength SiC/SiC, it is desirable to minimize the carbon content in the slurry to reduce exothermic reactions that can weaken the SiC fibers due to the resulting high temperatures.

5.2.4
Si Infiltration and Build-up of SiC Matrix

In the last process step, the porous, fiber reinforced preform is infiltrated with molten Si, at temperatures well above the melting point of Si (1420 °C) in vacuum. Thereby, the Si is sucked into the preforms porosity by capillary forces only. Simultaneous to the infiltration, the Si reacts with the C, supplied by the preform, forming the SiC matrix. In the preforms containing primarily SiC particulate the Si act as binder for the particulates.

Due to the specific properties of molten silicon (Table 5.1), such as low viscosity, high surface tension, and good wettability [17], fast filling of the preform porosity is obtained. The infiltration behavior can be simulated by a calculation model based on the capillary theory of Gibb and d'Arcy's law, describing the motion of fluid in the capillary, taking into account the influence of the capillary and reaction forces, initial porosity, and capillary diameter, as well as the time dependent capillary diameter caused by the build-up of SiC matrix in the capillaries. By numerical

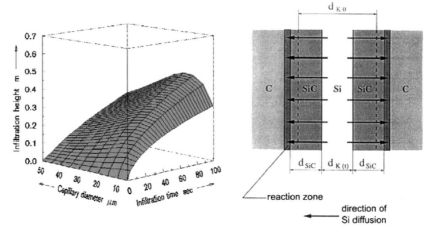

Figure 5.4 Infitration behavior of molten Si into a single capillary of the porous C/C preform [18] (left) and schematic view of the build-up of SiC matrix in a capillary (right). d_{K0} = capillary diameter before the infiltration of Si; $d_{K(t)}$ = capillary diameter in dependence of time t, d_{SiC} = thickness of SiC-layer.

solution of the resulting Navier Stokes equation [18], the possible infiltration height and time for common capillary diameters can be calculated (Figure 5.4). As a result, the time needed for the infiltration generally is very short and even large parts can be infiltrated completely within a few minutes.

For Reactive MI, parallel to the infiltration, the Si reacts with the carbon, forming the SiC matrix:

$$Si_{liquid} + C_{solid} \rightarrow SiC_{solid} [\Delta H = -68\,kJ\,mol^{-1}]$$

Due to the heavily exothermal reaction, SiC layers consisting of SiC granules are immediately built up on the walls of the capillaries and the reaction zone is located between the carbon and the SiC (Figure 5.4) [19]. Therefore, the subsequent chemical reaction is determined by the diffusion of Si-atoms through the SiC layer and is slowing down with increasing SiC-layer thickness. Due to the fact that the molar volume of the resulting SiC ($V_{SiC} = 12.45\,cm^3\,mol^{-1}$) is 30% smaller than the added volumes of the reacting C ($V_C = 6.53\,cm^3\,mol^{-1}$) and Si ($V_{Si} = 11.11\,cm^3\,mol^{-1}$), a single filling of the porosity is not sufficient to create a dense SiC matrix, but further Si has to be infiltrated continuously. However, the transformation of C to SiC leads to a volume increase of 91%. Thereby, the capillaries tend to be closed by the formation of SiC until blocking the pore channels.

At DLR, C/C materials typically are siliconized at T = 1650 °C and p < 3 mbar. Prior to heating up, a calculated amount of Si granulate is added to the C/C preform. The molten silicon is sucked into the micro-channel system of the preform reacting immediately with C-matrix and C-fibers at the channel walls,

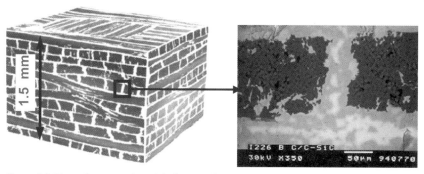

Figure 5.5 Three-dimensional model of a typical microstructure for fabric reinforced C/C-SiC basic material showing SiC layers (gray), wrapping the C/C bundles (dark gray). Right: SEM view of cross-section, showing free Si (light gray) as an intergranular phase inside the SiC layer (gray).

whereas the deuse C/C bundles are not infiltrated. Unlike C/SiC or SiC/SiC materials based on CVI fiber coating, where each fiber is individually embedded in the SiC matrix, the resulting CMC material is characterized by dense C/C bundles embedded in the SiC matrix and therefore consequentially is called C/C-SiC (Figure 5.5).

Depending on the maximum process temperature and duration, unreacted, free silicon remains as an intergranular phase in the SiC layers, especially in broad capillaries. For economic reasons, a compromise between process duration and conversion level has to be made, leading to Si contents in the range of 1 to 4 vol.-% in the final C/C-SiC basic material.

Typical process times are about 60 hours for large, thick-walled parts based on fabrics down to about 12 hours for small, short fiber reinforced parts with random fiber orientation.

One main advantage of the LSI process is the possibility to realize large and complex structures via near-net-shape manufacturing and *in-situ* joining. Thereby, the structures are divided up into basic components, for which individual CFRP preforms are manufactured and pyrolyzed. After assembling the single components by slight pressure fit, positive locking joints, or using a special joining paste, based on a mixture of polymer precursor and carbon particulates, the final ceramic joining is completed *in situ* by siliconizing the whole structure (Figure 5.6) [20]. The joining process is described in detail in a separate chapter of this book.

5.3
Properties

In CMC materials, the well-known properties of bulk ceramics, such as high thermal and chemical resistance, hardness, and wear resistance, can be combined with extremely unusual qualities such as thermal shock resistance, fracture tough-

Figure 5.6 Lightweight C/C-SiC intake ramp demonstrator, manufactured via near-net-shape and *in-situ* joining technique. Left: CFRP single elements manufactured via RTM. Middle: Assembled C/C preform (5 cell body structure, 10 end caps, 2 tubes). Right: Joint C/C-SiC structure after siliconization and final machining (275 mm × 265 mm × 35 mm).

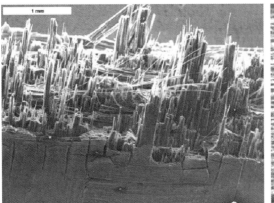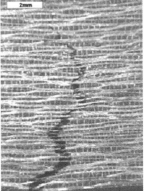

Figure 5.7 SEM images of typical fracture behavior of CMC characterized by fiber pull-out (left) and crack deflection (right).

ness, and quasi-ductile fracture behavior, leading to a totally new class of materials for high performance structural parts.

Carbon as well as SiC fibers offer significantly higher strength ($\sigma_t > 1800$ MPa) and, more importantly, strain level ($\varepsilon > 1.5\%$) compared to bulk ceramics (SiC: $\sigma_t = 400$ MPa; $\varepsilon < 0.05\%$). By integrating the fibers with a weak interface to the brittle ceramic matrix, the fibers can separate from the matrix. Therefore, cracks arising by local overstressing will be bridged by the fibers and the load will be transferred to the fibers. Additionally, cracks are stopped at the fiber matrix interface or at micro-cracks in the matrix. In contrast to bulk ceramics, where cracks are running through the entire part and cannot be stopped, extremely reduced brittleness and increased damage tolerance can be obtained in CMC due to the dissipation of crack energy by crack deflection, fiber breaking, and eventually fiber pull-out (Figure 5.7).

In summary, the material behavior of fiber reinforced CMCs is more similar to metals, like gray cast iron, than to monolithic ceramics and the material strength is not dependent on the volume of the part or structure as it is with "Weibull materials," such as bulk ceramics. Reliable CMC structures of large dimensions can be realized without increasing the failure risk.

For MI-CMC, a wide range of different process variations are used by several manufacturers. However, the availability of reliable data for material properties is limited. Additionally, the published values cannot be compared directly, due to different evaluation methods and lack of information about material composition and manufacturing details. Therefore, the given values in the following tables provide a rough orientation and cannot be used as design data directly, without consulting the material manufacturer. The mechanical properties are investigated on samples loaded parallel to the fabric layers or fibers. High temperature properties are determined in an inert gas atmosphere for C-fiber based CMC, and in ambient air for SiC/SiC material.

In Figures 5.8 and 5.9, as well as in Tables 5.2 and 5.3, typical microstructures and mechanical/thermal properties of two-dimensional-fabric and unidimensional

Table 5.2 Typical material properties of fabric and UD crossply (0°/90°; EADS) based C/SiC and C/C-SiC materials in dependence of the manufacturing method [21–23].

		CVI		LPI		LSI	
		C/SiC	C/SiC	C/SiC	C/C-SiC	C/C-SiC	C/SiC
Manufacture		SPS (SNECMA)	MT Aerospace	EADS	DLR	SKT	SGL
Density	g/cm³	2.1	2.1–2.2	1.8	1.9–2.0	>1.8	2
Porosity	%	10	10–15	10	2–5	–	2
Tensile strength	MPa	350	300–320	250	80–190	–	110
Strain to failure	%	0.9	0.6–0.9	0.5	0.15–0.35	0.23–0.3	0.3
Young's modulus	GPa	90–100	90–100	65	50–70	–	65
Compression strength	MPa	580–700	450–550	590	210–320	–	470
Flexural strength	MPa	500–700	450–500	500	160–300	130–240	190
ILSS	MPa	35	45–48	10	28–33	14–20	–
Fiber content	vol.-%	45	42–47	46	55–65	–	–
CTE coefficient ∥	$10^{-6} K^{-1}$	3[a]	3	1.16[d]	−1–2.5[b]	0.8–1.5[d]	−0.3
of thermal expansion ⊥		5[a]	5	4.06[d]	2.5–7[b]	5.5–6.5[d]	−0.03–1.36[e]
Thermal ∥	W/mK	14.3–20.6[a]	14	11.3–12.6[b]	17.0–22.6[c]	12–22	23–12[f]
conductivity ⊥		6.5–5.9[a]	7	5.3–5.5[b]	7.5–10.3[c]	28–35	–
Specific heat	J/kgK	620–1400	–	900–1600[b]	690–1550	–	–

∥ and ⊥ = fiber orientation.
- a RT-1000 °C.
- b RT-1500 °C.
- c 200–1650 °C.
- d RT-700 °C.
- e 200–1200 °C.
- f 20–1200 °C.

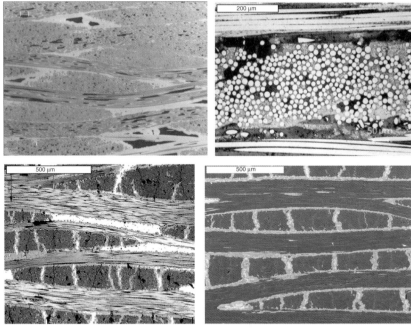

Figure 5.8 Typical microstructures of fabric reinforced C/SiC and C/C-SiC materials. Left above: CVI–C/SiC (MT Aerospace). Right above: LPI–C/SiC (EADS/Dornier). Both materials showing high porosity (black). Left bottom: LSI–C/SiC material (SGL). Right bottom: LSI–C/C-SiC XB (DLR). Both LSI/MI materials characterized by dense C/C bundles, embedded in SiSiC matrix (SiC: gray, Si: light gray) showing almost no porosity.

cross-ply (0–90 degrees) reinforced CMC materials, manufactured via CVI, LPI, and MI/LSI, are compared. The scattering of the values originates from standard material variants, based on different fiber types, fiber volume contents and matrix composition. In Figure 5.10 and Table 5.4, an overview of exemplary carbon fiber reinforced SiC materials is given, demonstrating the high variability of MI/LSI processes.

Table 5.3 Typical material properties of SiC/SiC materials in dependence of the manufacturing method [14, 24–26].

		Gasphase infiltration				Melt infiltration					
		SiC/SiC				Hypercomp PP-HN		Hypercomp SC-HN		N-24 B	
Manufacture method		CVI		CVI[a]		MI-Prepreg		MI-Slurry cast		MI-Slurry cast	
Manufacturer		Snecma Propulsion Solide (SPS)				General Electric (GE)				NASA	
Fiber type		Nicalon		Hi-Nicalon		Hi-Nicalon		Hi-Nicalon		Sylramic-iBN[b]	
Fiber content	vol.-%	40				20–25		35		36	
Temperature	°C	23	1400	23	1200	25	1200	23	1200	23	1315
Density	g/cm^3	2.5	2.5	2.3	–	2.8	–	2.7	–	2.85	–
Porosity	%	10	10	13	–	<2	–	6	–	2	–
Tensile strength (UTS)	MPa	200	150	315	–	321	224[c]	358	271	450	380
Proportional limit stress	MPa	–	–	–	–	167	165	120	130	170	160
Strain to failure	%	0.3	0.5	0.5	–	0.89	0.31	0.7	0.5	~0.55	–
Young's modulus	GPa	230	170	220		285	243	–	–	210	–
Interlam. shear strength	MPa	40	25	31	23	135	124	–	–	–	–
CTE ∥	10^{-6}K^{-1}	3	3[f]	–	–	3.57	3.73	3.74	4.34	–	–
CTE ⊥		1.7	3.4[f]	–	–	4.07	4.15	3.21	3.12	–	–
Thermal conductivity ∥	W/mK	19	15.2[f]	–	–	33.8	14.7	30.8[g]	14.8[g]	–	–
Thermal conductivity ⊥		9.5	5.7[f]	–	–	24.7	11.7	22.5	11.8	27[d]	10[e]
Specific heat	J/kgK	620	1200[f]	–	–	710	1140	700	2660[g]	–	–

∥ and ⊥ = fiber orientation.
a Si-B-C self healing matrix.
b COI Ceramics + NASA.
c Strain rate = 3 × 10^{-5}–10^{-4}; higher values are obtained at higher strain rates.
d 204 °C.
e 1204 °C.
f 1000 °C.
g Engineering estimates.

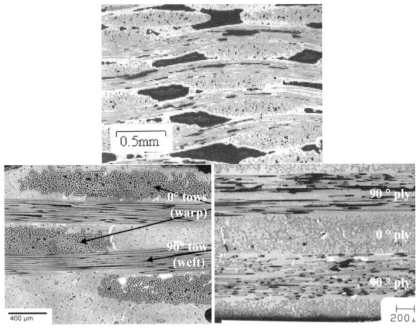

Figure 5.9 SEM micrographs of different SiC/SiC materials. Above: CVI-SiC/SiC (NASA), characterized by high open porosity (black). Bottom: Hipercomp (GE) MI–SiC/SiC materials based slurry casting (left) and on prepreg technology (right) [14].

Table 5.4 Material properties of fabric and short fiber reinforced C/C-SiC and C/SiC material variants, obtained by LSI (DLR, SGL, SKT).

		C/C-SiC				C/C-SiC	C/SiC
Material properties	Unit	XB	XT	XC	SF	FU 2952 P77	Sigrasic 6010 GNJ
Manufacturer	–	DLR	DLR	DLR	DLR	SKT	SGL
Fiber reinforcement	–	Fabric	Fabric	Fabric	Short fiber	Short fiber	Short fiber
Density	$g\,cm^{-3}$	1.9	1.92	2.05	2.03	>2	2.4
Open porosity	%	3.5	3.7	2.2	3.0	–	<1
Young's modulus[a]	GPa	60	100	–	50–70	–	20–30
Flexural strength	MPa	160	300	120	–	60–80	50–60
Tensile strength	MPa	80	190	–	–	–	20–30
Strain to failure	%	0.15	0.35	–	–	0.2–0.26	0.3
Thermal conductivity[g] $\parallel$	W/mK	18.5/17	22.6/20.8	–	–	18–23	40/20[b]
$\perp$		9.0/7.5	10.3/8.8	–	25–30[f]	28–33	
Specific heat (25 °C)	$J\,kgK^{-1}$	750	690	720	750	–	600–800
SiC content	vol.-%	21.2	19.8	30.4	26	–	60
Si content	vol.-%	5.4	4.1	5.2	1.3	–	10
C content	vol.-%	69.9	72.4	62.2	69.7	–	30
CTE $\parallel$	$10^{-6}K^{-1}$	−1/2.5[d]	−1/2.2[d]	–	0.5/3.5[e]	0.8–1.3[c]	1.8/3.0[a]
Ref. Temp. = 25 °C $\perp$		2.5/6.5[d]	2.5/7[d]		1.0/4.0[e]	5.5–6.0[c]	

$\parallel$ and $\perp$ = fiber orientation.
- a 0–300 °C/300–1200 °C.
- b 20/1200 °C.
- c 25–800 °C.
- d 100/1500 °C.
- e 25/1400 °C.
- f 50 °C.
- g 200/1600 °C.

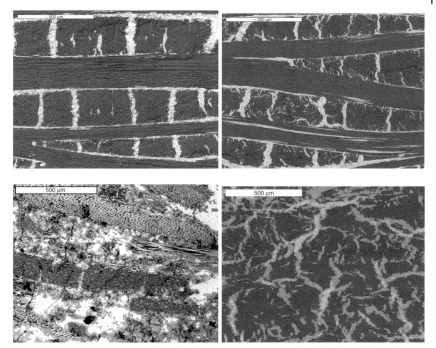

Figure 5.10 SEM micrographs of different C-fiber reinforced MI CMC. Left above: Basic C/C-SiC material based on HTA fiber fabrics (C/C-SiC XB; DLR). Right above: High tenacity C/C-SiC material based on thermally treated T 800 fiber fabrics (C/C-SiC XT; DLR). Left bottom: C/SiC material based on random oriented, short fibers (Sigrasic 6010 GNJ, SGL). Right bottom: C/C-SiC material based on random oriented, cut HTA fibers (C/C-SiC SF; DLR).

5.3.1
Material Composition

Regarding the material composition, the main characteristic of MI-CMC are low porosities and multiphase matrices. Due to process related and economical restrictions, the open porosity of CVI- and LPI-materials is usually higher than 10%, up to 15%. In contrast, the very effective MI process leads to relatively dense CMC materials with open porosities in the range of 2 to 6% (Figures 5.9 and 5.10). Whereas with CVI, LPI, or hot pressing, almost stoichiometric β-SiC matrices can be obtained, the multiphase matrices of MI-CMC are a mixture of β-SiC crystals formed by the reaction of Si with C, as well of non-reacted, residual C and Si. Additionally, crystalline SiC particulates are embedded in the matrix of SiC/SiC materials, built up via slurry casting, or prepreg technology.

Typical contents of non-reacted, residual silicon are in the range of 2 to 5 vol.-%, up to 17 vol.-% for C/SiC and C/C-SiC and 12 to 18 vol.-%, down to 5 vol.-% for

SiC/SiC, depending on the aimed application and the used raw materials and processes.

For fabric reinforced, quasi-ductile C/C-SiC materials, the main characteristic is the high amount of residual carbon matrix inside the C/C-bundles, used for embedding and protecting the carbon fibers, as well as a certain carbon matrix content in the crossing areas of the original fiber bundles. In total, a C-matrix content above 12 vol.-% and, by including the C-fibers, an overall C content in the range of 60 to 75 vol.-%, can be achieved. Due to the fact that the carbon fibers are not equipped with a protective coating individually, some of the fibers are transformed to SiC. However, the fiber contents of the final C/C-SiC materials are still high, because of the high fiber contents of the CFRP and C/C preform ($\phi_{F,C/Cpreform}$ = 55–65 vol.-%). The corresponding SiC contents are low, in the range of 20 to 30 vol.-%. In contrast, CVI or PIP derived C/SiC materials typically show a low overall carbon content (C-fiber and fiber coating) in the range of 40 to 50 vol.-% and much higher SiC contents of about 30 to 40 vol.-%. However, using short fiber reinforced CFRP preforms with low fiber contents, the SiC and Si content of MI-C/SiC can also be increased up to 60 vol.-% and 10 vol.-%, respectively [21].

The fiber content in Mi-SiC/SiC materials typically is low, in the range of 20 to 25 vol.-%, and about 35 vol.-% for prepreg and slurry based materials, respectively. For the fiber coatings, about 6 to 10 vol.-% are common. In slurry cast materials, the matrix is built up by 23 to 35 vol.-% CVI-SiC overcoating and 16 to 18 vol.-% SiC particulates, as well as by 12 to 18 vol.-% Si.

5.3.2
Mechanical Properties

The mechanical properties of CMC materials are heavily influenced by the fiber matrix bonding. Therefore MI, as well as CVI and PIP/LPI materials, show similar high ultimate strength and strain to failure capability, when highly complicent fibre coatings, for example, CVI applied C or BN are used. However, in the case of melt infiltrated C/C-SiC materials, where the carbon fibers are processed as delivered and no costly fiber coatings are used, tensile and flexural strength are significantly lower. One main advantage of MI-CMC is the generally higher interlaminar shear strength compared to CVI- and LPI/PIP–CMC, especially for SiC/SiC materials, resulting from their significantly lower porosity.

At elevated temperatures up to 1200 °C, in an inert gas atmosphere, the mechanical properties of C/C-SiC are slightly higher than at room temperature, similar to the behavior of C/C materials (Figure 5.11). However, at temperatures above 1350 °C in a vacuum, a certain decrease of tensile strength was observed. Due to oxidation at temperatures above 450 °C in air, the lifetime of C-fiber based CMC materials is limited. Oxygen can react with the C-fibres at CMC surfaces, where fiber ends are exposed, and through matrix cracks that extend to the CMC surface. These matrix cracks cannot only be generated by externally applied tensile loads on the CMC, but also during CMC fabrication because of the thermal expansion

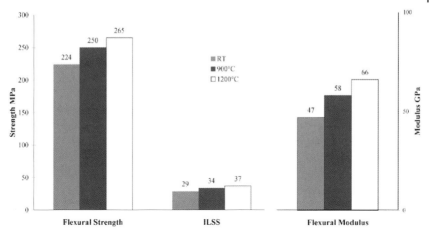

Figure 5.11 High temperature properties of fabric reinforced C/C-SiC manufactured via LSI.

mismatch that exists between the C-fibres and SiC matrix. Thus despite the fact that the SiC matrix and surface coatings significantly increase oxidation stability, C/SiC and C/C-SiC materials are not usable for long term applications, e.g. in gas turbines.

In contrast to C-fiber based CMC, the ultimate tensile strength (UTS) of SiC/SiC materials is decreasing at higher temperatures primarily due to slow-crack growth and creep cavitation in the reinforcing SiC fibers. However, the proportional limit stress, that is the maximum design stress level in the linear elastic regime, as well as strain to failure, or through thickness matrix cracking, are generally not influenced by temperature and remain stable even after 1000 to 4000 hours of exposure in air at 1200 °C [14, 27].

Extensive investigations of various SiC/SiC materials showed decreasing lifetime in air at temperatures ≤950 °C, with increased temperature and tensile loads above matrix crackings leading to enhanced access of oxygen into the material. Thereby lifetime was reduced from more than 1000 hours at 700 °C/250 MPa to 22 hours at 950 °C/250 MPa [28]. For this reason, recession of fiber coating and build-up of glassy interphases, resulting in localized fiber attack and high fiber matrix bonding, was assumed.

In the temperature range between 950 and 1150 °C, lifetime in air could be improved significantly to more than 1000 hours at 1150 °C/250 MPa. Thereby, the oxidation of SiC and Si led to a rapid formation of a protective SiO_2 sealing of the sample surfaces and the matrix cracks, generated by the applied stress.

At higher temperatures over 1200 °C up to 1400 °C and long exposure time of over 100 hours, tensile strength and strain to failure are rapidly decreasing due to fiber creep and oxygen ingress through matrix cracks, generated by high tensile stresses. In addition, above ~1300 °C, diffusion reactions of the residual Si in the matrix are restarting, attacking the fiber matrix interphase and the fiber itself [22, 27]. This again leads to severe degradation of the SiC fibers as well as to a brittle fracture behavior, due to significantly increased fiber matrix bonding. Therefore, very low Si content and almost stoichiometric SiC matrices are the focus of ongoing development of SiC/SiC materials, for use as high temperature components in gas turbines.

Additionally, SiC as well as SiO_2 are not stable in typical combustion environments of gas turbines, characterized by high partial pressure of water vapor. Silica is volatilizing by the formation of Si–O–H, leading to continuous attack and degradation of the SiC/SiC components. In sample tests, degradation of the wall thickness of 200 to 500 μm after 1000 hours exposure at 1200 to 1315 °C (p = 1 MPa, gas velocity v = 90 m s^{-1}) was observed [7].

In summary, the lifetime of SiC/SiC components can be increased by limiting applied stresses to avoid matrix cracking and to reduce the content of free Si as much as possible. Additionally, environmental barrier coatings (EBC) [29] based on multi-phase non-oxide or oxide ceramics are mandatory in corrosive environments. To avoid micro-cracks in the protective coating, the maximum allowable stress load of the composite is governed by the low strain to failure of the usually highly brittle coatings, and therefore has to be reduced significantly.

5.3.3
CTE and Thermal Conductivity

Thermal conductivity and Coefficients of Thermal Expansion (CTE) of MI CMC generally are low compared to metals and bulk ceramics. Compared to CVI and LPI CMC, significantly higher thermal conductivity, especially perpendicular to the fiber orientation, are offered by MI CMC, resulting from their lower porosity. Because C-fibers show very different, anisotropic thermal properties compared to the SiC matrix, the CTE and thermal conductivity of C/C-SiC and C/SiC materials can be tailored in a wider range than SiC/SiC materials, where the thermal properties of SiC fibers and SiC matrices are similar.

At low temperatures up to 200 °C, the CTE of C-fiber-based MI CMC parallel to the fiber direction is generally very low, due to the negative CTE of the C-fibers. In combination with the positive CTE of the SiC matrix, thermally stable ($CTE_{\parallel,RT} \approx 0 \times (10^{-6} K^{-1})$) CMC materials and components can be obtained by tailoring, for example, fiber content and fiber orientation. Perpendicular to the fibers, the CTE is dominated by the SiC matrix and therefore is significantly higher ($CTE_{\perp,RT} = 2-4 \times 10^{-6} K^{-1}$). Similar to monolithic SiC and C materials, the CTE values increase at high temperatures ($CTE_{\parallel,1200°C} = 2-4.5 \times 10^{-6} K^{-1}$; $CTE_{\perp,1200°C} = 7-8 \times 10^{-6} K^{-1}$). In contrast, the CTE of MI SiC/SiC materials show similar values parallel and perpendicular to the fibers, only slightly influenced by temperature. Typical values are in the range of 3.1 to $4.3 \times 10^{-6} K^{-1}$ (RT–1200 °C) [14].

C fiber-based MI CMC, with high contents of C-fiber and C matrix, show low thermal conductivities of $\lambda_\perp$ = 7 to 9 W/mK and $\lambda_\parallel$ = 12 to 22 W/mK at ambient as well as at high temperatures. However, these values can be increased significantly by using graphitized C-fibers or C/C performs, as well as by an increase of the volume content of SiC and Si. Using short fiber reinforced C/SiC materials with high contents of SiC and Si (ϕ_{SiC} = 60 vol.-%; ϕ_{Si} = 10 vol.-%), maximum thermal conductivities of up to λ_{II} = 40 W/mK (RT) and λ_{II} = 20 W/mK (1200 °C), comparable to gray cast iron, can be achieved.

SiC/SiC materials show about twice the thermal conductivity at ambient temperatures compared to C-fiber-based MI CMC. Similar to C/SiC and C/C-SiC, the in-plane thermal conductivity of SiC/SiC materials is about 30% higher than in the perpendicular direction ($\lambda_{\perp,RT}$ ≈ 23/25 W/mK and $\lambda_{II,RT}$ ≈ 31/34 W/mK for slurry cast/prepreg based materials) resulting from the anisotropic fiber architecture and the BN-coatings of the SiC fibers. At high temperatures, the thermal conductivity is reduced by half ($\lambda_{\perp,1200°C}$ ≈ 12 W/mK and $\lambda_{II,1200°C}$ ≈ 15 W/mK). However, the perpendicular conductivity could be increased to $\lambda_{\perp,RT}$ = 41 W/mK and $\lambda_{\perp,1200°C}$ = 17 W/mK by a high the CVI-SiC matrix content of 35 vol.-% and an additional thermal treatment prior to MI [24].

5.3.4
Frictional Properties

LSI-based C/SiC and C/C-SiC materials show unique properties for friction applications, such as high wear resistance, high temperature and thermal shock resistance, and a coefficient of friction, which can be tailored in a wide range for successful uses in brakes as well as in gliding systems. This application field is presented in detail in a separate chapter of this book.

5.4
Applications

C fiber reinforced CMC materials originally have been developed for lightweight thermal protection structures of spacecraft. Due to their unique thermal and mechanical properties, superior to all other available materials, such as metals or monolithic ceramics, even very high manufacturing costs were acceptable. However, by the development of cost-efficient MI processes, new application fields beyond aerospace could be opened. Some materials even went out of the laboratories into the market.

SiC/SiC materials have been developed for hot components in jet engines and stationary gas turbines, as well as for reactor structures in nuclear power generation and fusion technology. MI–SiC/SiC materials are commercially available but, to the knowledge of the author, are not yet used in commercial applications.

5.4.1
Space Applications

In TPS of reusable spacecraft, maximum temperatures of up to 1800 °C, high heating rates of several hundred K/s and high thermal gradients are obtained locally during the critical re-entry phase, lasting about 20 minutes. Therefore, in

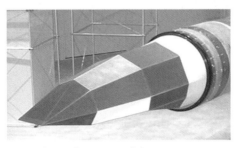

Figure 5.12 C/C-SiC structural parts for TPS. Top left: X 38 nose cap (ca. 740 × 640 × 170 mm^3; t ca. 6 mm; m ca. 7 kg, DLR), showing *in-situ* joint load bearing elements on the rear side. Top right: Front view of nose cap, mounted on the aluminium structural part of spacecraft. Bottom: Facetted TPS structure, built up by flat panels, mounted on a rocket system.

this application field, there is no lightweight alternative to carbon fiber reinforced CMC.

MI C/SiC materials were used for the first time for the nose cap in the TPS system of the Buran from the former Soviet Union [30]. The current status of MI C/C-SiC in TPS can be demonstrated on the nose cap for the NASA X-38 spacecraft (Figure 5.12), developed by DLR in the German TETRA program (1998–2002) [31], which was planned to be the future crew return vehicle for the International Space Station (ISS). This large, complex shaped and lightweight structure could be realized in near-net-shape technology using LSI and *in-situ* joining methods. Thereby the development of a well adapted mounting system, based on moveable C/C-SiC levers, ensured safe fixing of the nose cap to the metallic fuselage, without inducing inner stresses by the thermal expansion of the nose cap during re-entry. Recent developments for novel TPS systems concentrate on facetted structures, based on cost-efficient, flat panels [32]. During a re-entry test flight in 2005, a first C/C-SiC structure could withstand extreme thermal loads ($T_{max.} > 2000$ K) at the sharp nose tip, as well as at the edges of the structure.

Figure 5.13 C/C-SiC jet vane and sealing ring (left) and jet vane assembly in the exhaust nozzle extension (right).

5.4.2
Short-term Aeronautics

In military rocket motors, CMC materials offer high potential for use in lightweight structures at the most severe thermal, mechanical, and environmental loads known to date. However, due to the very short service time, usually a few seconds, oxidation is not critical in many cases so that carbon fiber based materials can be used.

C/C-SiC jet vanes for thrust vector control (TVC) systems have been introduced in military rocket motors (Figure 5.13). The moveable jet vanes are positioned in the exhaust jet stream, leading to high bending forces in the jet vane shaft, maximum temperatures of up to 3100 K, and heating rates of several thousand K/s in the leading edge, as well as to local temperature gradients up to 200 K mm^{-1} in the blade area. Additionally, the erosion of the leading edge, caused by Al_2O_3 particles impacting at velocities of up to 2000 m s^{-1}, can be limited to an acceptable level. Compared to metallic jet vanes, usually made of refractory metals such as tungsten, the use of C/C-SiC materials offers weight savings of up to 90%.

5.4.3
Long-term Aeronautics and Power Generation

Due to their higher temperature and thermal shock resistance, compared to metals and monolithic ceramics, CMC materials offer a high potential to increase gas turbine efficiency, leading to lower fuel consumption, higher output, and lower emission of NO_x, CO_2, and CO. This can be achieved by increased temperature capability of combustor liners and turbine components, allowing higher combustion, as well as turbine inlet temperatures. Novel, temperature stable CMC materials have a high potential to reduce the consumption of compressor air, usually needed to cool the hot components by half, as well as to increase hot wall temperature by 30%. Thereby, fuel consumption and pollutant emissions (156 Mio tons kerosene, 61 Mio tons CO_2, and 2 Mio tons NO_x for civil aviation worldwide in 2005) could be reduced by up to 15% and 80%, respectively.

SiC/SiC materials are well suited for use in gas turbine engines due to their unique properties, such as high temperature strength, creep resistance, and thermal conductivity, as well as low thermal expansion, porosity, and density.

In the United States, MI-SiC/SiC materials and components have been developed in multitude governmental funded programs since 1992. In 1997, field tests were started by the integration of exemplary components in small- and medium-sized stationary gas turbines from GE (GE 7 FA; 160 MW) and Solar Turbines (Centaur 50 S; 4.1 MW). The current status of these research activities is represented by successful field tests of combustor liners and turbine shrouds, as well as by commercially available SiC/SiC materials, so-called HyPerComp, offered by GE. For the field test, several SiC/SiC liners in original geometry (up to ⌀760 mm; l = 200 mm, t = 2–3 mm), with multilayer EBC based on SiC, BSAS, and mullite, have been tested for accumulated service times of more than 12 000 hours at $T_{max.}$ up to 1260 °C and a calculated stress load of about 76 MPa, induced by thermal gradients only [29].

However, the minimum required service time of 30 000 hours, which represents the typical target time between overhaul of small turbine engines, still presents a challenge. The most critical issue is the long-term stability of both, the SiC/SiC material and the EBC, in the highly corrosive combustor environment, characterized by high water vapor partial pressures, temperatures well above 1200 °C, and combustion gas velocities of up to 90 m s^{-1}. Current research is focused on SiC/SiC materials with Si-free matrices, reliable coatings, and lifetime calculation models.

5.4.4
Friction Systems

High performance brakes and clutches are one of the most important application areas for carbon fiber reinforced carbon (C/C) materials. However, due to high manufacturing costs and some unfavorable tribological properties, such as high wear and unstable friction behavior at low temperatures, C/C is limited to applications in aircrafts and racing cars. By the development of cost-efficient MI-C/SiC and C/C-SiC, these drawbacks could be overcome, and a wide range of potential application fields beyond aerospace could be opened. Additionally, the material properties could be adapted to a wide range of different applications, leading to serial production of high performance brake disks for automobiles, brake pads for high speed elevators, and even sliding elements for the high speed train, Transrapid. This important breakthrough of CMC technology is presented in a separate chapter in this book.

5.4.5
Low-Expansion Structures

The high thermal stability of C/SiC and C/C-SiC materials are favorably used in low-expansion structures, such as calibrating plates and telescope tubes for laser communication terminals (LCT).

Optical systems for satellite communication via laser beams provide high data rates, for example, 1 Gbps at a distance of 20 000 km, but require extremely precise structures for long service times of several years [33]. Figure 5.14 shows the optical unit of the LCTSX used on the Terrasar-X satellite, launched in 2007. Due to a tailored CTE in axial direction ($0 \pm 0.1 \times 10^{-6}\,K^{-1}$), a constant distance between the primary and secondary mirror, in the temperature range of −50 °C and +70 °C can be obtained, ensuring a safe data transfer without transmission losses. Compared to other low expansion materials, such as Cerodur, C/C-SiC offers lower density and higher fracture toughness, which enables the near-net-shape manufacture of thin-walled, lightweight structures. Additionally, the main drawbacks of CFRP materials, outgassing in vacuum and hygroscopic swelling, have been overcome.

Figure 5.14 *In-situ* joined C/C-SiC telescope tube (⌀ 140 mm, l = 160 mm, wall thickness = 3 mm) for the LCT in the satellite TerraSAR-X (Zeiss Optroniks).

5.4.6
Further Applications

Another very promising application for carbon fiber reinforced SiC is ballistic protection against high velocity or armor piercing projectiles. Compared to hardened armor steel, ceramic armor systems offer a weight reduction of about 50 to 70%, leading to areal weights of 30 to 40 kg m^{-2}. Well-known monolithic armor ceramics, such as Al_2O_3, SiC, and B_4C, offer high protection against single hits, whereas multiple hit requirements are difficult to meet, due to the typically high brittleness of bulk ceramics. More importantly, large SiC structures with complex geometries are difficult to manufacture via common technologies such as hot pressing, leading to high waste rates and production costs. Therefore, costly multi-tile protection systems, based on a large number of small tiles, for example, 50 mm by 50 mm, arranged precisely to large structures, are widely used for the protection of vehicles and aircraft.

The drawbacks of monolithic armor ceramics can be overcome by C/SiC materials, offering significantly reduced crack propagation and improved multiple hit performance [34], as well as reduced weight compared to monolithic Al_2O_3 and SiC. Additionally, complex-shaped, thin-walled structures can be manufactured in near-net-shape geometry via low cost MI/LSI processes (Figure 5.15).

In high temperature furnaces, carbon fiber reinforced SiC composites are used for charging devices (Figure 5.15) and as supporting structures for high temperature treatment of metallic parts, as well as for brazing processes. Compared to metals, C/SiC composites are significantly lighter and offer lower heat capacity, which is favorable for fast processes with high heating up and cooling down rates. Due to the low CTE of carbon fiber reinforced SiC, even unsymmetrical structures with sudden changes in wall thickness are geometrically stable at varying temperatures and do not show distortion even at elevated temperatures, offering higher service life and safer operation compared to metal structures.

 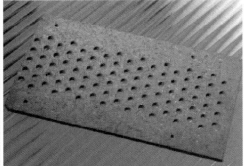

Figure 5.15 Examples for novel applications for LSI-C/SiC materials based on short fiber reinforcement (SGL) Left: Vest insert and bulletproof armor plate. Right: Perforated plate for high temperature furnace charging devices.

5.5
Summary

CMC materials offer a unique combination of properties at high as well as at medium and low temperatures, which cannot be achieved by any other materials, such as metals or monolithic ceramics. However, CMC materials are also expensive due to high raw material and process cost. These drawback can be partially cut by adding MI, as a more economical manufacturing process, to established CVI and PIP/LPI. MI processes are characterized by a fast build-up of Si/SiC matrix leading to short process times. The Si/SiC-matrix is built up almost simultaneously in the whole volume of the part in one single process step and no multiple infiltration cycles are needed to densify C as well as SiC fiber reinforced CMC. Therefore, even very large, complex-shaped and thin-walled lightweight structures, as well as very thick-walled components, can be realized in near-net-shape techniques.

Compared to CMC materials derived from CVI and LPI, MI-based C/SiC, C/C-SiC, and SiC/SiC materials offer significantly lower open porosities, leading to higher shear strength and thermal conductivity. However, lower tensile strength of C/SiC and C/C-SiC, as well as the limited lifetime of SiC/SiC materials caused by free Si in the matrix attacking the fibers in long-term use at high temperatures, are still a challenge.

After more than 20 years of development and gaining operational experience in aerospace applications, MI-CMC materials have just found their way out of the laboratory into high performance products for the industrial market. Typical application areas are TPS structures for spacecraft, hot structures for rocket propulsion, gas turbines and nuclear reactors, thermally stable structures, and friction materials. The introduction of C/SiC automotive brake disks in serial production is a breakthrough and an important milestone in CMC technology, offering high potential for further applications of C/SiC and C/C-SiC materials in other industrial areas.

The further development will not only focus on low cost processes and improvement of material properties and EBCs, but also on providing design and simulation tools in combination with convincing, non-destructive evaluation methods. With the latter, scientists and engineers should be able to reliably calculate lifetime as well as damage and failure behavior, which is a basic prerequisite for future applications of CMC components in very promising, but challenging, safety-critical applications fields.

References

1 Hillig, W.B. et al. (1975) Silicon/silicon carbide composites. *Ceramic Bulletin*, **54** (12), 1054–56.
2 Evans, C.C., Parmee, A.C. and Rainbow, R.W. (1974) *Silicon treatment of carbon fiber-carbon composites.* Proceedings of 4th London Conference on Carbon and Graphite, 231–35.
3 Fitzer, E. and Gadow, R. (1986) Fibre-reinforced silicon carbide. *American Ceramic Society Bulletin*, **65**, 368–72.

4. Krenkel, W. and Hald, H. (1988) *Liquid infiltrated C/SiC – an alternative material for hot space structures*. Proceedings of the ESA/ESTEC Conference on Spacecraft Structures and Mechanical Testing, Nordwijk, The Netherlands.
5. Corman, G.S., Luthra, K.L. and Brun, M.K. (2003) Silicon melt infiltrated ceramic composits_process and properties, in *Progress In Ceramic Gas Turbine Development, Vol. II, Ceramic Gas Turbine Component Development and Evolution, Fabrication, NDE, Testing and Life Prediction* (eds M. van Roode et al.), ASME Press, New York, USA.
6. Luthra, K.L., Singh, R.N. and Brun, M.K. (1993) Toughened silcomp composites – process and preliminary properties. *American Ceramic Society Bulletin*, **72** (7), 79–85.
7. DiCarlo, J.A. et al. (2002) *Progress in SiC/SiC ceramic composite development for gas turbine hot-section components under NASA EPM and UEET programs*. Proceedings of ASME Turbo Expo 2002, June 3–6, 2002, Amsterdam, The Netherlands.
8. Fareed, A.S. (2005) Silicon carbide and oxide fiber reinforced alumina matrix composites fabricated via direct metal oxidation, in *Handbook of Ceramic Composites* (ed. N.P. Bansal), Kluwer Academic Publishers, USA.
9. Winnacker, K. (2005) *Chemische Technik: Prozesse und Produkte*, Vol. 8 (eds R. Dittmeyer et al.), Wiley-VCH Verlag GmbH, Weinheim, Germany, pp. 1166–73.
10. DiCarlo, J.A. and Bansal, N.P. (1998) Fabrication Routes for Continuous Fibre-Reinforced Ceramic Composites (CFCC). NASA/TM-1998-208819.
11. Beyer, S. et al. (2006) *Advanced composite materials for current and future propulsion and industrial applications*. Proceedings of CIMTEC.
12. Krenkel, W. and Fabig, J. (1995) *Tailoring of microstructure in C/C-SiC composites*. Proceedings of the 10th International Conference on Composite Materials, ICCM-10, Whistler, Canada, Vol. 4, 601–9.
13. Lee, S.P., Yoon, H.K., Park, J.S., Katoh, Y. and Kohyama, A. (2003) Processing and properties of SiC and SiC/SiC composite materials by melt infiltration process. *International Journal of Modern Physics B*, **17** (8–9), 1833–38.
14. Corman, G.S. and Luthra, K.L. (2005) Silicon melt infiltrated ceramic composites (HiPerComp™), in *Handbook of Ceramic Composites* (ed. N.P. Bansal), Kluwer Academic Publishers, USA.
15. Schulte-Fischedick, J. et al. (1998) Untersuchung der Entstehung des Rissmusters während der Pyrolyse von CFK Vorkörpern zur Herstellung von C/C-Werkstoffen. Werkstoffwoche Materialica 98, Munich, Germany.
16. Suyama, S., Itoh, Y., Nakagawa, S., Kohyama, N. and Katoh, Y. (1999) Effect of Residual Silicon Phase on Reaction Sintered Silicon Carbide. The 3rd International Agency Workshop on SiC/SiC ceramic composites for fusion structural application, 108–12.
17. Fitzer, E. and Gadow, R. (1986) Fibre-reinforced silicon carbide. *American Ceramic Society Bulletin*, **65**, 368–72.
18. Gern, F. (1995) Kapillarität und Infiltrationsverhaltenbei der Flüssigsilicierung von Carbon/Carbon-Bauteilen, Doctoral Thesis. University of Stuttgart, DLR-FB 95-26.
19. Hillig, W.B. (1994) Making ceramic composites by melt infiltration. *American Ceramic Society Bulletin*, **73**, 56–62.
20. Krenkel, W., Henke, T. and Mason, N. (1996) In situ *joined CMC components*. Int. Conference on Ceramic and Metal Matrix Composites CMMC, San Sebastian, Spain.
21. Data sheets from SGL Carbon Group (2005) SIGRASIC 6010 GNJ – Short Fiber Reinforced Ceramics, SIGRASIC 1500J Fabric Reinforced Ceramics, www.sglcarbon.de (accessed 03.03.2007).
22. Kochendörfer, R. (2003) Möglichkeiten und Grenzen faserverstärkter Keramiken, in *Keramische Verbundwerkstoffe* (ed. W. Krenkel), Wiley-VCH Verlag GmbH, Weinheim, Germany, pp. 1–22.
23. Weiss, R. (2001) Carbon Fiber Reinforced CMCs: Manufacture, Properties, Oxidation Protection, in *High Temperature Ceramic Matrix Composites*, (eds W. Krenkel, R. Naslain and H. Schneider), Wiley-VCH Verlag GmbH, Weinheim, Germany, pp. 440–56.

24 DiCarlo, J.A. *et al.* (2005) SiC/SiC composites for 1200°C and above, in *Handbook of Ceramic Composites* (ed. N.P. Bansal), Kluwer Academic Publishers, USA.

25 Lamon, J. (2005) Chemical vapor infiltrated SiC/SiC composites (CVI SiC/SiC), in *Handbook of Ceramic Composites* (ed. N.P. Bansal), Kluwer Academic Publishers, USA.

26 General Electric Data Sheet: Melt Infiltration Products (2007) www.gepower.com (accessed 03.03.2007).

27 Bhatt, R.T., McCue, T.R. and DiCarlo, J.A. (2003) Thermal stability of melt infiltrated SiC/SiC Composites. *Ceramic Engineering and Science Proceedings*, **24** (3), 295–300.

28 Lin, H.T., Becher, P.F. and Singh, M. (1999) Lifetime Response of a HI-Nicalon TM Fiber–Reinforced Melt-Infiltrated SiC Matrix Composite. ORNL/CP-102833.

29 van-Roode, M. et al. (2007) Ceramic matrix composite combustor liners: a summary of field evaluations. *Journal of Engineering for Gas Turbine and Power*, **129**, 21–30. ASME.

30 Trefilov, V.I. (1995) *Ceramic- and Carbon-Matrix Composites*, Chapman and Hall.

31 Hald, H. *et al.* (1999) *Developmentof a nose cap system for X 38*. Proceedings of International Symposium Atmospheric Reentry Vehicles and Systems, Arcachon, France.

32 Weihs, H., Reimer, T. and Laux, T. (2004) *Mechanical architecture and status of the flight unit of the sharp edge flight experiment SHEFEX*. IAF Congress October 2004, I301, Vancouver, Canada.

33 Schöppach, A. *et al.* (2000) *Use of ceramic matrix composites in high precision laser communication optics*. European Conference on Spacecraft Structures, Materials and Mechanical Testing, ESTEC, Nordwijk, The Netherlands.

34 Heidenreich, B., Gahr, M., Straßburger, E. and Lutz, E. (2006) *Biomorphic SISIC-materials for lightweight armour*. Proceedings of 30th International Conference on Advanced Ceramics & Composites, Cocoa Beach, Florida, USA.

6
Chemical Vapor Infiltration Processes for Ceramic Matrix Composites: Manufacturing, Properties, Applications
Martin Leuchs

6.1
Introduction

Ceramic matrix composites (CMCs) are a class of materials, which–while keeping most of the properties of conventional ceramics–overcome their main disadvantage: their tendency to brittle fracture and the problem of sudden failure of components in applications. The greatly improved crack resistance of CMCs is caused by energy consuming mechanisms at the fiber/matrix interface, which divert, bridge, or stop the propagating cracks [1, 2]. Essential for this mechanism is weak bonding between fibers and matrix, as only this can allow the crack bridging mechanism and the effective use of strength and elongation capabilities of the fibers (usually >2000 MPa and 2% (Chapters 1 and 2)). With a strong bond between fibers and matrix, the fibers would have to sustain infinite elongation capability to bridge a crack (Figure 6.1). With this weak bonding mechanism, resulting material is damage tolerant and demonstrates deformations and elongations, which are impossible in conventional ceramics. Pictures of crack surfaces of CMC materials demonstrate this weak bonding, by showing the so-called fiber pull-out (Figure 6.2).

Technically, this weak bonding is achieved via a thin carbon or boron nitride [3] coating of the fibers, which allows the matrix to glide along the fibers (Chapter 3).

The properties of this material, with respect to reliability and behavior under dynamic loads and thermal shocks, reach the level of metallic components.

Presently, besides carbon/carbon composites (Chapter 4), only CMCs using carbon or silicon carbide fibers, together with a silicon carbide matrix, have reached major importance in industrial development and manufacturing processes. Some importance of oxide CMCs [4] can also be observed (Chapters 8 and 9).

The manufacturing processes, presently relevant on an industrial level, differ in their methods to infiltrate the so-called fiber preform with the specific matrix material. In oxide CMCs, the matrix between standard fibers of ceramic oxides is obtained by low temperature sintering of special organic liquids (sol/gel) mixed with oxide powders. For a SiC matrix, essentially three processes are introduced.

Ceramic Matrix Composites. Edited by Walter Krenkel
Copyright © 2008 WILEY-VCH Verlag GmbH & Co. KGaA, Weinheim
ISBN: 978-3-527-31361-7

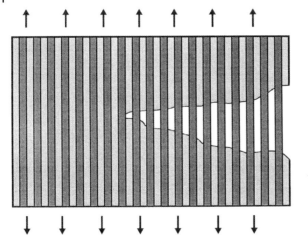

Figure 6.1 Scheme of crack formation and fiber pull-out mechanism.

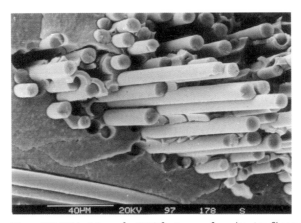

Figure 6.2 SEM picture of a CMC fracture surface showing fiber pull-out.

The chemical vapor infiltration (CVI) process, discussed below, the liquid polymer infiltration (LPI, sometimes called the polymer infiltration and pyrolysis(PIP) process (Chapter 7), and the liquid silicon infiltration (LSI) process (Chapter 5). These different manufacturing techniques lead, of course, to different materials with different properties, manufacturing costs, and fields of application.

All these manufacturing techniques start their processes by fixing the fibers into a shape, which – as in the manufacture of carbon fiber reinforced plastic materials – is as close as possible to the final shape of the intended product. The design of the fiber preform has to choose the fiber orientation in a way that considers the expected loads of the component. The second step is the matrix infiltration and the third one, if required, usually consists of machining (Chapter 12) the component to its final dimensions. Sometimes an additional coating or impregnation of the residual porosity is applied.

In this chapter, material derived from CVI will be presented with its essential manufacturing steps, properties, and some specific fields of application.

6.2
CVI Manufacturing Process for CMCs

The process for SiC infiltration is mostly identical to the coating process called chemical vapor deposition (CVD). In a furnace at temperatures above 800 °C, the fiber preform is exposed to a process gas (vapor). For the deposition of SiC, this gas is usually a mixture of hydrogen and methyl-trichloro-silan (MTS, CH_3SiCl_3). Because the gas also reaches the inner layers of the preform, the process is not called the CVD-process, but instead the CVI-process. The overall chemistry of this process follows the simple scheme:

$$CH_3SiCl_3 \xrightarrow{\text{in } H_2} SiC + 3\, HCl$$

and shows that hydrogen is working as a catalyst. Beta-SiC is generated in this process.

To ensure the deposition of SiC to a good and near stoichiometric quality, three parameters have to be observed, namely pressure, temperature, and volume ratio of hydrogen and MTS. According to [5], lower pressures and temperatures require a higher H_2 to MTS ratio to reach stoichiometric SiC deposition. Lower ratios lead to carbon-rich deposition and too much hydrogen yields silicon-rich matrix material. Higher temperatures and pressures increase the deposition rate and thus reduce the processing time.

To ensure the penetration of gas into deeper layers of the preform, the penetration depth of gas into pores has to be taken into account. Figure 6.3 shows the

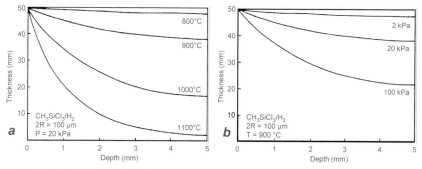

Figure 6.3 (a) Thickness of SiC deposition in a pore with a diameter of 100 microns for different temperatures at 20 kPa pressure; (b) for different pressures at 900 °C.

calculated effect of temperature and pressure on the penetration depth of gas into a cylindrical pore of 100 micron diameter [6]. It can be seen that higher temperatures and pressures substantially reduce the penetration depth, because the deposition close to the surface is closing the entrance to the pore faster under these conditions. Lower temperatures and pressures increase the mean free path of the gas molecules, so they can penetrate deeper into the preform channels.

Basically, two types of CVI processes are relevant in the field of CVI techniques for industrial manufacturing of SiC-matrix composites with either carbon or silicon carbide fibers: isothermal-isobaric infiltration and temperature-pressure gradient infiltration.

6.2.1
Isothermal-Isobaric Infiltration

In the isothermal-isobaric process [7], the fiber preform is placed in a special furnace (Figure 6.4) and heated to the process temperature somewhere below 1000 °C. Then, at a well defined pressure, typically around 100 hPa, a mixture of hydrogen and MTS is introduced into the furnace at a well defined flow rate. The process conditions have to be chosen in a way that avoids premature closure of the preform surface by the growing SiC. Since the movement of gas molecules is controlled by diffusion only, a higher mean path length is necessary. Therefore, comparatively low process pressures and temperatures are necessary, which in turn lead to process times of several weeks. Furthermore, the wall thickness of preforms is limited to a few (~3–5) millimeters. Components with higher wall thickness have to be infiltrated in two or more infiltration steps, with an intermediate grinding-off of the tight surface layer.

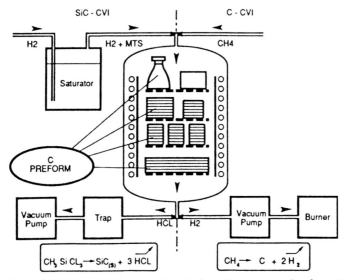

Figure 6.4 Scheme of an isothermal and isobaric CVI process, taken from [6].

6.2.2
Gradient Infiltration

In the gradient CVI-process [8, 9], the same process gas is penetrating the preform, not by diffusion but by being forced through via a pressure gradient. In this method it is necessary to place the preform in a device that seals the gas input side from the output side (Figure 6.5). To avoid a faster SiC deposition on the input side, a thermal gradient is applied over the wall thickness. The gas input side is kept at a lower temperature level. Therefore, at the beginning of the process, the SiC deposition rate is higher on the hotter output side. With increasing infiltration, the heat conductivity of the preform grows and the deposition rate on the input side increases. At the same time, it decreases on the output side, because the process gas is mostly used up before reaching there. During the infiltration process the matrix grows from the hotter side of the preform to the cooler side.

If temperature gradient and gas flow are properly controlled, the SiC matrix is evenly distributed over the wall thickness. Because no diffusion mechanism has to be taken into account, the temperature and pressure level can be raised to more than 1150 °C and up to ambient pressure. Both measures increase the chemical reaction rate and the supply of MTS molecules. The necessary infiltration time in the gradient CVI process therefore is reduced to a few days. Furthermore, due to forced gas flow, fresh process gas is always present at the site of the deposition

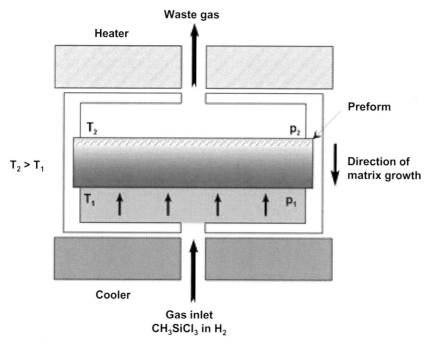

Figure 6.5 Scheme of a gradient CVI process.

and less intermediate gas products of the MTS decomposition are poisoning the process.

6.2.3
Discussion of the Two CVI-processes

For the industrial production of CMC, usually a programmable and automatic system is installed, which controls furnace temperatures, gas flows, and process pressure in a reproducible way, depending on the type and size of the preforms. The coating of the fibers with carbon or boron nitride tends to be done in the same furnace.

The isothermal CVI process has the advantage of great flexibility with respect to the shape of the preform. Furthermore, in large furnaces, many preforms, even with different shapes, can be infiltrated in one shot, which reduces specific costs for the CMC components. The problem with this process is the long infiltration time due to the diffusion controlled infiltration mechanism, which in turn leads to higher component costs and increased length of design cycles for developing components. The limitation of the wall thickness to about 5 mm is another drawback of this process, if a one-shot process is chosen.

The gradient CVI process has the advantage of short infiltration times, allowing quicker reaction to design needs, and the capability to infiltrate a preform wall thickness up to about 30 mm. The problem with this process lies in the need for tools to cool the preform and force the gas flow through. Therefore, only simple shaped components, such as tubes with round and rectangular profiles and plates, have been infiltrated in this process. More complex parts need to be built by joining techniques from such elementary components.

In the present situation of industrial manufacturing, both types of processes show advantages in specific situations. Also, the small market for CMC components does not allow a final judgement about a commercial or technical advantage of one or the other method.

It should also be mentioned that other types of CVI processes have been studied, for example, a simple forced flow process with pressure gradient and without temperature gradient [10] and a pressure pulse process [11]. However, they have not as yet reached an industrial level of production. The same holds for the technology to infiltrate the fiber preform via an electrophoretic infiltration process [12, 13].

6.3
Properties of CVI Derived CMCs

6.3.1
General Remarks

The SiC-matrix deposited with a CVI process is a homogeneous beta-SiC with very fine crystalline structures (Figure 6.6). They grow in radial orientation around the

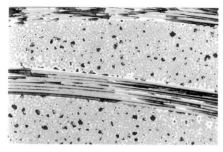 

Figure 6.6 Micrographs of SiC fiber reinforced SiC, showing the SiC matrix as light areas between the gray fibers. Pores are black and pores between the layers of fabric have an extension up to about 0.5 mm. Many of the small pores in between the filaments seen on the left are closed.

Figure 6.7 With three-dimensional fiber structures; the problem of weak interlaminar strength can be solved.

filaments of the fibers. The carbon or boron nitride coating of the filaments has a graphite structure with the planes parallel to the filament surface. This allows the matrix to glide along the filaments.

The use of fiber fabrics, three-dimensional structures (Figure 6.7), or other structures of fibers, for example, filament wound preforms, must have channels wide enough to allow the gas to penetrate into the deeper layers of the preform before being closed too soon by the SiC deposition. Thus, at the end of an isothermal infiltration process, the material has a closed porosity if a tight surface has grown on the preform. A mixture of closed and open porosity is the result of a gradient-CVI process.

A general feature of all types of CMC material properties, besides their residual porosity, is their anisotropy, given by the pattern of the fiber orientation. CMC material with two-dimensional fiber structures, often realized by conventional lay-up of fabrics, shows low strength values perpendicular to the fabrics. Interlaminar shear strength and interlaminar crack resistance are low, which has to be considered at the design stage, when bending loads for the component are expected.

A special difference between CMC using carbon and silicon carbide fibers should be mentioned. Since carbon and silicon carbide have a different coefficient of thermal expansion and the matrix forming process takes place at high temperatures, the silicon carbide shrinks more than the carbon fibers when cooling down to ambient temperature. This leads to a pattern of micro-cracks in the matrix of all C/SiC materials, while SiC/SiC material has no micro-cracks in its fabricated state.

Furthermore, the fiber content has to be greater than a certain minimum to carry the tensile load, which at the moment of crack formation immediately has to be sustained only by the reduced cross-section of the fibers in load direction. The matrix no longer contributes after the crack has formed.

Finally, material data of CMC usually show a scatter in the range of 10 to 20%, and the type of CVI process has no significant influence on the properties.

6.3.2
Mechanical Properties

6.3.2.1 Fracture Mechanism and Toughness

The main feature of CMCs is their elongation capability and fracture behavior. The latter is usually characterized by the fracture toughness or the stress intensity factor, K_{Ic}. For the exact measurement of this quantity, it is necessary to determine the size of the fracture surface. The complex pattern of cracks, which are formed in the matrix and between fibers and matrix, does not allow this measurement. Therefore, measurements of samples for single edge notch bend (SENB) tests [14, 15] were done using the initial notch surface as the value of the fracture surface, yielding lower data than with the true, but unknown and larger surface. The calculation based on this value was defined as the "formal stress intensity factor" (SIF). The data obtained are not values characterizing the material in the conventional sense, but allow a comparison of different types of CMC materials, if the samples have the same geometry. The results shown in Figure 6.8 demonstrate that different types of CMC material have different behavior concerning the crack resistance. The liquid silicon infiltration process is yielding the lowest peak values, because this process cannot ensure the single filament pull-out mechanism, but rather only something like a filament bundle pull-out. Furthermore, compared with conventional silicon carbide, the energy for pushing the crack through the sample is a multiple for all CMC types (the area under the curve is a measure for the energy it takes for complete crack formation).

The peak value of the curves gives information about the force, which is necessary to make the crack propagate through the sample. Materials produced by CVI and LPI techniques are superior to material of the other process types, mainly because in CVI and LPI materials, the filament pull-out mechanism makes best

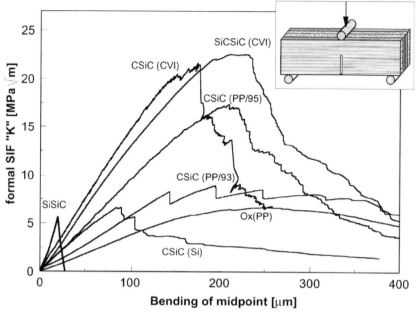

Figure 6.8 The result of single-edge-notch-bend (SENB) test demonstrates the variety of different CMC materials with respect to crack resistance. CSiC(CVI) and SiCSiC(CVI): CVI materials with C- and SiC-fibers; CSiC(PP/95) and CSiC(PP/93): two types of LPI material; Ox(PP) is oxide CMC and CSiC(Si) is LSI-derived material.

use of the fiber and filament properties. The result of this property is the insensitivity toward notches and bearing loads [16, 17].

It should be noted that often the damage tolerance of CMC material is demonstrated by measurements of the bending strength, the documented curves which usually show some quasi-plastic behavior; but these measurements do not indicate the toughness level. The quantitative measurement of the stress intensity factor of any material cannot be performed by bending tests.

6.3.2.2 Stress-Strain Behavior

Following the toughness behavior, the elongation capability of CVI material is more than 10-fold compared to conventional SSiC. Figure 6.9 and the data in Table 6.1 demonstrate this behavior, which again shows that the CVI process technique is using the fiber potential in an effective way. In Figure 6.9, SiC/SiC has an exact linear elastic behavior up to a load of about 90 MPa. Beyond this load the matrix starts to build up a pattern of micro-cracks, which are bridged by the filaments. After a first load, this linear elastic behavior is no longer observable and the Young's modulus is reduced (Figure 6.10).

To completely characterize CMC materials, it is advisable to look at three test results: bending strength, tensile strength, and toughness. CMC material with

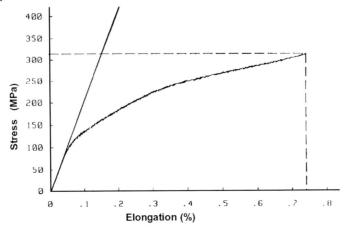

Figure 6.9 The stress-strain curve of SiC/SiC material, derived via CVI, shows a linear elastic range during the first loading (the straight line is an extra line created by the computer to determine the slope and the Young's modulus).

Table 6.1 Data of CVI derived C/SiC and SiC/SiC compared with those of conventional silicon carbide.

	Unit	CVI-SiC/SiC	CVI-C/SiC	SSiC
Fiber fraction	%	42–47	42–47	–
Density	g cm^{-3}	2.3–2.5	2.1–2.2	3.1
Porosity	%	10–15	10–15	<1
Tensile strength	MPa	280–340	300–320	100
Strain to failure	%	0.5–0.7	0.6–0.9	0.05
Young's modulus	GPa	190–210	90–100	400
Bending strength	MPa	450–550	450–500	400
Compressive strength	MPa	600–650	450–550	2200
Interlaminar shear strength	MPa	45–48	45–48	–
Coefficient of thermal expansion at r. t.	10^{-6} K^{-1}	4 (pl) / 4 (ve)	3 (pl) / 5 (ve)	4.1
Heat conductivity at r. t.	W/mK	20 (pl) / 10 (ve)	14 (pl) / 7 (ve)	110

"pl" and "ve" denote parallel and vertical orientation to the fabric, respectively.

high fiber content and only a small fraction of matrix material will show a high level of tensile strength, because the test is close to a fiber tensile test; the same material will show poor bending result, because bending a fiber bundle does not require much force.

On the other hand, material with low fiber content and a high share of matrix material will show high bending strength values close to those of conventional

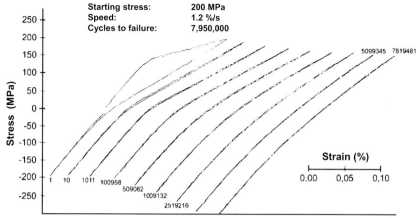

Figure 6.10 In strain controlled low-cycle-fatigue (LCF) tests, CVI-SiC/SiC reaches about 8 million cycles before rupture. The stress level at the beginning has been 80% of the material strength. On the tensile side, the Young's modulus is changing after the first load.

ceramics, and poor elongation capabilities, because there are not enough fibers to carry the load in the cracks.

6.3.2.3 Dynamic Loads

The crack bridging and pull-out mechanism allow the use of high-quality CMC materials under dynamic loads. Figure 6.10 shows the test results of a strain controlled low-cycle-fatigue test with SiC/SiC material derived via a gradient CVI process. With a starting load of 80% of the tensile strength, about 8 million cycles between the tensile and compressive strain position were reached before failure. On the tensile side, the fatigue effect on the stress level is small; on the compressive side, a higher increase of the stress level is observed, because the surfaces of the cracks do not match perfectly when pressed together. It can also be seen that the stress strain curve after the first load changes its slope (the Young's modulus) to a lower value.

6.3.2.4 High Temperature Properties and Corrosion

Mechanical reliability under mechanical and thermal stress, as well as thermal shock resistance and mechanical strength in a temperature range between 750 and 1300 °C, and sometime above, have been the major incentive to develop CMC materials and components for hot structures for space re-entry vehicles and energy plants (e.g. heat exchangers, combustion chambers, and components for gas turbines). The properties of CVI derived SiC/SiC and C/SiC material under high temperatures depend heavily on the details of the conditions such as load, ambient gas composition, and speed of the gas. Several publications, including modeling, deal with the topic of high temperature corrosion of silicon carbide based CMC materials, using either carbon or silicon carbide fiber material under various conditions and including some measures of protection [18–23]. The basic properties can be summarized as follows:

The behavior under thermal shock can be characterized as being excellent. Figure 6.11 shows a test result, where small pieces of $25 \times 25 \times 3$ mm have been exposed to an acetylene-oxygen flame and cold air in cycles lasting 30 s. Within these cycles the temperature changed between 300 and 1100 °C. No macroscopic damage was observed with SiC/SiC after 1000 cycles, zirconia was destroyed after 3 cycles, and SiSiC after 10 cycles.

SiC matrix material derived by CVI can stand temperatures over 2300 °C in vacuum or an inert gas atmosphere without problems, and this is also the case for carbon fibers. Standard silicon carbide fibers start to reorganize their amorphous structure at temperatures above 1200 °C and lose their mechanical properties, even in vacuum [24]. Meanwhile, more stoichiometric and crystalline SiC-fibers are available from Nippon Carbon (Hi-Nicalon S) and Ube Industries (Tyranno SA), showing improved high temperature stability [25, 26].

Most important for high temperature applications is the behavior in the presence of oxygen. Carbon fibers start to convert into CO_2 at temperatures above 500 °C and the amorphous silicon carbide fibers also start reacting with oxygen at about the same temperature, due to their content of free carbon. Crystalline silicon carbide starts to form SiO_2 to a significant extent at temperatures above 850 °C. The formation of this glassy coating gives a passivation effect and slows down further penetration of the oxidation process. At temperatures above 1600 °C, and in an oxygen atmosphere with a low partial pressure of about 5 kPa, lack of oxygen leads to the formation of gaseous SiO [5]. Under this so-called process of active oxidation, the silicon carbide rapidly disappears.

Therefore, it is clear that in components for the envisaged heat shields of space re-entry vehicles or gas turbines, an anti-oxidation protection system had to be developed for the silicon carbide based CMC materials. Such systems, based on CVD coating with SiC, have been qualified for components of space

Figure 6.11 SiC/SiC (left), SiSiC and ZrO2 (right) samples, $25 \times 25 \times 3$ mm, after thermal shock tests in cycles of 30 s between 300 and 1100 °C.

re-entry vehicles, where a qualification for less than 100 hours of lifetime is sufficient. For longer lifetimes, coating systems based on oxides are under development [27, 28]. These are of limited success so far, except for an oxide based system, which has successfully been applied to gas turbine combustors [29] (Section 6.4.2).

In other environments, other than in hot oxygen containing gases, corrosion problems are less severe. Silicon carbide and carbon are among the most corrosion resistant materials that exist. Only strong bases can corrode silicon carbide, and carbon fibers show corrosion not only with oxygen at high temperatures above 500 °C, but also with strongly oxidizing agents such as concentrated nitric acid [5].

6.3.2.5 Thermal and Electrical Properties

Due to the anisotropic orientation of fibers, thermal and electrical properties also are dependent on the direction of measurement. The data are influenced and characterized by the type of fiber, the level of porosity, and the anisotropy of the material. Silicon carbide is a semi-conductor, where the electrical resistance is reduced with rising temperature, and carbon fibers have a much better conductivity, both electrical and thermal, than silicon carbide fibers. Due to the locally inhomogeneous material, the scatter of data is usually in the range of 10 to 20%. The scatter is increasing if the sample size is reduced, because then the local variations of the porosity become more important.

6.4
Applications and Main Developments

6.4.1
Hot Structures in Space

Hot structures for space vehicles, where the material for reusable components has to survive in several thermal and mechanical load cycles for a total of less than 100 hours, have been qualified. In the European program, HERMES, several components for heat shields and thermal insulation systems have been developed [30, 31]. In a German development program for hypersonic airplane technology, (HYTEX), a component for the hot air intake ramp (Figure 6.12) has been designed and manufactured on the basis of CVI and LPI technology. In this program, for the first time screws and nuts made from CVI-C/SiC material were used (Figure 6.13). Under mechanical load of 32 kN at 1600 °C, this component showed the deformation pre-calculated by design [32].

In a further program financed by ESA, the German and Bavarian government, the rear body flaps (Figure 6.14) of the NASA crew return vehicle X-38 were designed, manufactured in C/SiC material, and qualified [33]. In total, three such flaps were produced. One was used for extensive qualification tests. A mechanical deformation test at IABG, Munich (Germany), again demonstrated the correct design of the stiffness and deformation. The central bearing was tested at DLR in

Figure 6.12 C/SiC component of an air intake ramp developed in the HYTEX program for hypersonic jet engines.

Figure 6.13 C/SiC screws and nuts, which can be used due to their damage tolerance and notch insensitivity.

Figure 6.14 Pair of C/SiC body flaps for flight control during the re-entry phase of the space vehicle X-38 of NASA. Each flap is about 1.6 × 1.5 × 0.15 m and has a weight of 68 kg. It is probably the largest component that is CMC constructed, built and qualified for flight so far.

Figure 6.15 Development of a new type of C/SiC flight control flap, where the momentum is introduced via an axis. The length of this test sample is about 450 mm.

Stuttgart (Germany) under near to real re-entry conditions. Under a mechanical load of 4 tons and simultaneous ±8 degree movement at 1600 °C in air with controlled oxygen concentration, 5 re-entries were successfully simulated [34]. After acceptance by NASA, two flaps were mounted in Houston, Texas, and final mechanical acceptance tests for the flap control were performed. Due to financing and space shuttle problems, NASA cancelled the test flight, in which a space shuttle would have brought the unmanned X-38 into space for a first realistic re-entry event with one of the largest CMC components built so far. More than 400 screws and nuts have been used for the joining of the 4 segments of each flap.

In European programs (e.g. the Future Launcher Preparatory Programme FLPP, SHEFEX, Pre-X) of the European Space Agency (ESA), further development work has commenced [35], where different concepts for controlling the re-entry phase with moveable rudders and other hot re-entry control surfaces are to be tested. Figure 6.15 shows one example, which concentrates on innovative design details and new preform and joining techniques.

6.4.2
Gas Turbines

The expected possible increase of efficiency is the motivation to develop CMC material as hot gas turbine components. With respect to mechanical load, the combustion chamber can be considered as the component with the lowest thermomechanical requirements. To realize efficiency increase, the whole system of combustor, stator vanes, and rotor vanes has to sustain the increased temperature level. The thermal shock resistance of SiC/SiC material has been demonstrated in combustion chamber tests [36] lasting a total of 150 hours, with peak temperatures up to 1400 °C and several start/stop cycles.

In the United States, the problem of oxidation of ceramic fibers and SiC matrix by residual oxygen and hot water vapor was attacked by using crystalline silicon carbide fibers and an oxidation protection coating based on a system of oxide layers [37]. A test with a combustor has reached more than 15 000 hours [38] at a conventional temperature level.

Further progress is under development with respect to improved fibers based on a SiBNC composition, which are expected to show further improvement of properties in temperatures up to 1500 °C and in the presence of oxygen (Chapters 1 and 2).

For the turbine engines of military jet planes CMC components, weight saving flaps for thrust control, have been developed and tested in France [39, 40]. The temperature load of these components under normal flight conditions is low (~600 °C). Therefore, several hundred hours of flight time, including shorter periods of higher temperature loads at about 900 °C, can be achieved.

6.4.3
Material for Fusion Reactors

The high temperature properties and the first results of irradiation tests have made CMC materials based on crystalline SiC fibers and matrix interesting candidates for components in future fusion reactor concepts [41–46]. The main reason is because crystalline SiC has shown promising behavior when irradiated with neutrons. Therefore, the crystalline silicon carbide fibers have to be chosen for this application. Tests of material with Tyranno SA fibers from UB.E. Industries and Hi-Nicalon Type S fibers from Nippon Carbon, as well as Sylramic fibers developed by Dow Corning together with CVI derived SiC matrix, are on their way for application oriented testing. Within the frame of the EFDA (European Fusion Development Agreement), for example, two- and three-dimensional (Figure 6.7) samples have been supplied to check the effect on through-thickness data.

The differences, with respect to through-thickness strength and heat conductivity, are listed in Table 6.2.

6.4.4
Components for Journal Bearings

The use of sintered silicon carbide (SSiC) in journal bearings of pumps started in the 1980s [47]. In these bearings, pumped liquids such as water or hazardous

Table 6.2 Comparison of through-thickness data of two- and three-dimensional SiC/SiC based on Tyranno SA fibers from Ube Industries.

	2D-SiC/SiC	3D-SiC/SiC
Strength (MPa)	6.8	16
Heat conductivity (W/mK)	17.0	24.3

chemicals are used as lubricants. Examples are magnetic coupled pumps or pumps for seawater desalination plants, water gates, or waterworks.

The use of CMC material improves the reliability of such bearings in cases where conventional ceramics, especially for large pumps running under heavy load conditions, are not working reliably enough.

To qualify such material, the tribological behavior of CVI-derived C/SiC and SiC/SiC was studied in ring-on-disc tests [48, 49]

The surface of a machined CMC (Figure 6.16) demonstrates that the tribological properties of the material and the tribological system can be influenced by:

- the type of fibers used (carbon or silicon carbide);
- the orientation of the fibers with respect to the surface;
- the filling of the porosity with some proper material.

Lubricant, temperature, and tribological partner are further parameters of the system.

Results of the ring-on-disc tests are presented in Table 6.3. They were obtained following the German standard DIN 58835, in which a ring with 14 mm inner and 20 mm outer diameter is pressed onto a rotating disc. The rubbing conditions of $2\,m\,s^{-1}$ and 5 MPa pressure realize mixed friction.

It can be seen that tribological systems can be complex: with SSiC as rubbing partner, no influence of the fiber orientation has been observed (line 2 of the table), while under dry conditions or with aluminium oxide as partner in water, the results with fibers of vertical orientation were much better than with parallel orientation (line 3 and 1 of the table). Looking at the effect of the type of fiber, with parallel fiber orientation with SSiC as partner, there was also no significant difference in water, while in air and in water with aluminium oxide as partner, there is a significant difference (lines 4, 5, and 6 of the table).

Figure 6.16 The surface of SiC/SiC demonstrates the mixture of open pores, areas with fibers of different orientation, and matrix material.

Table 6.3 Tribological data of ring-on-disk test.

No.	Disk	Ring	Medium	Wear in mm^3 h^{-1}		Coeff. of friction
				Ring	Disk	
1	SiC/SiC pl	Al$_2$O$_3$	Water	0.080	0.150	n.a.
	SiC/SiC ve	"	"	0.014	0.065	n.a.
2	SiC/SiC pl	SSiC	Water	0.005	0.005	0.04
	SiC/SiC ve	"	"	0.004	0.005	0.03
3	SiC/SiC pl	SSiC	Air	4.0	36.0	0.66
	SiC/SiC ve	"	"	1.8	17.3	0.66
4	SiC/SiC pl	Al$_2$O$_3$	Water	0.080	0.15	n.a.
	C/SiC pl	"	"	0.005	0.03	n.a.
5	SiC/SiC pl	SSiC	Water	0.005	0.005	0.03
	C/SiC pl	"	"	0.003	0.010	0.10
6	SiC/SiC pl	SSiC	Air	4.0	36.0	0.66
	C/SiC pl	"	"	0.013	1.16	0.12

"pl" denominates samples with fiber orientation parallel to the friction surface, "ve" with vertical orientation.

Further tests in water with zirconia, tungsten carbide, and chromium oxide as partner materials, showed higher wear and friction coefficients and the pair of SSiC and C/SiC or SiC/SiC as partner turned out to be the best. Subsequent tests on radial and axial journal bearings at the pump company KSB in Frankenthal, Germany, showed:

- The open porosity of CVI-derived CMC material has to be impregnated to allow the build-up of a hydrodynamic lubricating film.

- The pair of SiC/SiC running against SSiC allowed a factor of 2.5 to 3 higher pressure times velocity (so-called PV) values, compared to many other tested pairs of materials (Figure 6.17).

In these tests, speed or load of the ring on disk rig were increased and the limiting speed or load was measured with 80 °C water as the lubricant. For many systems, the product, the so-called PV value, is found in area 1 of Figure 6.17. The system of SiC/SiC running against SSiC, shown in curve 2, demonstrates the capability of taking almost the triple load of all other tested systems.

With this result it was possible to furnish large pumps with shaft diameters of 100 mm or more (up to >300 mm) with all-ceramic journal bearings, where CMC would be used as shaft sleeve material (Figure 6.18), shrink fitted on the shaft component, and conventional SSiC would be the bearing, also shrink fitted into its metallic environment [50, 51]. In this case, SSiC is under compressive stress and thus shows sufficient reliability. Since 1994, several hundred pumps have been equipped with this bearing system, which in some cases has proven more than 40 000 successful running hours. For small pumps, the reliability of bearings

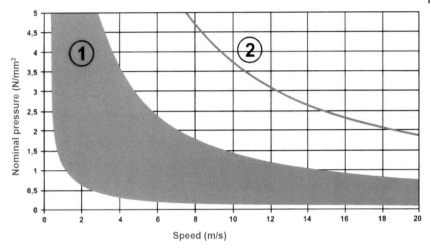

Figure 6.17 Result of tests, where the load limit of the tribological system of different pairs of material in water of 80 °C was determined. Area 1 is a region where the load limits of several pairs of conventional materials, including ceramics, have been found; line 2 gives the limit of PV values of the pair SSiC running against SiC/SiC material.

Figure 6.18 SiC/SiC shaft sleeves for water lubricated journal bearings of large pumps. The outer diameter of these sleeves lies between 100 and 305 mm.

using conventional SSiC is technically sufficient and for economic reasons CMC material is not used.

On the basis of these results, development programs are testing the possibility of applying this bearing concept for liquid oxygen (LOx) turbo pumps of space rocket engines. In a first step, the compatibility of C/SiC and SiC/SiC with LOx was checked. Aging tests, an impact test (ASTM Standard D2512–95), and auto-ignition tests (French standard NF 29–763) were successfully performed. No deterioration of strength or traces of ignition on impacts with 100 J of energy were observed. Auto-ignition of fine powder of all constituents of CMC (carbon fiber, silicon carbide fiber, silicon carbide CVI-matrix) was tested at 120 bar of pure oxygen with temperatures up to 525 °C. Only the powder of carbon fibers showed auto-ignition at 437 °C.

In pin-on-disc tests, the tribological behavior under mixed friction, a series of several loads and speeds was measured. Pins with a diameter of 8 mm were pressed onto a rotating disk and the pin center formed a circle of 70 mm diameter on the disk. Wear and friction coefficient of the test with a load of 2.8 MPa and a speed of 4 m/s, the highest tested values, are shown in Figure 6.19. They demonstrate a friction coefficient reduced to about 50% and a wear rate reduced by roughly a factor of 50 compared with a metallic (steel 440C) system. They also show that under these conditions there is no difference between C/SiC and SiC/SiC materials, and that for SiC/SiC the fiber orientation on the rubbing surface has no real effect [52].

In a subsequent test series of a hydrostatic journal bearing with a CVI-SiC/SiC shaft sleeve and an SSiC bush (Figure 6.20), a promising result was obtained. Several hours at a speed of 10 000 rpm (shaft diameter 70 mm) with changing radial loads between 120 and 1150 N and a series of 49 start/stop-tests with various load profiles, showed no significant wear and good behavior with respect to stiffness and LOx consumption. The inner surface of the bush became shiny and its roughness was reduced from an R_a of 0.3 to 0.04 microns. Due to a slight misalignment in one area, the wear was 3 microns [53].

In case of a successful further qualification, longer life times and reusability of such pumps together with better damping properties, simpler construction, and reduced costs for mounting and maintenance for such bearings, can be expected [54].

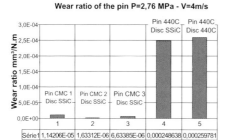
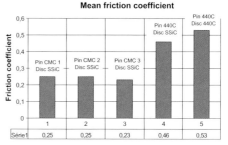

Figure 6.19 The diagrams show wear (left) and coefficient of friction (right) of pin-on-disk tests in liquid oxygen (LOx). The pin diameter has been 8 mm and the circle diameter on the disk 70 mm. The data shown are for a load of 2.76 MPa and a speed of 4 m/s. CMC1: SiC/SiC pin with fabric perpendicular to friction surface. CMC2: SiC/SiC pin with fabric parallel to friction surface. CMC3: C/SiC pin with fabric perpendicular to friction surface. 440C: type of steel, which can be used in LOx. SSiC: sintered silicon carbide.

Figure 6.20 Journal bearing system for application in liquid oxygen. The shrink fit on a steel component still holds at LOx temperatures; this means that at room temperature the CMC sleeve has a permanent hoop stress of about 160 MPa.

6.5
Outlook

The class of the different CMC materials is still new and starting to appear on the industrial market. The situation is comparable to the fiber reinforced plastic materials of the 1960s or 1970s. Therefore, applications concentrate on components, which would mean high efforts, risks, and costs with conventional ceramics. Normally such components have a specific function in a larger system and, although costly, create a substantial benefit.

Several applications of CVI-derived CMC components are under field observation or development, beyond the ones already mentioned above. Examples are valve components, which have to sustain temperature and pressure shocks, heat shields for racing cars, casings for rocket engines, and fins for supersonic rockets.

However, knowledge about these innovative materials is still limited and not yet a significant part of any education curriculum for engineers. Furthermore, it is still necessary to collect additional data, and their statistical behavior, and particularly information about the performance under more or less well defined conditions in various applications. Communication between research and development engineers also has to be intensified.

The main focus of future development work probably will be:

- Improvement of thermal (>1200 °C) stability of ceramic fibers in corrosive environments, such as air or exhaust gases. The incentives for these activities are applications of CMC components in high temperature heat exchangers and gas turbines. The long-term stability will have to face the problem of high temperature behavior of inhomogeneous materials, which tend to change their structure via diffusion according to the physical laws of phase diagrams.

- Reduction of manufacturing costs for silicon carbide fibers and the ceramic composite, which is the cost for the matrix creation. This will allow increasing markets, especially since several applications have been technically successful, but are not yet applied because of high costs.

References

The following publications are only a small fraction of those published in the last 30 years, which are dealing with different CVI process technologies, their modeling, the material properties and models, and the various experimental results of application oriented tests, for example, in the field of corrosion or tribology. For more and deeper detail, a literature survey via the Internet in accessible databases or other sources is necessary.

1 Phillips, D.C. (1983) Fibre reinforced ceramics, in *Handbook of Composites*, Vol. 4 (eds A. Kelly and S.T. Mileiko), Elsevier Science Publishers, p. 373.

2 Marshall, D.B. and Evans, A.G. (1985) Failure mechanisms in ceramic-fiber/ceramic-matrix composites. *Journal of the American Ceramic Society*, **68** (5), 225.

3 Rebillat, R. et al. (2000) The concept of a strong interface applied to SiC/SiC composites with a BN interphase. *Acta Materialica*, **48** (18/19), 4609.

4 Pritzkow, W. et al. (2007) *Failure effects in oxide ceramic matrix composites with defects due to the laminating*. 10th International Conference and Exhibition of the European Ceramic Society, Berlin, paper E-689.

5 Schröder, F. et al. (ed.) (1986) *Gmelin Handbook of Inorganic Chemistry, Si–Silicon, Supplement Volume B3, Silicon Carbide, Part 2*, 8th edn, Springer Verlag Berlin, Heidelberg, New York, Tokyo, pp. 194, 322–97.

6 Naslain, R.A. and Langlais, F. (1986) CVD-processing of ceramic-ceramic composite materials. *Material Science Research*, **20**, 145–64.

7 Naslain, R.A. (1989) The CVI-processing of ceramic matrix composites. *Journal de Physique Collogue C5*, Supp. au No. **5**, 191.

8 Stinton, D.P. et al. (1986) Synthesis of fiber-reinforced SiC composites by chemical vapor infiltration. *Journal of the American Ceramic Society Bulletin*, **65** (2), 347.

9 Mühlratzer, A. (1999) Production, properties and applications of ceramic matrix composites, *cfi/Ber. Deutsche Keramische Gesellschaft*, **76** (4), 30–5.

10 Roman, Y.G. (1994) *Forced Flow Chemical Vapour Infiltration*, doctor thesis. Addix, Wijk, Netherlands, ISBN 90-9007243-8.

11 Bertrand, S., Lavaud, J.F. et al. (1998) The thermal gradient-pulse flow CVI process: a new chemical vapor infiltration technique for the densification of fibre performs. *Journal of the European Ceramic Society*, **18** (7), 857.

12 Stoll, E. et al. (2005) Progress in the electrophoretic deposition technique to infiltrate oxide fibre materials for fabrication of ceramic matrix composites. Proceedings of the 2nd International Conference on Electrophoretic Deposition: Fundamentals and Applications, Castellvecchio Pascoli, Italy, 2005. *Key Engineering Materials*, **314**, 195.

13 Damjanovic, T. et al. (2005) Electrophoretic deposition of mullite based oxygen diffusion systems on C/C-Si-SiC composites. Proceedings of the 2nd International Conference on Electrophoretic Deposition: Fundamentals and Applications, Castellvecchio Pascoli, Italy, 2005. *Key Engineering Materials*, **314**, 201.

14 Kuntz, M. (1996) *Risswiderstand keramischer Faserverbundwerkstoffe*, Dissertation Universität Karlsruhe, Shaker Verlag.

15 Kuntz, M., Horvath, J. and Grathwohl, G. (2001) High temperature fracture toughness of a C/SiC (CVI) composite as used for screw joints in re-entry vehicles,

in *4th Int. Conf. On High Temperature Ceramic Matrix Composites (HT-CMC4)*, October 2001, Munich, Wiley-VCH Verlag GmbH, ISBN 3-527-30320-0, p. 469.

16 Sygulla, D. et al. (1993) *Integrated Approach in Modelling, Testing and Design of Gradient-CVI Derived CMC Components.* AGARD Report 795, Introduction of Ceramics into Aerospace Structural Composites, Neuilly sur Seine, France, chapter 14.

17 Mühlratzer, A. et al. (1996) Design of gradient-CVI derived CMC components. *Industrial Ceramics*, **16** (2), 111–17

18 Crossland, C.E. et al. (1997) Thermochemistry of corrosion of ceramic hot gas filters in service. Proceedings of the 2nd International Workshop on Corrosion in Advanced Power Plants. *Materials at High Temperatures*, **14** (3), 365.

19 Darzens, S. et al. (2002) Understanding of the creep behavior of SiC/SiBC composites. *Scripta Materialica*, **47** (7), 433.

20 Sauder, C. et al. (2004) The tensile behaviour of carbon fibers at high temperatures up to 2400 degree C. *Carbon*, **42** (4), 715.

21 Wu, S. et al. (2006) Oxidation behaviour of SiC-Al2O3-Mullit coated carbon/carbon. *Composites Part A*, **37** (9), 1396.

22 Mall, S. et al. (2006) Fatigue and stress-rupture behaviour of SiC/SiC composite under humid environment at elevated temperature. *Composite Science and Technology*, **66** (15), 2925.

23 Quemard, L. et al. (2007) Degradation mechanisms of a SiC fiber reinforced self-sealing matrix composite in simulated combustor environments. *Journal of the European Ceramic Society*, **27** (1), 377.

24 Kurtenbach, D. (2000) *Untersuchung zur Steuerung des Kristallisationsverhaltens von SiC aus amorphen Vorstufen*. Doctoral thesis Bergakademie Freiberg, Shaker Verlag, Aachen.

25 Havel, M. and Colomban, Ph. (2005) Raman and Rayleigh mapping of corrosion and mechanical ageing of SiC fibres. *Composite Science and Technology*, **65** (3–4), 353.

26 Takeda, M., Urano, A., Sakamoto, J. and Imai, Y. (1998) Microstructure and oxidative behaviour of silicon carbide fibre Hi-Nicalon Type S. *Journal of Nuclear Materials*, **258–263** (2), 1594.

27 Huang, J.-F. et al. (2005) Oxidation behaviour of SiC-Al2O3-mullit coated carbon/carbon composites at high temperature. *Carbon*, **43** (7), 1580.

28 Huang J.-F. et al. (2006) Preparation and oxidation kinetics mechanism of three-layer multi-layer-coatings-coated carbon/carbon composite. *Surface and Coating Technology*, **200** (18–19), 5379.

29 More, K.L. et al. (2002) *Evaluating Environmental Barrier Coatings on Ceramic Matrix Composites after Engine and Laboratory Exposures*, ASME Turbo Expo Amsterdam, paper GT-2002-30630.

30 Mühlratzer, A. et al. (1995) *Development of a new cost-effective ceramic composite for re-entry heat shield applications.* 46th International Astronautical Congress 2–6 October, 1995/Oslo, Norway, paper IAF-95-I.3.01.

31 Sygulla, D. et al. (1996) *Experiences in design and testing of ceramic shingle thermal protection systems – a review.* ESA/ESTEC Conference on Spacecraft Structures, Materials and Mechanical Testing, March 1996, Nordwijk, Netherlands.

32 Mühlratzer, A., Köberle, H. and Sygulla, D. (1996) *Hypersonic air intake ramp made of C/SiC ceramic composites.* 3rd International Conference on Composites Engineering, New Orleans, LA, USA, July 1996 (ed. D. Hue), p. 609.

33 Dogigli, M. and Weihs, H. et al. (2000) *New high-temperature ceramic bearing for space vehicles.* 51st International Astronautical Congress Oct. 2000, Rio de Janeiro, IAF-paper 00-I.3.04.

34 Pfeiffer, H. (2001) *Ceramic body flap for X-38 and CRV.* 2nd International Symposium on Atmospheric Re-entry Vehicles and Systems, Arcachon, March 2001.

35 Baiocco, P. et al. (2007) The pre-X atmospheric re-entry experimental lifting body: program status and system synthesis. *Acta Astronautica*, **61** (1), 459.

36 Filsinger, D. et al. (1997) *Experimental assessment of fiber reinforced ceramics for combustor walls.* International Gas Turbine

and Aeroengine Congress and Exhibition, Orlando, Florida, 2–5 June 1997, paper 97-GT-154.
37 van Roode, M. et al. (2001) Design and testing of ceramic components for industrial gas turbines, in *7th International Symposium on Ceramic Materials and Components for Engines*, Goslar, June 2000, Wiley-VCH Verlag GmbH, p. 261.
38 Miriyala, N. et al. (2002) *The evaluation of CFCC liners after field testing in a gas turbine–III*, ASME Turbo Expo 2002, Amsterdam, paper GT-2002-30585.
39 Spriet, P.C. and Habarou, G. (1996) *Applications of continuous fiber reinforced ceramic matrix composites in military turbojet engines*. International Gas Turbine and Aeroengine Congress, Birmingham, 1996, paper 96-GT-284.
40 Boullon, P., Habarou, G., Spriet, P.C., Lecordix, J.L., Ojard, G.C., Linsey, G.D. and Feindel, D.T. (2002) *Characterization and Nozzle Test Experience of a Self Sealing Ceramic Matrix Composite for Gas Turbine Applications*, ASME Turbo Expo, Amsterdam, paper GT-2002-30458.
41 Sharafat, S. et al. (1995) Status and prospects of SiC/SiC composite materials development for fusion applications. *Fusion Engineering and Design*, **29**, 411.
42 Jones, R.H. et al. (1998) Recent advances in the development of SiC/SiC as a Fusion Structural Material. *Fusion Engineering and Design*, **41**, 15.
43 Riccardi, B. et al. (2000) Status of the European R&D activities on SiC_f/SiC composites for fusion reactors. *Fusion Engineering and Design*, **51**, 11.
44 Hinoki, T. (2002) Effect of fiber properties on neutron irradiated SiC/SiC composites. *Materials Transactions JIM*, **43** (4), 617.
45 Riccardi, B. et al. (2004) Issues and advances in SiC_f/SiC composites development for fusion reactors. *Journal of Nuclear Materials*, **329–333**, 56.
46 Tost, S. et al. (2007) Nitrogen and helium permeation tests of SiC/SiC composites. *Fusion Engineering and Design*, **87**, 317.
47 Prechtl, W. (1993) Technische Keramik für Pumpen und Armaturen (Technical Ceramics for Pumps and Fittings). *Keramische Zeitschrift*, **45** (4), 197.
48 Leuchs, M. and Prechtl, W. (1996) *Faserverstärkte Keramik als tribologischer Werkstoff im Maschinenbau*, DGM-Tagung Reibung und Verschleiß, Bad Nauheim, März.
49 Leuchs, M. (2002) *Ceramic Matrix Composite Material in Highly Loaded Journal Bearings*, ASME Turbo Expo, Amsterdam, paper GT-2002-30460.
50 Gaffal, K. et al. (1997) *Neue Werkstoffe ermöglichen innovative Pumpenkonzepte für die Speisewasserförderung in Kesselanlagen*, VDI Berichte 1331 Innovationen für Gleitlager, Wälzlager, Dichtungen und Führungen, VDI-Verlag, Düsseldorf, pp. 275–89.
51 Kochanowski, W. and Tillack, P. (1998) *New Pump Bearing Materials Prevent Damage to Tubular Casing Pumps*. VDI-Berichte No. 1421, pp. 227–42.
52 Bozet, J.L., Nelis, M., Leuchs, M. and Bickel, M. (2001) *Tribology in liquid oxygen of SiC/SiC ceramic matrix composites in connection with the design of hydrostatic bearing*. 9th European Symposium on Space Mechanisms and Tribology, Liège, Belgium, September 2001, ESA Publication SP-480, p. 35.
53 Bickel, M., Leuchs, M., Lange, H., Nelis, M. and Bozet, J.L. (2002) *Ceramic journal bearings in cryogenic turbo-pumps*. 4th International Conference on Launcher Technology–Space Launcher Liquid Propulsion, Liège, Belgium, paper 129.
54 Henderson, T.W. and Scharrer, J.K. (1992) *Hydrostatic bearing selection for the STME hydrogen turbopump*. 28th Joint Propulsion Conference and Exhibition, Nashville, TN, USA, paper AIAA 92-3283.

7
The PIP-process: Precursor Properties and Applications

Günter Motz, Stephan Schmidt, and Steffen Beyer

7.1
Si-based Precursors

7.1.1
Introduction

Organometallic compounds (precursors) based on silicon have attracted considerable interest in recent years, owing to their promising potential for the formation of ceramics with tailored properties [1–4], matrices in composites [5–7], ceramic fibers [8–14], and polymer as well as ceramic surface coatings [15–19]. Since the first synthesis of the very simple silicone $[(H_5C_6)_2SiO]_x$ in 1901 by Kipping [20], pioneering works of Fink [21, 22] in the 1960s, and Verbeek and Winter [23] in the mid-1970s, a wide variety of precursors have been developed for the preparation of different polymer derived ceramics [24–28] (PDC). The major advantages of such polymer based materials are their intrinsic homogeneity at the atomic level, low processing temperatures, since the precursors can be transformed into amorphous covalent ceramics at temperatures between 800 and 1000 °C, and the applicability of established polymer processing techniques. In general, processing of ceramic materials via precursor technology involves the synthesis of the precursor from monomer units, followed by cross-linking into an unmeltable, pre-ceramic polymer network (thermoset), and finally the pyrolysis at elevated temperatures. The pyrolysis initiates the organic-inorganic transition and results in amorphous ceramics, which crystallises at temperatures exceeding 1000 °C (Figure 7.1).

The requirements for the precursor are given by the appropriate application and differ from a cross-linkable liquid to meltable and curable, or unmeltable but soluble solids. For the broader distribution of these precursors, they should be processable by conventional polymer processing techniques. Furthermore, the starting material should be readily available, relatively cheap, and the synthesis should be uncomplicated.

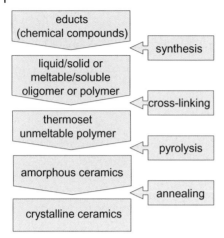

Figure 7.1 General processing of ceramics based on organometallic oligomers or polymers.

7.1.2
Precursor Systems and Properties

Many different precursor systems are used in industry for the processing of ceramic materials (Figure 7.2). Small and simple, gaseous precursor molecules are applied for the deposition of hard ceramic coatings such as TiN, Si_3N_4, or SiC via the CVD process (chemical vapor deposition) on different substrates. But also the generation of a ceramic matrix (e.g. SiC) in porous structures is possible by using chemical vapor infiltration (CVI). This process can be very time-consuming, depending on the thickness of the specimen.

The Sol-Gel-technique is applied to manufacture oxide ceramic coatings, for example, SiO_2, Al_2O_3, TiO_2, or ITO (indium-tin-oxide). The solved molecular compound, such as TEOS (tetraethoxysilane), reacts after the addition of an acid to the gel under elimination of an alcohol. The residual organic groups separate during the subsequent heat treatment. The ceramic yield is relatively low due to the high content of the necessary solvent and the organic groups.

In terms of the chemical composition, oligomer or polymer precursors are generally divided into oxygen containing and oxygen free materials. The class of

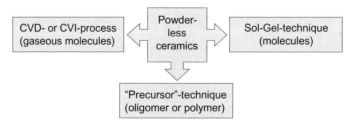

Figure 7.2 Classification of techniques for the powderless processing of ceramics.

silicones (Si-O-C precursors) is well established and commercialized. A great variety of different products is available in large quantities from companies such as Dow Corning (United States), Wacker Chemie AG (Germany), and Bluestar Silicones (China) (former Rhodia). But only a very small fraction is used for processing of ceramics, despite the relatively low costs and the good availability of silicones. This is due to the reduced thermal stability of the resulting SiCO ceramics in comparison to non-oxide systems.

Since the 1960s, numerous different non-oxide precursors for processing of ceramics have been developed for several applications. Polysilazanes without any organic group lead to SiN ceramics. Polysilazanes functionalized with organic groups, polycarbosilazanes, and polysilylcarbodiimides, are used to obtain SiCN ceramics [1–4, 25]. Polysilanes and polycarbosilanes serve as a source for the formation of SiC. The reaction of carbosilazanes with boron compounds lead to polyborocarbosilazanes, which are attractive for processing of high temperature stable SiBCN ceramics [29, 30]. But also Si-free precursors are known. Pure BN ceramics are available by pyrolysis of, for example, poly(methylamino)borazine [31–33].

Whereas the different chemical composition of the pre-ceramic polymers influences properties such as moisture sensitivity of the precursor or high temperature stability, corrosion, and oxidation behavior, as well as the tendency for crystallization of the resulting ceramics, the structure and the functional groups of the polymers determine the processability and the ceramic yield.

Especially the viscosity, the cross-linking behavior of the pre-ceramic polymer and the ceramic yield are important properties in the processing of CMC-materials via the polymer infiltration and pyrolysis process (PIP-process). Table 7.1 shows, by means of selected self-synthesized, examples of how the precursor architecture influences their properties. The silazane (Si-N) units of the so-called VN-precursor are linearly linked with each other and cycle to small rings with low average molecular weight because of steric influence of the vinyl groups. The higher molecular weight and resulting increase in viscosity of the HPS-precursor are due to the substitution of every second vinyl-group at the silicon atoms by hydrogen. The very low steric demand of the H-atom reduces the tendency to form small cycles. The HVNG-precursor contains a third NH-function at every second silicon atom. This additional three-dimensional bonding leads to a higher average molecular weight and a remarkable increase in viscosity in comparison to the VN- and HPS-precursors.

7.1.3
Cross-Linking Behavior of Precursors

The cross-linking of the precursors to form an unmeltable thermoset is possible by addition as well as condensation reactions and also depends on special functional groups (Figure 7.3). Most of the commercialized non-oxide precursors, such as silazanes and carbosilanes, contain Si-H and Si-vinyl functionalities, respectively.

Table 7.1 Molecular structure and resulting properties of different silazanes.

Precursor	Simplified molecular structure	Average molecular weight (g mol^{-1})	Viscosity at 20°C (Pas)	Ceramic yield (%) (1000°C N$_2$-atm.)
VN	[Si(CH$_3$)(CH=CH$_2$)—NH]$_n$, n > 2	255	0.004	62[a]
HPS	[HN—Si(CH=CH$_2$)(CH$_3$)]$_n$ [HN—SiH(CH$_3$)]$_n$	440	0.05	73[a]
HVNG	[HN—Si(CH=CH$_2$)(NH)]$_n$ [HN—SiH(CH$_3$)]$_n$	920	29	82[a]

a By using of a radical initiator (e.g. dicumylperoxide [DCP]) for cross-linking.

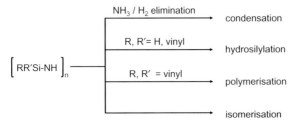

Figure 7.3 Different kind of cross-linking reactions for silazanes.

Both condensation and addition reactions, to form a thermoset, usually start at temperatures above 200 °C A good thermal insulation is required. Furthermore, low molecular precursors, such as VN or HPS can already boil at this temperature. Cross-linking is only possible by using an autoclave under high pressure conditions. By using a suitable radical initiator or catalyst, it is possible to reduce the cross-linking temperature and to promote addition reactions. The desired hydrosilylation (Si-H + Si-vinyl) and polymerization (Si-vinyl + Si-vinyl) reactions already start at 100 °C, which avoids the evaporation of oligomers. This approach lowers

Figure 7.4 Polysilazane cross-linked at 300 °C (N₂): (a) only thermally; (b) with initiator.

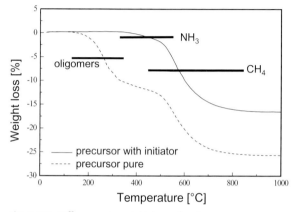

Figure 7.5 Differences in weight loss of a silazane precursor with and without an initiator investigated by FTIR-coupled TG-measurements.

the apparent effort for thermal insulation, provides for a bubble-free thermoset (Figure 7.4) and leads to an increased ceramic yield after pyrolysis.

Figure 7.5 shows typical differences in the pyrolysis behavior of a low molecular pure and initiator added SiCN-precursor, respectively. The pure precursor shows a significant weight loss during cross-linking in the temperature range between 100 and 250 °C, mainly based on the evaporation of low molecular oligomers. In contrast, the use of an initiator leads to connection of smaller molecules with larger molecules or with each other below the boiling point. A remarkable weight loss starts at temperatures higher than 350 °C due to further cross-linking by condensation of gaseous ammonia.

7.1.4
Pyrolysis Behavior of Precursors

Depending on the used precursor system, the pyrolysis starts at temperatures higher than 400 °C, characterized by the elimination of organic groups evaporated as methane, ethylene, and hydrogen (Figure 7.5). The polymer network will be destroyed during this complex process and the amorphous ceramic is formed. The composition of the resulting ceramic depends on the used pre-ceramic polymer and the pyrolysis atmosphere.

The transformation of the precursor polymer into ceramics is also coupled with remarkable shrinkage of the material (Figure 7.6). This is one of the most important drawbacks of the precursor technology in comparison to the sintering of ceramic powders (up to 20 vol.-%). The curvature in Figure 7.6 indicates up to 400 °C the typical thermal expansion of the polymer with increasing temperature. Simultaneous to the first weight loss at 400 °C (Figure 7.5), the material starts to shrink rapidly during the transformation from the polymer state to amorphous ceramics. This shrinkage is based on two effects. On the one hand, organic or inorganic groups at the polymer backbone with certain sterical demand were separated off. This weight loss results in shrinkage. On the other hand, many bonds were broken and a denser but random network composed of, for example, Si, C, and N atoms, were formed. This effect dominates the further shrinkage at temperatures above 800 °C, because only small amounts of residual hydrogen release the amorphous system. With increasing temperature, the atoms can be adjusted in an increasingly regular manner, which leads to a further shrinkage of about 10%. Finally, at higher temperatures, crystallization is initiated, depending on both the used elementary composition as well as the pyrolysis atmosphere.

The shrinkage of the precursor during pyrolysis and the formation of gaseous species avoid the processing of greater ceramic parts directly from the pre-ceramic polymer. Therefore, the applications are limited to specimens with small dimensions, such as ceramic fibers and coatings or on CMC materials, where the porosity is filled by multiple infiltration/pyrolysis steps.

In accordance with the results from the TG-measurements and dilatometry, the pyrolysis leads to a significant change in hardness and density (Figure 7.7). First

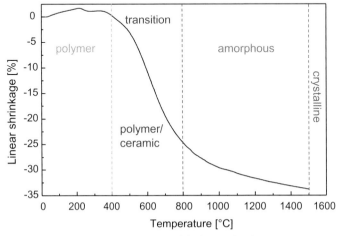

Figure 7.6 Dilatometric measurement of the linear shrinkage of cross-linked SiCN material during pyrolysis.

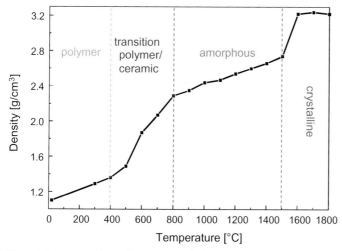

Figure 7.7 Density values of SiCN material after pyrolysis at different temperatures.

increase in density was observed up to 400 °C because of a higher degree of cross-linking. The strongest increase in density was measured during the transition from the polymer into the ceramics in the temperature range from 400 up to 800 °C. The further compaction effect up to 1500 °C is due to rearrangements in the amorphous state. The last increase in density at temperatures higher than 1500 °C in Figure 7.7 is based on the decomposition of the SiCN ceramic by the loss of nitrogen and the formation of crystalline SiC.

7.1.5
Commercial Available Non-oxide Precursors

For a long time the use of non-oxide precursors in many markets was strongly restricted by their relatively high prices, the variability, and the availability. In the meantime, not only the techniques for processing of precursors were improved but also a great variability of pre-ceramic polymers with different properties (Table 7.2) for several applications was developed. Recently, KiO.N. Corp. (Clariant GmbH, Germany) publicized a new method for the manufacture of polysilazanes in great quantities at reasonable prices [4].

7.2
The Polymer Impregnation and Pyrolysis Process (PIP)

7.2.1
Introduction

CMCs are a class of technologically advanced materials, which have been developed in recent years for different industrial applications such as aerospace, energy,

Table 7.2 Commercial available precursors (not complete).

Precursor name	Resulting ceramic system	Important properties[a]	Company
KiON HTT 1800	SiCN	Liquid	Clariant GmbH[b]
KiON HTA 1500 slow or rapid cure	SiCN	Liquid	Clariant GmbH[b]
KiON ML33	SiCN	Liquid	Clariant GmbH[b]
KiON ML66	SiCN(O)	Liquid	Clariant GmbH[b]
KiON ML20	SiCN(O)	Liquid	Clariant GmbH[b]
Ceraset PURS 20	SiCN	Liquid	Clariant GmbH[b]
Ceraset PSZ 20	SiCN	Liquid	Clariant GmbH[b]
Ceraset Ultra	SiCN	Meltable solid	Clariant GmbH[b]
NL 120A-20	SiN	Solution	Clariant GmbH[b]
NN 120–20	SiN	Solution	Clariant GmbH[b]
SMP-10	SiC	Liquid	Starfire Systems[c]
MCs	SiC	Moldable	Starfire Systems[c]
CVD-2000	SiC	Solution	Starfire Systems[c]
CVD-4000	SiC	Solution	Starfire Systems[c]
SOC-A35	SiCO	Meltable solid	Starfire Systems[c]

a   For further information see data sheets.
b   Clariant Produkte GmbH, Am Unisys Park 1, 65843 Sulzbach, Germany.
c   Starfire Systems, 10 Hermes Road, Malta, NY 12020, USA.

and the automobile industry, due to their excellent mechanical strength, chemical inertness, and heat resistance, along with their further features as lightweight materials. Currently, studies for the application of CMCs in the design of future controlled thermonuclear fusion reactors are being conducted.

As far as the basic materials are concerned, several types of ceramic matrices, ceramic fibers, and also manufacturing technologies may be used for the production of CMCs. The most important currently available CMC-qualities (fiber/matrix) are:

- Nextel/Al_2O_3 or mullite (Al_2O_3-SiO_2)   usable up to 1100 °C*
- SiC/SiC   usable up to 1300 °C*
- C/SiC   usable up to 1600 °C*,**
- C/C   usable up to >2000 °C**

* long term use, ** with oxidation protection.

In terms of processing, only multi-filament fiber bundles with small diameter mono-filaments (5–20 µm) can be used in the manufacture of long-fiber reinforced ceramics. Commercially available fibers are:

- carbon fibers;
- oxide fiber types based on Al_2O_3 (Nextel); and
- non-oxide fibers based on SiC (Nicalon, Tyranno).

In principle, the working temperatures of the SiC-based reinforcing fibers are between 1000 and 1300 °C of the oxide fibers at 1100 °C and one of the carbon fibers up to more than 2000 °C (in an inert atmosphere). These temperatures depend significantly on the particular working time.

There are a variety of methods used to create the composite material. Often two or more of these methods are combined. The methods used depend on architecture of the fibers and matrix precursor chemistry. Problems include making a fully dense matrix, reduction of cracking during processing, and producing the required matrix at a low enough temperature so that the fibers are not damaged or even destroyed. To manufacture CMC components, an infiltration process is required to build up the matrix in between the fibers. Depending on the infiltration media, it is a so-called gas phase or liquid phase infiltration process. All processes have one thing in common: the infiltration media does not have the final composition of the matrix material, which is generated in chemical reactions during processing [34–37].

A particular process, the so-called CVI, is known and used on an industrial scale for the manufacture of CMC materials. In addition, melt infiltration processes, such as the liquid silicon infiltration (LSI-process), are used to build up CMC-structures. These processes are described in detail in the other chapters in this book.

7.2.2
Manufacturing Technology

An additional process, not yet used on an industrial scale, is being investigated and developed. It is based on the wet infiltration of fiber rovings or preformed fiber materials with not gaseous, but liquid, ceramic polymeric precursors in a melted state or in solution. The so-called PIP process, used to manufacture CMCs, is widely recognized as a versatile method to fabricate large, complex-shaped structures. Compared with other ceramic composite fabrication processes, the PIP process offers significantly greater flexibility. By utilizing low temperature forming and molding steps typically used for manufacturing polymer matrix composites, the PIP approach allows use of the same existing equipment and processing technologies. The infiltration of a matrix preform with a liquid ceramic precursor gives a high-density matrix and permits many possible reinforcement geometries [38, 39].

7.2.2.1 Preform Manufacturing
The ability to fabricate preforms having the desired shape and architecture is of primary importance in the CMC industry. There are a number of processes that can fabricate such a complex shape (Figure 7.8), including:

- filament winding;
- hand lay-up of prepreg;
- braiding (in combination with stitching to generate three-dimensional material);
- three-dimensional weaving [40].

Figure 7.8 Preform manufacturing by: (a) filament winding; (b) hand lay up of prepreg; (c) two-dimensional braiding; (d) three-dimensional braiding; (e) two-dimensional stitching, (f) three-dimensional stitching (EADS).

7.2 The Polymer Impregnation and Pyrolysis Process (PIP)

Each of these techniques, combined with further processing, has its own particular benefits such as low cost, fiber orientation and architecture, and size, depending on the design and stress requirements of the particular application. However, the high temperature treatments involved can lead to reactions between matrix and fibers, which can adversely affect the strength of the composite.

7.2.2.2 Manufacturing of CMC

A carbon-fiber coating with thin protective layers is of crucial importance with regard to the damage-tolerant fracture behavior of several fiber-reinforced ceramics. For a C/SiC material, a coating of pyrolytic carbon (pyC) leads to an optimized fiber/matrix interface to guarantee sufficient debonding of the fibers in the matrix. This leads to the typical fiber pull-out or to a crack deflection in the interface between fiber and matrix, and so to excellent mechanical characteristics of the material (Figures 7.9 and 7.10) [41, 42].

The viscosity tends to be high in the ceramic precursors, making complete infiltration and wetting of reinforcing fibers difficult. The infiltration can be done by filament winding or vacuum impregnation using a polymeric matrix precursor. Analogous to the manufacturing technique with carbon fiber reinforced plastics

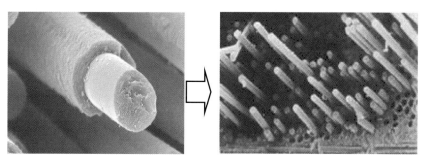

Figure 7.9 Fiber coating and fiber pull-out (EADS).

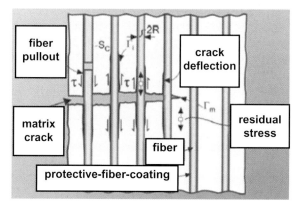

Figure 7.10 Mechanism of "quasi-ductilizing" of a fiber-coated PIP-CMC [37].

(CFRP), the structure is laminated either via pre-produced prepregs, filament winding, or textile technologies, then compacted and finally cross-linked in an autoclave at moderate temperatures between 100 and 300 °C and pressures of 10 to 20 bars.

For the production of C/SiC material, a polycarbosilane with a high ceramic yield of SiC is an allyl hydridopolycarbosilane, consisting of a chain of alternating C- and Si-atoms, such as:

$$-SiH_2-CH_2-SiHR-CH_2-SiH_2-$$

Here R is equal to an allyl-group:

$$-CH_2-CH=CH_2.$$

The length of the chain, as well as the frequency of the allyl-groups, defines the viscosity of the polymer system. Because of the double-bond of the allyl-group, all molecules get irreversibly cross-linked in the curing process in the autoclave. In addition to the polycarbosilane precursor, different precursor polymers can be used, which are converted into the ceramic matrix during the following high temperature process (Figure 7.11).

Both the initial shaping and fabrication of the polymer composite are carried out with low-temperature processing equipment. The subsequent transformation of the pre-ceramic polymer into the ceramic matrix is provided by a high temperature treatment (pyrolysis) between 1100 and 1700 °C, resulting in an amorphous or nano-crystalline structure according to the pyrolysis conditions.

One problem of the PIP process appears during pyrolysis, which results in a significant loss of weight of the precursor-polymer (up to 40%, depending on the polymer). This is based on the evaporation of oligomers as well as solvents and

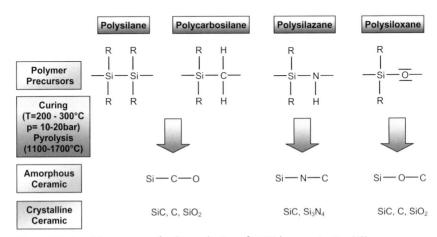

Figure 7.11 Possible precursors for the production of C/SiC by ceramization [43].

7.2 The Polymer Impregnation and Pyrolysis Process (PIP)

structural rearranging with evolution of gases, such hydrocarbons and hydrogen. Simultaneously, the transformation of the polymeric material having a typical density of about $1\,g\,cm^{-3}$ into a ceramic material of higher density (2.5–$2.7\,g\,cm^{-3}$ of SiC precursors) implies a bulk contraction (volume reduction of the matrix) with production of porosity, such as macro- and micro-pores in the matrix being formed. Nevertheless, the microstructure is characterized by large interbundle pores, which show a negative impact on the mechanical properties. Even microcracks appear during the high temperature formation of the ceramic matrix.

The porosity can be reduced by repeating the polymer infiltration, the so-called re-infiltration (or re-impregnation) and the pyrolysis processes several times. This leads to significant reduction of the porosity and a simultaneously hardening of the matrix. Mechanical characteristics of the composite, such as tensile strength or shear strength, are improved with the number of re-infiltration steps until reaching a maximum of strength. Unfortunately, almost four to eight or more cycles are necessary to reduce the porosity and fill up the matrix to a satisfying degree.

In order to overcome such a problem, a further known solution is to use, along with the pre-ceramic polymer, inert "fillers" such as ceramic powders, for example, SiC, which do not undergo bulk contraction during the thermal treatment and are compatible with the matrix. Such powders are used because they allow the voids in the fiber materials and among the material layers to be better filled to reduce the total number of the necessary infiltration and pyrolysis cycles in the whole process.

For C/SiC materials, the currently used inert fillers are commercially available SiC powders of micrometric granulometry. However, such powders suffer from further problems. These problems include the non-homogeneous distribution of the powder, as the powder particles barely infiltrate the fiber bundles (typically formed of 500–1000 fibers of 5–15 µm in diameter). The gaps among the fibers are by two orders of magnitude narrower than the size of the fibers [43, 44].

Moreover, the amorphous or nanocrystalline nature of the matrix generated by the pyrolysis of pre-ceramic polymers can be affected by the presence of crystalline ceramic particles. This is especially so of particles with larger size, which influence the mechanical characteristics of the resulting CMC. Recent developments have dealt with the use of ceramic fillers of a nano-size diameter in combination with the pre-ceramic polymer. In this case, the size of the ceramic particles is much lower than the mean distance among the ceramic fibers. It also allows the gaps between the single fibers to be filled in a generally homogeneous manner. Also the amorphous or nano-crystalline structure of the pre-ceramic polymer matrix is not substantially affected by the presence of particles that remain nano-metric sized after the pyrolysis. The homogeneous dispersion of such nano-metric powders into an initially amorphous matrix can act as a crystallization nucleus for the matrix formation.

The advantages of using nano-metric powders are even more apparent if the technique of liquid infiltration and pyrolysis is applied to three-dimensional fiber preforms produced by textile technologies, such as braiding or weaving in combination with stitching. Only particles of extremely reduced size can be uniformly distributed in three-dimensional fiber preforms.

In addition, the active filler controlled pyrolysis (AFCOP process [43, 44]) can be used to create a matrix product with an increased volume during ceramization. In the ideal case, the shrinkage of the polymer system and the increase of volume due to the active filler component can be compensated for (Figure 7.12). It is relatively easy and can be used to produce matrices at low cost.

During the ceramization process, the component geometry constantly relies on shrinkage so that all the geometries are feasible, which can be realized with classical CFRP technology. Optionally, the component may be finished by, for example, turning, grinding, and coating with an anti-oxidation layer in a final production step.

The overall PIP-process consists of the following single steps (Figure 7.13):

1. coating of fibers to create a weak fiber-matrix interface (ductilizing of fiber matrix bonding);
2. infiltration of the fibers/fiber architecture with a pure polymer-system or a powder-filled slurry system, e.g. via fiber winding, lay up of prepregs, or infiltration of textile fiber preforms by RTM (resin transfer moulding);
3. lamination of prepregs or joining of infiltrated parts, as well as forming if necessary;
4. curing via temperature and pressure in the autoclave;
5. ceramization via pyrolysis (high temperature process in vacuum or inert atmosphere);
6. multiple re-infiltration with a pre-ceramic polymer and following pyrolysis;
7. optional coating with an external oxidation protection system.

As can be seen, a CMC having improved mechanical characteristics, particularly as far as density and homogeneity of the matrix is concerned, is still desired. It is

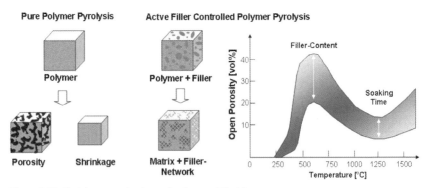

Figure 7.12 Shrinkage mechanism of polymers [43, 44].

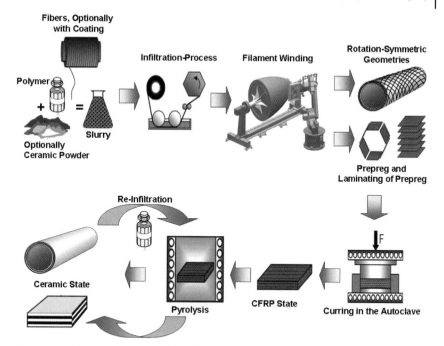

Figure 7.13 Schematic flowchart of the PIP process.

also important for industry that such a material can be produced within industrially practical times and costs, as well as with large enough thicknesses to be used, for instance, as structural components in large plants.

The PIP process provides the following advantages:

- lower cost of the one-step PIP process route, compared to the multistage CVI process;
- low schedule risk due to comparable short processing time;
- feasible for integral structure design due to the possibility of *in-situ* joining of different pre-hardened single parts within the autoclave process;
- similarity in design, analysis, and manufacturing with conventional CFRP standards and low tool costs lead to cost efficiency.

However, one preferred CMC is certainly the C/SiC (carbon fiber reinforced silicon carbide) because of its excellent thermal and mechanical properties in high temperature applications for a limited lifetime. C/SiC is currently used for space applications, such as heat shields or propulsion parts as well, as in the military field because of the following characteristics [45]:

Material Features

- ultra-light weight
- high chemical and erosion resistance
- fast and low-cost near-net-shaping

Table 7.3 Exemplary material properties of SICARBON in different fiber architecture.

SI CARBON			Temperature range (°C)	Fiber orientation	Data
Thermal expansion	CTE	10^{-6} 1/K	RT–1500	⊥/∥	5/2
Thermal conductivity	λ	W/mK	RT–1500	⊥/∥	6/14
Spec. heat capacity	c_p	kJ/kgK	RT–1500	–	0.6
Density	ρ	$g\,cm^{-3}$	RT–1500	UD/0° 90°	1.8
Porosity		vol%	RT–1500	UD/0° 90°	8
Tensile strength		MPa	RT–1500	UD/0° 90°	470/260
E-modulus	E	GPa	RT–1500	UD/0° 90°	80–90
Interlaminar-shear-strength		MPa	RT–1500	UD/0° 90°	11–20

- high temperature resistance (thermal shock)
- high flexibility in structural design
- potential reduction of component weight up to 60% (in comparison to the particular metal component).

Material Properties

A wide range of properties is adjustable, due to the free variation of fiber orientation:

- permissible material temperature: → long-time use/short-time use
- thermal conductivity
- thermal expansion
- high stiffness
- high tensile strength.

For EADS-Astrium's C/SiC material, produced via a PIP-process named SICARBON, the following exemplary material properties can be obtained (Table 7.3).

7.3
Applications of the PIP-process

7.3.1
Launcher Propulsion

Due to advantages inherent in ceramic composites, currently engine manufacturers and research institutes are stepping up their activities geared toward the use of ceramics in rocket engine thrust chamber components. In view of the extreme thermo-mechanical loads in the combustion chamber of liquid-propellant rocket engines, previous developments mostly concentrated on the use of ceramic composites in the less thermally loaded nozzle extensions. At EADS-Astrium, subscale nozzles on the ratio of 1 : 5 for the Ariane 5 main engine "Vulcain" was made of

PIP-C/SiC and also successfully tested on the research test bench P8 at DLR in Lampoldshausen [46–48].

The objectives to use CMCs for rocket nozzle extension are:

- simplification of cooling design (active cooling → radiation cooling);
- reduction of component mass (high thrust-to-mass ratio);
- potential to increase the payload capabilities;
- substitution of metal materials to expand the operation temperature;
- increase the performance and reliability.

The high mechanical loads during the operation sequence of the rocket motor require additional design measures, such as local C/SiC stiffeners at the outer nozzle shell. Figure 7.14 shows two subscale test nozzles made by EADS-Astrium's PIP-C/SiC.

Figure 7.15 shows the Vulcain subscale nozzle during a hot test; temperatures of up to 2300 K with extreme thermal gradients were measured.

The leap from subscale to large structures, such as the Ariane 5 upper-stage engine Aestus nozzle, represented a particular problem. The process-induced component shrinkage occurring during manufacture and as a function of the fiber orientation had to be solved during production development, with a special empha-

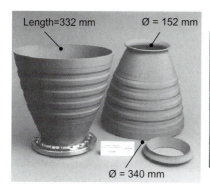
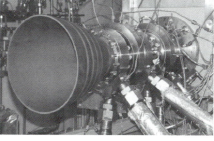

Figure 7.14 Vulcain subscale PIP-C/SiC nozzle extensions (ratio 1:5).

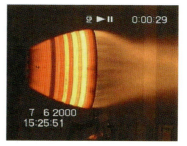

Figure 7.15 Hot test of Vulcain subscale PIP-C/SiC nozzle extensions.

Figure 7.16 Ceramic nozzle on test bench (left), during the vacuum hot-firing test (middle) and during vibration test (right).

sis on adherence to the geometrical tolerances. Based on the FEM and thermal analyses, the required angle and the wall-thickness progression of the Aestus nozzle were set via the winding technique in near-net-shape and lightweight design, as for the Vulcain subscale nozzle. The loads occurring on launching the Ariane necessitate providing a stiffening ring at the end of the nozzle.

Within the framework of the test program, the structural integrity and the thermal load on the C/SiC nozzle extension were verified in a sine-load vibration and vacuum hot-firing test. The test was performed at the DLR's test facility in Lampoldshausen (Figure 7.16).

7.3.2
Satellite Propulsion

Combustion chambers for apogee- and attitude-control engines for satellites are currently made of refractory heavy metals such as rhenium, iridium, and platinum, due to their high resistance to chemical attack and the high service temperature. Besides high material and manufacturing costs, as well as the substantial use of raw materials, heavy metals exhibit high density, amounting to more than $21\,g\,cm^{-3}$. The potential offered by CMC as a structural material for small thrusters, in addition to the benefits already mentioned as objectives in the nozzle extension programs, lies in the clearly lower manufacturing costs compared to metal construction. Further advantages are:

- simplification of the construction method by reducing the individual components (single-piece construction), hence reduced test effort;
- increase in the permissible wall temperatures of currently 1900 to ~2200 K (with suitable layer system), hence increase of engine performance.
- reduction of engine mass of 30 to 50%.

In 1998, in order to study the use of composite ceramics (PIP-C/SiC) for small thrusters, the first hot-firing tests were carried out at sea level with different C/SiC

Figure 7.17 PIP-C/SiC combustion chamber (left) and during hot test (right).

Figure 7.18 EAM PIP-SICARBON production model.

combustion chambers. The propellant compatibility (MMH/N_2O_4), diverse joining concepts, and investigation of Environmental Barrier Coatings (EBCs) with regard to long-term deployment and leakage, comprised the main areas of effort. Figure 7.17 shows the combustion chamber design and also during the hot-firing operation.

Due to the low specific weight, the high specific strength over a large temperature range, the low CTE, the good chemical and erosion resistance with hypergolic propellants, their good damage tolerance, and the feasibility of manufacturing a single piece reinforced ceramic combustion chamber combined with a nozzle extension, the Astrium's LPI-C/SiC was defined as baseline material for the development of the new **E**uropean **A**pogee **M**otor EAM.

The natural micro-cracks and pores of reinforced ceramics and the associated gas leakage has to be mastered. Figure 7.18 shows the prototype of the single piece ceramic component.

In addition to propulsion applications, advanced two-dimensional-C/SiC based, as well as oxide/oxide CMC materials, can be applied to hot structures and thermal protection systems, which are suitable for future re-useable space transportation vehicles. The particular PIP process of these products will allow the design and

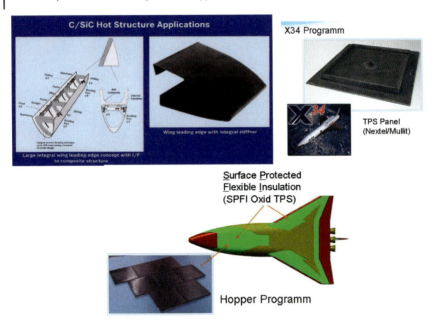

Figure 7.19 TPS components produced by PIP- = process.

manufacturing of complex and highly integrated components, for example, leading edges, stiffened panels, nose caps, flaps and rudders, engine protection shields, and so forth. (Figure 7.19) [47].

7.4
Summary

CMCs fabricated via the polymer infiltration and pyrolysis process are attractive construction materials for lightweight applications at temperatures higher than 1000 °C. It is possible to produce large-sized, and if required, complex as well as thin-walled, high-temperature structures via the PIP process with a following ceramization. C/SiC is especially suitable for high temperature aerospace applications, such as thermal protection systems (heat-shields) and propulsion components. Current investigations have the objective of increased long-term resistance and significant reduction of production costs.

References

1 Birot, M., Pillot, J.-P. and Dunoguès, J. (1995) Comprehensive chemistry of polycarbosilanes, polysilazanes and polycarbosilazanes as precursors for ceramics. *Chemical Reviews*, **95**, 1443–77.

2 Laine, R.M., Blum, Y.D., Tse, D. and Glaser, R. (1988) Synthetic routes to

oligosilazanes and polysilazanes, Chapter 10, in *Inorganic and Organometallic Polymers*, American Chemical Society.

3. Motz, G., Hacker, J. and Ziegler, G. (2001) Design of SiCN-precursors for various applications. *Ceramic Materials and Components for Engines*, 581–5.

4. Lukacs, A. (2007) Polysilazane precursors to advanced ceramics. *The American Ceramic Society Bulletin*, **86** (1), 9301–6.

5. Miller, D.V., Pommell, D.L. and Schiroky, G.H. (1997) Fabrication and properties of SiC/SiC composites derived from Ceraset™ SN preceramic polymer, *Ceramic Engineering and Science Proceedings*, **18**.

6. Ziegler, G., Richter, I. and Suttor, D. (1999) Fiber reinforced composites with polymer derived matrix: processing, matrix formation and properties. *Composites Part A*, **30**, 411–17.

7. Rak, Z.S. (2001) A process for Cf/SiC composites using liquid polymer infiltration. *Journal of the American Ceramic Society*, **84** (10), 2235–9.

8. Yajima, S., Okamura, K., Hayashi, J. and Omori, M. (1976) Synthesis of continuous SiC fibers with high tensile strength. *Journal of the American Ceramic Society*, **59** (7–8), 324–7.

9. Baldus, H.-P., Jansen, M. and Sporn, D. (1999) Ceramic fibers for matrix composites in high-temperature engine applications. *Science*, **285**, 699–703.

10. Arai, M., Funayama, O., Nishii, H. and Isoda, T. (1989) High-Purity Silicon Nitride Fibers, Patent US 4818611.

11. Hacker, J., Motz, G. and Ziegler, G. (2001) *Novel Ceramic SiCN-Fibers from the Polycarbosilazane ABSE, High Temperature Ceramic Matrix Composites*, Wiley-VCH Verlag GmbH, Weinheim, pp. 52–5.

12. Motz, G., Hacker, J. and Ziegler, G. (2002) New SiCN fibers from the ABSE polycarbosilazane. *Ceramic Engineering and Science Proceeding*, **23** (3), 255–60.

13. Motz, G. (2006) Synthesis of SiCN-precursors for fibres and matrices. *Advances in Science and Technology (Advanced Inorganic Fibrous Composites for Structural Applications)*, **50**, 24–30.

14. Kokott, S. and Motz, G. (2007) Cross-linking via electron beam treatment of a tailored polysilazane (ABSE) for processing of ceramic SiCN-Fibers. *Soft Materials*, **4** (2–4), 165–74.

15. Mucalo, M.R. and Milestone, N.B. (1994) Preparation of ceramic coatings from pre-ceramic precursors. *Journal of Materials Science*, **29**, 5934–46.

16. Cross, T.J., Raj, R., Prasad, S.V. and Tallant, D.R. (2006) Synthesis and tribological behavior of silicon oxycarbonitride thin films derived from poly(urea)methyl vinyl silazane. *International Journal of Applied Ceramic Technology*, **3** (2), 113–26.

17. Motz, G. and Ziegler, G. (2002) Simple processibility of precursor-derived SiCN coatings by optimised precursors. *Key Engineering Materials*, **206–213**, 475–8.

18. Motz, G., Kabelitz, T. and Ziegler, G. (2004) Polymeric and ceramic-like SiCN coatings for protection of (light) metals against oxidation and corrosion. *Key Engineering Materials*, **264–268**, 481–4.

19. Günthner, M., Albrecht, Y. and Motz, G. (2007) Polymeric and ceramic-like coatings on the basis of SIN(C) precursors for protection of metals against corrosion and oxidation. *Ceramic Engineering and Science Proceeding*, **27** (3), 277–84.

21. Fink, W. (1966) Beiträge zur Chemie der Si-N Bindung, VIII. Silylierungen an 1.2- und 1.3-Diaminen. *Chemische Berichte*, **99**, 2267–74.

22. Fink, W. (1967) Beiträge zur Chemie der Si-N Bindung, XI. polymere 1,3-diaza-2-sila-cyclopentane. *Helvetica Chimica Acta*, **50**, 1144–53.

23. Verbeek, W. and Winter, G. (1974) German Patent 2236078.

24. Riedel, R., Kienzle, A., Dressler, W., Ruwisch, L., Bill, J. and Aldinger, F. (1996) A silicoboron carbonitride ceramic stable to 2000 °C. *Nature*, **382**, 796–98.

25. Kroke, E., Li, Y.L., Konetschny, C., Lecompte, E., Fasel, C. and Riedel, R. (2000) Silazane derived ceramics and related materials. *Materials Science and Engineering, R: Reports*, **26**, 97–199.

26. Seyferth, D. and Wiseman, G.H. (1984) High-yield synthesis of Si3N4/SiC ceramic materials by pyrolysis of a novel polyorganosilazane. *J. Am. Ceram. Soc.*, **67**, C132–C133.

27 Peuckert, M., Vaahs, T. and Brück, M. (1990) Ceramics from organometallic polymers. *Advanced Materials*, 2(9), 398–404.

28 Whitmarsh, K. and Interrante, L.V. (1991) Synthesis and structure of a highly branched polycarbosilane derived from (chloromethyl)trichlorosilane. *Organometallics*, 10, 1336–44.

29 Riedel, R., Kienzle, A., Dressler, W., Ruwisch, L., Bill, J. and Aldinger, F. (1996) A silicoboron carbonitride ceramic stable to 2000 °C. *Nature*, 382, 796–8.

30 Bernard, S., Weinmann, M., Gerstel, P., Miele, P. and Aldinger, F. (2005) Boron-modified polysilazane as a novel single-source precursor for SiBCN: synthesis, melt-spinning, curing and ceramic conversion. *Journal of Materials Chemistry*, 15, 289–99.

31 Paine, R.T. and Narula, C.K. (1990) Synthetic routes to boron nitride. *Chemical Reviews*, 90, 73–91.

32 Paciorek, K.J.L. and Kratzer, H. (1992) Boron nitride preceramic polymer studies. *European Journal of Solid State and Inorganic Chemistry*, 29, 101–12.

33 Toury, B., Miele, P., Cornu, D., Vincent, H. and Bouix, J. (2002) Boron nitride fibers prepared from symmetric and asymmetric alkylaminoborazine. *Advanced Functional Materials*, 12, 228–34.

34 Claussen, N. (1992) *Verstärkung keramischer Werkstoffe*, DGM Informationsgesellschaft Verlag.

35 Hüttinger, K. and Greil, P. (1992) Keramische Verbundwerkstoffe für Höchsttemperaturanwendungen, Cfi/ Ber. DKG 96, No. 11/12.

36 Lehman, R., Rahaiby, S. and Wachtman, J. (1995) *Handbook of Continuous Fibre-Reinforced Ceramic Matrix Composites, Ceramics*. Information Analysis Center, Purdue University.

37 Evans, A. G., Zok, F.W. and Davis, J. (1991) The role of interfaces in fibre-reinforced brittle matrix composites. *Composites Science and Technology*, 24, 3–24.

38 Vogel, W.D. and Spelz, U. (1995) Cost effective production techniques for continuous fiber reinforced ceramic matrix composites. *Ceramic Processing Science and Technology*, 51, 225–59.

39 Mühlratzer, A. (1990) *Entwicklung zur kosteneffizienten Herstellung von Faserverbundwerkstoffen mit keramischer Matrix*. Proceedings Verbundwerkstoffe Wiesbaden, 22.1–22.39.

40 Brandt, J., Drechsler, K. and Meistring, R. (1990) *The application of three-dimensional fiber preforms for aerospace composite structures*. Proc. ESA Symp.: Space Applications of Advanced Structural Materials, ESTEC Noordwijk/NL, 71–7.

41 Naslain, R. (1995) The concept of layered interphases in SiC/SiC. *High-Temperature Ceramic-Matrix Composites II, Ceramic Transactions*, 58, 23–38.

42 Helmer, T. (1992) Einfluß einer Faserbeschichtung auf die mechanischen Eigenschaften von Endlosfasern und C/SiC-Verbundwerkstoffen, Dissertation Universität Stuttgart.

43 Greil, P. (1998) Near-net-shape manufacturing of polymer derived ceramics. *Journal of the European Ceramic Society*, 18, 1905–14.

44 Greil, P. (2000) Polymer derived engineering materials. *Advanced Engineering Materials*, 2, 339–48.

45 Beyer, S., Knabe, H., Schmidt, S., Immich, H., Meistring, R. and Gessler, A. (2002) *Advanced ceramic matrix composite materials for current and future technology application*. 4th International Conference on Launcher Technology Space Launcher Liquid Propulsion, 3–6 December 2002, Liege (Belgium).

46 Schmidt, S., Beyer, S., Knabe, H., Immich, H., Meistring, R. and Gessler, A. (2004) Advanced ceramic matrix composite materials for current and future propulsion technology applications. *Acta Astronautica*, 55 (3–9), 409–20.

47 Schmidt, S., Beyer, S., Immich, H., Knabe, H., Meistring, R. and Gessler, A. (2005) Ceramic matrix composites: a challenge in space-propulsion technology applications. *International Journal of Applied Ceramic Technology*, 2 (2), 85–96.

8
Oxide/Oxide Composites with Fiber Coatings

George Jefferson, Kristin A. Keller, Randall S. Hay, and Ronald J. Kerans

8.1
Introduction

Advanced aerospace applications require structural materials that are lighter, durable, and capable of operation at higher temperatures. Ceramic matrix composites (CMCs), also called ceramic fiber-matrix composites (CFMCs) and continuous fiber ceramic composites (CFCCs), have structural capability at temperatures above those for advanced intermetallics, and damage tolerance far superior to monolithic ceramics [1, 2]. The most advanced CMCs use SiC fibers and matrices, but are limited in lifetime by oxidation, particularly in combustion environments and other environments with high pH_2O, despite the use of environmental barrier coatings (EBCs) [3–5]. Superior high-temperature environmental stability can be achieved by using oxides for all constituents: fibers, matrices, and fiber coatings. Oxide/oxide CMCs also tend to be less expensive than SiC-SiC CMCs. However, reduced creep resistance and strength at high temperatures limit the application of oxide/oxide CMCs. This is a direct consequence of the creep resistance and high temperature strength of the current polycrystalline oxide fibers (Table 8.1) [6, 7, 20]. Oxide/oxide CMCs also have higher thermal expansions (CTE) and lower thermal conductivities than SiC. Component design for high thermal gradients and transients (thermal shock) is therefore more difficult [2].

The toughness of current commercially available oxide/oxide CMCs is a consequence of the matrix fracture energy dissipation, by diffuse micro-cracking in a porous matrix. The properties of these types of CMCs have been extensively reviewed (Table 8.2, [21, 28–34]) and additional information is given in the following chapter. Impressive mechanical properties are reported, but there are limitations to the matrix-dominated properties as a consequence of the matrix porosity. Poor interlaminar properties [22, 35–37], low thermal conductivity [38–40], matrix permeability, and fretting and wear [41] are substantial concerns. The temperature capability of porous-matrix CMCs is limited by the matrix sintering temperatures; coarsening or densification of the matrix degrades CMC toughness [23].

Ceramic Matrix Composites. Edited by Walter Krenkel
Copyright © 2008 WILEY-VCH Verlag GmbH & Co. KGaA, Weinheim
ISBN: 978-3-527-31361-7

Table 8.1 Constituent properties.

	Comp. (wt %)		Strength (Mpa)		Modulus (GPa)	Dens (g/cc)	Dia. (μm)	Creep limit[d] (°C)	Expansion (ppm/°C)
	Al_2O_3	SiO_2	Filament[c]	Tow[c]					
Fibers									
Nextel 610	>99	<0.3	3100	1600	380	3.9	10–12	1000	8.0
Nextel 650	89	–	2550	–	358	4.1	10–12	1100	8.0
Nextel 720	85	15[b]	2100	800	260	3.4	10–12	1200	6.0
Sapphire[a]	100	0	3100	2250	435	3.8	70–250	–	–
Matrix constituents (fully dense properties)									
Alumina	100	0	–	–	380	3.96	–		8.1
Mullite	75.5	24[b]	–	–	220	3.16	–		5.0

a Saphikon inc, c-axis single-crystal.
b In crystalline mullite phase.
c 2.5 mm gauge length.
d 1% strain after 1000 hours @70 MPa, single filament; Refs: fibers [6–16]; mullite/alumina: [6, 7, 14–19].

Table 8.2 Room temperature mechanical properties of porous matrix oxide/oxide CMCs.

Designation	Fiber direction properties			Off axis (matrix dominated) strength			
	E (Gpa)	Prop. limit (MPa)	UTS (MPa)	±45° (MPa)	In-plane shear (MPa)	Interlaminar shear (MPa)	Trans-thickness (MPa)
COI-610/AS	124	–	366	–	48.3	15.2	–
GE-610/GEN-IV	70	100	205	54	27	–	7.1
UCSB-610/M	95	–	215	–	–	–	–
UCSB-720/M	60	80	150	28	–	10	–
COI-720/AS	75.6	–	220	–	31	11.7	2.7
COI-720/A-1	75.1	50	177	–	–	–	–

S. Butner, Personal Communictaion; References [21–27].

The logical approach to improving the matrix-dominated properties is through the use of a dense matrix with an engineered fiber-matrix interface [29, 42]. The toughness of dense-matrix oxide/oxide CMCs is a consequence of matrix crack detection at the fiber-matrix interface, and subsequent fiber pull-out [29, 42, 43]. Typically, this is accomplished by engineering the fiber-matrix interface through the use of fiber coatings.

The objective of this chapter is to review recent progress in oxide/oxide CMCs that utilize fiber coatings. The review will focus on high-temperature (>1000 °C), dense-matrix oxide/oxide CMCs, which excludes CMCs with glass or glass-ceramic matrices and fibers.

8.2
Applications

The improved interlaminar properties, hermeticity, wear resistance, and other matrix-dominated properties of dense matrix oxide/oxide CMCs, are expected to eventually drive their transition to some of the applications currently under development for the more mature porous-matrix CMCs. Recent field tests of a combustor outer liner for over 25 000 hours in a *Centaur* 50S (Solar Turbines) gas turbine, demonstrate the potential of porous-matrix oxide/oxide CMCs (Figure 8.1) [44–46]. Exhaust components of interest include nozzle flaps for a fighter aircraft [38] and lightweight helicopter exhaust ducts that insulate and protect the surrounding structure [47]. Other components include the seal-casing shroud and stationary vanes [48] in a turbine engine. Rotating components are currently not under consideration because of the modest high temperature strength and creep resistance of current oxide fibers.

In current oxide CMC applications, the primary part loading is in-plane, which is in the fiber direction and therefore the best-strength orientation for the CMC. However, the interlaminar and off-fiber-axis strength of current CMCs is so much lower than the fiber-direction strength (in-plane) (Table 8.2), that interlaminar failure is a design limiting issue [49, 50]. It is this limitation that is driving the development of dense matrix composites.

8.3
CMC Fiber-Matrix Interfaces

A tough dense matrix CMC requires crack deflection at or near the fiber-matrix interface, and fiber pull-out afterwards. Therefore, the fiber-matrix interface must

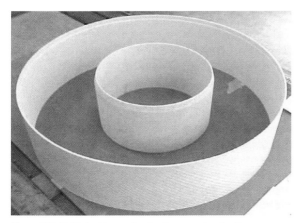

Figure 8.1 Oxide/oxide combustor inner and outer liners fabricated by ATK-COI Ceramics (Photo courtesy of ATK-COI Ceramics).

be weak relative to the fiber, which is important because virtually all fabrication processes degrade the fiber strength to some degree. Fiber-matrix interface issues are briefly reviewed in this section. A comprehensive review of CMC fiber-matrix interfaces is available elsewhere [42].

8.3.1
Interface Control

Dense matrix CMCs use a fiber-matrix interface that is engineered to be weak (Figure 8.2a). A matrix crack that approaches a coated fiber is deflected either at the matrix-coating interface, within the coating layer, or at the coating-fiber inter-

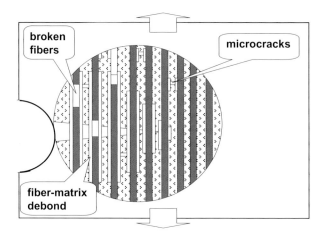

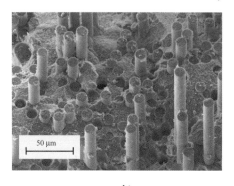

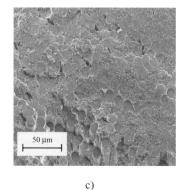

b)   c)

Figure 8.2 (a) Schematic showing toughening through crack deflection at the fiber-matrix interface; (b) SEM micrograph of fracture surface of Nextel 610/monazite/alumina composite, heat treated at 1200°C for 5 hours, showing extensive fiber pull-out; (c) fracture surface of Nextel 650/alumina composite with no interface coating after heat treatment at 1200°C. Absence of crack deflection mechanism led to brittle failure in (c).

face. The fiber ultimately fractures at a point out of the matrix crack plane, and pulls out of the matrix with continued loading. Fiber pull-out requires some load to overcome friction, which in turn can be affected by the fiber surface roughness. A fracture surface showing fiber pull-out in a Nextel 610/monazite/alumina CMC is shown in Figure 8.2b. For comparison, a fracture surface of a brittle Nextel 650/alumina composite, with almost no fiber pull-out, is shown in Figure 8.2c.

The optimal interface properties, such as coating thickness, friction, and pull-out lengths, depend on the CMC constituents. General fracture mechanics guidelines for crack deflection usually refer to the He and Hutchinson criteria; the ratio of the interface and fiber fracture energies must be less than 0.25 when the fiber and matrix are of similar elastic modulus [51, 52]. A more comprehensive discussion of these criteria is given elsewhere [42, 53, 54].

The design and evaluation of oxide fiber coatings for controlled interface properties is complex. Carbon (C) and boron nitride (BN) are the usual fiber coatings, but they have poor oxidation resistance. The accumulated understanding of CMCs based on C and BN coatings provides a good foundation for oxide coating design. However, oxide coatings can also be expected to have major differences with C and BN. Oxide coatings are less compliant, so the optimal coating thicknesses are likely to be thicker [55]. Even with thick coatings, the low friction achievable with C and BN is probably unattainable.

Despite the challenges, monazite coatings ($LaPO_4$) have been definitively demonstrated to deflect matrix cracks and substantially improve alumina-based CMC properties [56, 57]. Use of monazite fiber coatings in Nextel 610 reinforced alumina matrix CMCs increased life by over two orders of magnitude compared to a CMC with no coating [57]. These composites did have a porous matrix so that, while the effectiveness of the coatings was established, the expected improvement in matrix-dominated properties, to be gained with a dense matrix, was not demonstrated.

8.3.2
Fiber Coating Methods

Coatings have been deposited on ceramic fibers by a variety of methods, including solution/sol/slurry precursors [58–60], chemical vapor deposition (CVD) [61–64], and electrophoretic deposition (EPD) [65, 66]. The process choice is often dictated by the coating material. CVD has been successful for the production of C and BN coatings [67–70]. However, it is difficult to continuously deposit fiber coatings of multi-component oxides with the correct stoichiometry using this process. EPD requires a conductive substrate, and is better suited to batch coating of cloth and preforms.

Liquid precursor (solution/sol/slurry) coating techniques have been the most successful for the deposition of multi-component oxides on continuous oxide fiber tows [59, 71–73]. A typical fiber coater is shown in Figure 8.3a [59, 74]. An immiscible liquid floating on the coating precursor can aid in removal of excess sol, allowing discrete coating of individual filaments, with minimal bridging of the coating between filaments and crusting of the coating around the tow perimeter

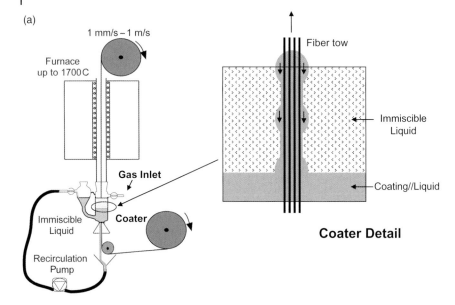

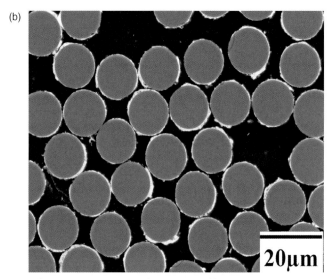

Figure 8.3 (a) Fiber coating apparatus utilizing immiscible liquid solution [72]; (b) typical monazite (LaPO$_4$) coatings produced using the high volume fiber coater. The coatings are relatively uniform with very little bridging between the fiber tows.

(Figure 8.3b). The sol chemistry must be carefully controlled to avoid fiber degradation [72, 73, 75]. Multiple coating passes (2–5) are generally needed for deposition of coatings greater than 100 nm in thickness. A high volume fiber coater has recently been developed by UES, Inc. that deposits multiple coating layers in one coating run. This decreases both the coating cost and time. The use of an immiscible liquid is unnecessary because the fiber tow passes over many rollers that break up coating bridges and fins.

Bridging and crusting of the coating is also a major problem in dip coating of fiber cloth and preforms. Fiber tows can be coated prior to weaving into cloth [76]. There is a possibility that the coating will be damaged during weaving, but this can be minimized by appropriate fiber sizing. The deposition of monazite coatings on cloth has recently been demonstrated by a precipitation process (Figure 8.4) [77]. These coatings are much more uniform than those made by the continuous coating process. If acceptable coated fiber strengths are achieved, it is expected that this process will be very attractive for cloth and preform coating.

Another recent approach to cloth and preform coatings uses a hybrid technique [29, 78]. The fabric is initially coated with carbon using a solution. After matrix processing, the carbon is burned away, leaving a slight gap at the fiber-matrix interface. The composite is then infiltrated with a monazite precursor solution, which fills the gaps.

Coated fibers are generally characterized for coating thickness, uniformity, composition, microstructure (including porosity), and single fiber strength. Coating characterization is challenging and often requires both SEM and TEM. TEM sample preparation requires specialized procedures [79, 80]. Coating thickness uniformity can be quantified for some coatings using SEM image analysis and statistical methods [81].

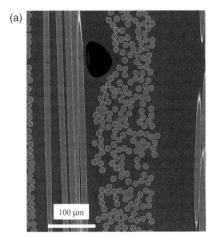 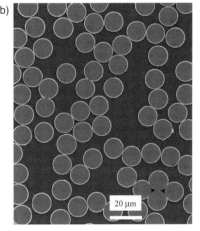

Figure 8.4 Polished cross-sections showing discrete monazite coatings deposited onto Nextel fabric. The coatings, shown as brighter halos around the fibers, are uniform with very little bridging.

The strength of the coated fiber is critical, since significant degradation can occur during coating [71–73, 75, 82]. Tow testing and single filament testing are both used [17, 83]. Smaller strength variations can be measured by single filament testing. Fiber coatings may also protect fibers from damage during processing.

8.3.3
CMC Processing

As oxide fiber coating technology matures, processes to produce dense oxide matrices must be developed to make best use of the coated fibers. Achieving high matrix density in CMCs is complicated by the continuous fibers. The one-dimensional constraint introduced by fibers inhibits densification [84, 85]. Furthermore, temperatures, pressures, and chemistries are limited by the need to avoid damaging fibers. This situation is further aggravated by fiber cloth (two-dimensional) and three-dimensional fiber reinforcement. Fiber cloth imposes densification constraints in two dimensions, but allows some shrinkage in the third, while three-dimensional fiber preforms do not allow shrinkage in any direction. Since some porosity is usually unavoidable, processing options are restricted to deciding how best to distribute the void volume. This benefits the processing of porous matrix composites, but improved processing methods are required for dense-matrix CMCs with engineered fiber-matrix interfaces.

Oxide/oxide CMC processing typically involves the following steps:

1. coating fibers (Section 8.3.2)
2. arranging fibers in the desired architecture
3. matrix green body formation
4. consolidating the matrix
5. final machining

If fiber preforms are used, step 2 precedes step 1. Matrix forming techniques (step 3) include pressure infiltration, pre-pregging followed by vacuum bagging/ autoclaving and EPD. After green body formation, CMCs are consolidated (step 4) using pressureless sintering, hot pressing, or hot isostatic pressing (HIPing).

Hot pressing or HIPing can provide a dense oxide matrix composite. For example, fully dense Al_2O_3-$LaPO_4$ matrices have been created by hot pressing, as a demonstration, using sapphire fibers [86]. However, the current engineering grade oxide fibers (Nextel series, Table 8.1) will not survive the required 1400 °C process temperature needed for full density. Therefore, lower processing temperatures are required. In most cases, the oxide composite is pressureless sintered and then re-infiltrated with a sol, solution, slurry, or a gaseous oxide precursor to increase the matrix density.

Re-infiltration with a liquid ceramic precursor typically requires many infiltration steps, since only small increments in densification result from each step. This is directly related to the rheology and ceramic yield of the liquid precursor. For example, alumina sols are limited to about 8 vol.-% solid yield per step before becoming too viscous for infiltration [87]. Furthermore, the ultimate densification is limited as the matrix porosity becomes increasingly difficult to infiltrate with each step [87, 88]. Recently, a novel infiltration technique, using an aqueous CrO_3 solution, has been investigated [89, 90]. During heating, the solution deposits as solid Cr_2O_3 on the porous matrix surfaces. The aqueous CrO_3 solution has a lower viscosity and a higher solid yield per infiltration step than other ceramic sols. Composite matrix porosity as low as 15% has been achieved using this method. However, it should be noted that toxicity concerns may limit the practical use of chromia as a composite matrix [91].

Other processing techniques include directed metal oxidation and reaction bonding. Some fiber coatings, such as rare earth orthophosphates, are carbothermally reduced at about 750 °C [92] and require processing methods that avoid low pO_2. These methods are reviewed in more detail elsewhere [34, 93].

8.3.4
Fiber-Matrix Interfaces

Interpretation of results on interface control is complicated. It is difficult to distinguish between the absence of fiber-matrix crack deflection and the strength degradation of fibers [94]. It can also be difficult to distinguish between fiber-matrix crack deflection and porous-matrix distributive damage mechanisms. These factors make careful choice of control specimens (with no fiber coatings) and tests important, to properly evaluate the effect of fiber coatings on CMC mechanical properties.

8.3.4.1 Weak Oxides
Monazite and xenotime rare earth orthophosphates satisfy the crack deflection and fiber pull-out criteria for CMCs, and have been definitively demonstrated in high fiber volume fraction CMCs [57, 70, 95, 96]. Monazite ($LaPO_4$) is the most widely studied, having a high melting point (>2000 °C) and is stable with oxides such as alumina [95, 97, 98]. It also bonds weakly with many oxides, particularly alumina, is relatively soft for a refractory material, and is machinable [95, 99]. Morgan and Marshall first demonstrated the functionality of monazite as a fiber-matrix interface [95]. The importance of plastic deformation and the relative softness of monazite have been demonstrated by characterization of pushed-out monazite-coated single-crystal fibers. The monazite coating was intensely plastically deformed, but the surrounding matrix and fiber were not deformed [100]. Five deformation twin modes have been identified in monazite [101, 102] and other interesting plastic deformation mechanisms, including dislocation climb at room temperature, have been described [103, 104].

Mechanical testing, either tensile or flexural, of CMCs with high volume fractions of monazite-coated fibers are listed in Table 8.3. Monazite coatings increased the strength and use temperatures for Nextel 610, porous alumina matrix CMCs [57]. CMCs with monazite fiber-matrix interfaces lost about 30% of their tensile strength after 100 hours and 1000 hours at 1200 °C; control samples lost more than 70% of their strength after only 5 hours at 1200 °C [57]. Fractography showed

Table 8.3 Oxide/oxide composites with weak oxide interface coatings.

Fiber[a] Coating	Matrix	Fiber arch.	Vf (%)	Comp. density (%)	Test type	Test condition[b]	UTS (MPa)	Ref.
LaPO$_4$	Al$_2$O$_3$	0°	20[c]	70–75	Tensile	RT	198 ± 12	[57]
						RT, after 1200 °C, 100 h	143 ± 7	
Uncoated	Al$_2$O$_3$	0°	20[c]	70–75		RT	45 ± 20	
LaPO$_4$	Al$_2$O$_3$	0°	20[c]	70–75	Tensile	RT	168 ± 10.5	[57]
						1200 °C	167 ± 2.5	
Uncoated	Al$_2$O$_3$	0°	20[c]	70–75		RT	93 ± 6.4	
						1200 °C	100 ± 26	
LaPO$_4$	Al$_2$O$_3$	8HS	40	~87	3-pt. bend	RT (3pt. bend)	226	[105]
LaPO$_4$	AS[d]	8HS	–	–	Tensile	RT	~140	[106]
Uncoated	AS[d]	8HS				RT	~210	
In-situ LaPO$_4$	Al$_2$O$_3$-LaPO$_4$	8HS	40	~80	Tensile	RT	200–250	[86, 109]
CaWO$_4$	Al$_2$O$_3$-CaWO$_4$	0°		~80	Tensile	RT	~350	[56, 110]
NdPO$_4$ (Nextel 720)	Mullite + 5 wt.-% ZrO$_2$	8HS	35	~86	4-pt. bend	RT	235 ± 32	[107, 108]
						1000 °C	234 ± 17	
						1200 °C	233 ± 33	
						1300 °C	230 ± 41	

a All fibers Nextel 610, except NdPO$_4$.
b RT = room temperature test; HT = high temperature test.
c Composite strengths normalized to a V$_f$ = 20%. Actual V$_f$ ranged from 18–35%, although most ~18–25%.
d Aluminosilicate matrix from COI.

extensive fiber pull-out of monazite coated fibers, with evidence of monazite smearing, diagnostic of plastic deformation (Figure 8.5) [100]. Tensile strengths at 1200 °C were similar to those at room temperature. Control samples were brittle, with no fiber pull-out.

Monazite coating functionality was confirmed for monazite-coated Nextel 610 and Almax fabrics in relatively dense matrix CMCs [96, 105]. By hot pressing at high temperatures (≥1200 °C), composite densities between 80 and 90% were attained. LaPO$_4$-monazite coatings were not successfully evaluated in Nextel 312 and Nextel 610/aluminosilicate matrix CMCs [106]. Fiber strength degradation during fiber coating may be responsible for this result. CMCs with NdPO$_4$-monazite fiber-matrix interfaces have also been evaluated [107, 108]. The CMCs were about 86% dense and tested using 4-pt. flexure at high temperature. The strengths were about 235 MPa at room temperature; this was retained up to 1300 °C. The strength decreased to less than 15% after 300 thermal cycles to 1150 °C.

An in-situ monazite coating was developed for porous matrix Nextel 610-based CMCs [86, 109]. The coating was formed from an alumina matrix precursor mixed

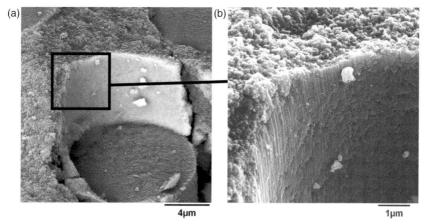

Figure 8.5 SEM micrographs showing Nextel 610/monazite/alumina fracture surface: (a) The monazite coating (lighter phase) can be seen between the fiber and matrix; (b) Deformation of the monazite in the trough appears as smearing of the coating.

with about 10 to 20 vol.-% monazite precursor during CMC fabrication. The CMC had higher damage tolerance and reduced notch sensitivity compared to control samples. Monazite was also dispersed throughout the matrix; the effect of this dispersion on matrix-dominated properties was not characterized.

Scheelite ($CaWO_4$) fiber-matrix interfaces were demonstrated for Nextel 610/alumina CMCs [56, 110]. Relatively dense (>80% dense) unidirectional CMCs had tensile strengths approaching 350 MPa and strain-to-failure of about 0.25%. Some debonding at the fiber-matrix interface was evident from fractography.

8.3.4.2 Porous Coatings and Fugitive Coatings

A porous fiber coating can, in principle, function like a porous matrix, but with less effect on the matrix-dominated properties. There are several limited studies of porous coatings, including zirconia-silica [111], ZrO_2 [112], and rare-earth aluminates [113]. The term fugitive is used for fiber coatings that are removed after CMC fabrication. Carbon is the obvious choice. Carbon coated fibers can be incorporated in oxide/oxide CMCs and then oxidized, leaving a gap at the fiber-matrix interface [114]. Functional fugitive coatings were shown for sapphire-reinforced yttrium-aluminium garnet and for Nextel 720/calcium aluminosilicate glass-ceramic CMCs [114]. Push-out testing was used for both CMCs. Tensile testing of the Nextel 720 CMCs found higher strengths (92 ± 42 MPa) in comparison to control samples without a fugitive interface (22 ± 3 MPa). The initial carbon coating thickness was found to be very important. Fugitive carbon coatings were also examined for the small diameter Nextel 610 fibers in a mullite matrix [37]. These coatings were not successful due to incomplete coverage of the fibers. Duplex fugitive carbon + ZrO_2 fiber coatings have also been reported [64, 115, 116]. CMCs with fugitive fiber coatings may sacrifice strength, stiffness, and thermal conductivity in directions perpendicular to fibers, as compared to CMCs with dense fiber-matrix interfaces and matrices. This may inhibit selection of such CMCs for some applications.

8.3.4.3 Other Coatings

Research has been conducted on other fiber-matrix interface concepts. Other coatings include easy-cleaving coatings, such as phyllosilicates [117, 118], hexaluminates [119], and layered perovskites [120] and ductile coatings such as platinum [121, 122]. A commercial product, "Soft-Cera," is available from Mitsui-Mining Materials Corp. It consists of 22 vol.-% Almax fabric (Al_2O_3) with a ZrO_2 coating and a porous alumina matrix [123, 124]. Initial reports indicate relatively low strengths of 50 to 60 MPa and little effect of gauge length on the composite strength.

8.4
Summary and Future Work

Porous matrix oxide/oxide CMCs, useful at temperatures up to about 1200 °C, are commercially available. The matrix-dominated properties of these CMCs are modest, and limit application of such CMCs. Dense matrix oxide/oxide CMCs with engineered fiber-matrix interfaces and improved matrix-dominated properties have greater application potential. The functionality of rare earth orthophosphate fiber-matrix interfaces has been definitively demonstrated. These fiber coatings may also benefit porous matrix oxide/oxide CMCs if the fiber is subject to environmental attack in a combustion environment. Hermetic monazite ($LaPO_4$) coatings are being developed to protect fibers from aggressive environments [125]. Insertion into many applications may also require increased sophistication in CMC design, such as component-specific fiber architectures with reproducible economical processing, and improved processing methods for three-dimensional densification constraints.

Current oxide/oxide CMCs are limited in use temperature by the capabilities of commercially available oxide fibers. There are clear approaches to the development of higher temperature, fine diameter oxide fibers. Yttrium-aluminium garnet and mullite have creep resistance and micro-structural stability superior to current alumina based fibers. Preliminary research demonstrates the potential of such fibers [6, 7, 126–128]. Commercialization of such fibers will improve the maximum use temperatures and will allow for matrix densification processing at higher temperatures.

References

1 Elfstrom, B.O. (1998) The role of advanced materials in aircraft engines. *Nouvelle Revue D'Aeronautique et D'Astronautique*, **2**, 81–5.

2 Richerson, D.W., Ferber, M.K. and Roode, M.V. (2003) the ceramic gas turbine – retrospective, current status and prognosis, in *Ceramic Gas Turbine Component Development and Characterization* (eds M.V. Roode, M.K. Ferber and D.W. Richerson), ASME Press, New York, pp. 696–741.

3 Robinson, R.C. and Smialek, J.L. (1999) SiC recession caused by SiO_2 scale volatility under combustion conditions: I, experimental results and empirical model. *Journal of the American Ceramic Society*, **82**, 1817–25.

4 Jacobson, N.S. (1993) Corrosion of silicon-based ceramics in combustion environments. *Journal of the American Ceramic Society*, **76**, 3–28.

5 Jacobson, N.S., Opila, E.J. and Lee, K.N. (2001) Oxidation and corrosion of ceramics and ceramic matrix composites. *Current Opinion in Solid State and Materials Science*, **5**, 301–9.

6 Wilson, D.M. (2001) New high temperature oxide fibers, in *High Temperature Ceramic Matrix Composites* (eds W. Krenkel, R. Naslain and H. Schneider), Wiley-VCH Verlag GmbH, Weinheim, Germany, pp. 1–12.

7 Wilson, D.M. and Visser, L.R. (2001) High performance oxide fibers for metal and ceramic composites. *Composites A*, **A32**, 1143–53.

8 Parthasarathy, T.A., Boakye, E., Cinibulk, M.K. and Petry, M.D. (1999) Fabrication and testing of oxide/oxide microcomposites with monazite and hibonite as interlayers. *Journal of the American Ceramic Society*, **82**, 3575–83.

9 3M (2001) *3M Nextel Textiles Product Description Brochure*, 3M, St. Paul, MN.

10 Richerson, D.W. (1997) Ceramic matrix composites, in *Composites Engineering Handbook* (ed. P.K. Mallick), Marcel Dekker, New York, pp. 983–1038.

11 Saphikon (2003) *Properties and Benefits of Sapphire: A Quick Reference Guide*, Saphikon Inc., Milford, NH.

12 Cinibulk, M.K., Parthasarathy, T.A., Keller, K.A. and Mah, T.-I. (2002) Porous yttrium aluminum garnet fiber coatings for oxide composites. *Journal of the American Ceramic Society*, **85**, 2703–10.

13 Parthasarathy, T.A., Boakye, E., Keller, K.A. and Hay, R.S. (2001) Evaluation of porous ZrO_2-SiO_2 and monazite coatings using Nextel™ 720 fiber-reinforced blackglas (TM)minicomposites. *Journal of the American Ceramic Society*, **84**, 1526–32.

14 Wilson, D.M., Lunenburg, D.C. and Lieder, S.L. (1993) High temperature properties of Nextel 610 and alumina-based nanocomposite fibers. *Ceramic Engineering and Science Proceedings*, **14**, 609–21.

15 Wilson, D.M., Lieder, S.L. and Lunenburg, D.C. (1995) Microstructure and high temperature properties of Nextel 720 fibers. *Ceramic Engineering and Science Proceedings*, **16**, 1005–14.

16 Wilson, D.M. (1997) Statistical tensile strength of Nextel 610 and Nextel 720 fibres. *Journal of Materials Science*, **32**, 2535–42.

17 Petry, M.D. and Mah, T-I. (1999) Effect of thermal exposures on the strength of Nextel 550 and 720 filaments. *Journal of the American Ceramic Society*, **82**, 2801–7.

18 Kingdon, A.I., Davis, R.F. and Thackeray, M.M. (1991) *Engineering Properties of Multicomponent and Multiphase Oxides. Engineering Materials Handbook, Ceramics and Glasses*, ASM International, Metals City, OH, pp. 758–74.

19 Miyayama, M., Koumoto, K. and Yanagida, H. (1991) *Engineering Properties of Multicomponent and Multiphase Oxides. Engineerin Materials Handbook, Ceramics and Glasses*, ASM International, Metals City, OH, pp. 748–57.

20 Berger, M.H., Lavaste, V. and Bunsell A.R. (1999) Properties and microstructure of small-diameter alumina-based fibers, in *Fine Ceramic Fibers* (eds A.R. Bunsell and M.H. Berger), Marcel Dekker, New York, pp. 111–64.

21 Zawada, L.P., Hay, R.S., Staehler, J. and Lee, S.S. (2003) Characterization and high temperature mechanical behavior of an oxide/oxide composite. *Journal of the American Ceramic Society*, **86**, 981–90.

22 Mattoni, M.A., Yang, J.Y., Levi, C.G. and Zok, F.W. (2001) Effects of matrix porosity on the mechanical properties of a porous-matrix, all-oxide ceramic

23 Carelli, E.A.V., Fujita, H., Yang, J.Y. and Zok F.W. (2002) Effects of thermal aging on the mechanical properties of a porous-matrix composite. *Journal of the American Ceramic Society*, **85**, 595–602.

24 Jurf, R.A. and Butner, S.C. (2000) Advances in all-oxide CMC. *Journal of Engineering for Gas Turbine and Power*, **122**, 202–5.

25 Tandon, G.P., Buchanan, D.J., Pagano, N.J. and John, R. (2001) Analytical and experimental characterization of thermomechanical properties of a damaged woven oxide-oxide composite. *Ceramic Engineering and Science Proceedings*, **22**, 687–94.

26 COI Ceramics I (2003) Oxide-Oxide CMC Data Sheets, San Diego, CA.

27 Heathcote, J.A., Gong, X.-Y., Yang, J.Y., Ramamurty, U. and Zok, F.W. (1999) In-plane mechanical properties of an all-oxide ceramic composite. *Journal of the American Ceramic Society*, **82**, 2721–30.

28 Levi, C.G. (1999) Microstructural design of stable porous matrices for all-oxide ceramic composites. *Zeitschrift für Metallkunde*, **90**, 1037–47.

29 Zok, F.W. (2006) Developments in oxide fiber composites. *Journal of the American Ceramic Society*, **89**, 3309–24.

30 Marshall, D.B. and Davis, J.B. (2001) Ceramics for future power generation technology: fiber reinforced oxide composites. *Current Opinion in Solid State and Materials Science*, **5**, 283–9.

31 Evans, A.G., Marshall, D.B., Zok, F. and Levi, C. (1999) Recent advances in oxide-oxide composite technology. *Advanced Composite Materials*, **8**, 17–23.

32 Kerans, R.J., Hay, R.S. and Parthasarathy, T.A. Structural ceramic composites. *Current Opinion in Solid State and Materials Science*, **4**, 445–51.

33 Keller, K.A., Jefferson, G. and Kerans, R.J. (2005) Progress in oxide composites. *Annales de Chimie: Science des Materiaux*, **30**, 547–63.

34 Keller, K.A., Jefferson, G. and Kerans, R.J. (2005) Oxide-oxide composites, in *Handbook of Ceramic Composites* (ed. N. Bansal), Kluwer Academic, Boston, pp. 377–422.

35 Zawada, L.P. and Jenkins, M. (2002) Specimen size effects and round robin results for transthickness tensile strength of N720/AS-1. *Ceramic Engineering and Science Proceedings*, **23**, 637–45.

36 Simon, R.A. and Danzer, R. (2006) Oxide fiber composites with promising properties for high-temperature structural applications. *Advanced Engineering Materials*, **8**, 1129–34.

37 Peters, P.W.M., Daniels, B., Clemens, F. and Vogel, W.D. (2000) Mechanical characterization of mullite-based ceramic matrix composites at test temperatures up to 1200 °C. *Journal of the European Ceramic Society*, **20**, 531–5.

38 John, R., Zawada, L.P. and Kroupa, J.L. (1999) Stresses due to temperature gradients in ceramic-matrix-composite aerospace components. *Journal of the American Ceramic Society*, **82**, 161–8.

39 Kanka, B.J., Goring, J., Schmucker, M. and Schneider, H. (2001) Processing microstructure, and properties of Nextel™ 610, 650, and 720 fiber/porous mullite matrix composites. *Ceramic Engineering and Science Proceedings*, **22**, 703–10.

40 Ma, X.-D., Shiono, K., Fukino, Y., Hayashi, S. and Nakagawa, Z. (2003) Fabrication of fiber-reinforced porous ceramic composite setter produced by organic flocculating method and its thermal shock behavior. *Journal of the Ceramic Society of Japan*, **111**, 37–41.

41 Staehler, J.M. and Zawada, L.P. (2000) Performance of four ceramic-matrix composite divergent flap inserts following ground testing on an F110 turbofan engine. *Journal of the American Ceramic Society*, **83**, 1727–38.

42 Kerans, R.J., Hay, R.S., Parthasarathy, T.A. and Cinibulk, M.K. (2002) Interface design for oxidation resistant ceramic composites. *Journal of the American Ceramic Society*, **85**, 2599–632.

43 Evans, A.G. and Zok, F.W. (1994) Review: the physics and mechanics of fibre-reinforced brittle matrix composites. *Journal of Materials Science*, **29**, 3857–96.

44 DiCarlo, J.A. and Roode, M.V. (2006) *Ceramic Composite Development for Gas*

Turbine Engine Hot Section Components. ASME Turbo Expo 2006: Power for Land, Sea, and Air, ASME, Barcelona, Spain.
45. Lane, J. (2007) *Oxide-Based CMCs for Combustion Turbines*. 31st International Cocoa Beach Conference and Exposition on Advanced Ceramics and Composites, American Ceramic Society, Daytona Beach, FL.
46. Szweda, A. (2005) *Development and Evaluation of Hybrid Oxide/Oxide Ceramic Matrix Composite Combustor Liners*. ASME Turbo Expo 2005: Power for Land, Sea, and Air, ASME, Reno, NV, pp. 315–21.
47. Jurf, B. (2003) *Fabrication and Test of Insulated CMC Exhaust Pipe*. Aeromat 2003. Dayton, OH.
48. Morrison, J.A. and Krauth, K.M. (1998) Design and analysis of a CMC turbine blade tip seal for a land-based power turbine. *Ceramic Engineering and Science Proceedings*, **19**, 249–56.
49. Mattoni, M.A., Yang, J.Y., Levi, C.G., Zok, F.W. and Zawada, L.P. (2005) Effects of combustor rig exposure on a porous-matrix oxide composite. *International Journal of Applied Ceramic Technology*, **2**, 133–40.
50. Parthasarathy, T.A., Zawada, L.P., John, R., Cinibulk, M.K., Kerans, R.J. and Zelina, J. (2005) Evaluation of oxide-oxide composites in a novel combustor wall application. *International Journal of Applied Ceramic Technology*, **2**, 122–32.
51. He, M.Y. and Hutchinson, J.W. (1989) Crack deflection at the interface between dissimilar materials. *International Journal of Solids and Structures*, **25**, 1053–67.
52. He, M.Y., Hutchinson, J.W. and Evans, A.G. (1994) Crack deflection at an interface between dissimilar elastic materials: role of residual stresses. *International Journal of Solids and Structures*, **31**, 3443–55.
53. Kerans, R.J. and Parthasarathy, T.A. (1999) Crack deflection in ceramic composites and fiber coating design criteria. *Composites Part A*, **30A**, 521–24.
54. Faber, K.T. (1997) Ceramic composite interfaces: properties and design. *Annual Review of Materials Science*, **27**, 499–524.
55. Kerans, R.J. (1995) The role of coating compliance and fiber/matrix interfacial topography on debonding in ceramic composites. *Scripta Metallurgica et Materiala*, **32**, 505–9.
56. Goettler R.W., Sambasivan S., Dravid, V. and Kim, S. (1999) Interfaces in oxide fiber–oxide matrix ceramic composites, in *Computer Aided Design of High Temperature Materials* (eds A. Pechenik, R. Kalia and P. Vashishta), Oxford University Press, pp. 333–49.
57. Keller, K.A., Mah, T., Parthasarathy, T.A., Boakye, E.E., Mogilevsky, P. and Cinibulk, M.K. (2003) Effectiveness of monazite coatings in oxide/oxide composites after long term exposure at high temperature. *Journal of the American Ceramic Society*, **86**, 325–32.
58. Gundel, D.B., Taylor, P.J. and Wawner, F.E. (1994) Fabrication of thin oxide coatings on ceramic fibres by a sol-gel technique. *Journal of Materials Science*, **29**, 1795.
59. Hay, R.S. and Hermes, E.E. (1990) Sol-gel coatings on continuous ceramic fibers. *Ceramic Engineering and Science Proceedings*, **11**, 1526–32.
60. Dislich, H. and Hussmann, E. (1981) Amorphous and crystalline dip coatings obtained from organometallic solutions: procedures, chemical processes, and products. *Thin Solid Films*, **77**, 129–39.
61. Jero, P.D., Rebillat, F., Kent, D.J. and Jones, J.G. (1998) Crystallization of lanthanum hexaluminate from MOCVD precursors. *Ceramic Engineering and Science Proceedings*, **19**, 359–60.
62. Haynes, J.A., Cooley, K.M., Stinton, D.P., Lowden, R.A. and Lee, W.Y. (1999) Corrosion resistant CVD mullite coatings for Si_3N_4. *Ceramic Engineering and Science Proceedings B*, **20**, 355–62.
63. Chayka, P.V. (1997) Liquid MOCVD precursors and their application to fiber interface coatings. *Ceramic Engineering and Science Proceedings*, **18**, 287–94.
64. Nubian, K., Saruhan, B., Kanka, B., Schmucker, M., Schneider, H. and Wahl, G. (2000) Chemical vapor deposition of ZrO_2 and C/ZrO_2 on mullite fibers for interfaces in mullite/aluminosilicate fiber-reinforced composites. *Journal of the European Ceramic Society*, **20**, 537–44.

65 Brown, P.W. (1995) Electrophoretic deposition of mullite in a continuous fashion utilizing non-aqueous polymeric sols, in *Ceramic Transactions*, Vol **56** (ed. K.V. Logan), American Ceramic Society, Columbus, OH, pp. 369–76.

66 Illston, T.J., Ponton, C.B., Marquis, P.M. and Butler, E.G. (1994) Electrophoretic deposition of silica/alumina colloids for the manufacture of CMCs. *Ceramic Engineering and Science Proceedings*, **15**, 1052–9.

67 Rice, R.W. (1987) BN Coating of Ceramic Fibers for Ceramic Fiber Composites. US patent 4,642,271.

68 Naslain, R., Dugne, O., Guette, A., Sevely, J., Brosse, C.R., Rocher, J.P. and Cotteret, J. (1991) Boron nitride interphase in ceramic matrix composites. *Journal of the American Ceramic Society*, **74**, 2482–8.

69 Griffin, C.J. and Kieschke, R.R. (1995) CVD processing of fiber coatings for CMCs. *Ceramic Engineering and Science Proceedings*, **16**, 425–32.

70 Lewis, M.H., Tye, A., Butler, E.G. and Doleman, P.A. (2000) Oxide CMCs: interphase synthesis and novel fiber development. *Journal of the European Ceramic Society*, **20**, 639–44.

71 Hay, R.S. and Boakye, E. (2001) Monazite coatings on fibers: I, effect of temperature and alumina doping on coated fiber tensile strength. *Journal of the American Ceramic Society*, **84**, 2783–92.

72 Boakye, E., Hay, R.S. and Petry, M.D. (1999) Continuous coating of oxide fiber tows using liquid precursors: monazite coatings on Nextel 720. *Journal of the American Ceramic Society*, **82**, 2321–31.

73 Boakye, E.E., Hay, R.S., Mogilevsky, P. and Douglas, L.M. (2001) Monazite coatings on fibers: II, coating without strength degradation. *Journal of the American Ceramic Society*, **84**, 2793–801.

74 Hay, R.S. and Hermes, E.E. (1993) Coating Apparatus for Continuous Fibers. US patent 5,217,533.

75 Hay, R.S., Boakye, E. and Petry, M.D. (1996) Fiber strength with coatings from sols and solutions. *Ceramic Engineering and Science Proceedings*, **17**, 43–53.

76 Boakye, E.E., Mah, T., Cooke, C.M., Keller, K. and Kerans, R.J. (2004) Initial assessment of the weavability of monazite-coated oxide fibers. *Journal of the American Ceramic Society*, **87**, 1775–8.

77 Fair, G.E., Hay, R.S. and Boakye, E.E. (2007) Precipitation coating of monazite on woven ceramic fibers–I. Feasibility. *Journal of the American Ceramic Society*, **90**, 448–55.

78 Weaver, J.H., Yang, J., Levi, C.G., Zok, F.W. and Davis, J.B. (2007) A Method for Coating Fibers in Oxide Composites. *Journal of the American Ceramic Society*, **90**, 1331–3.

79 Cinibulk, M.K., Welch, J.R. and Hay, R.S. (1996) Preparation of thin sections of coated fibers for characterization by transmission electron microscopy. *Journal of the American Ceramic Society*, **79**, 2481–4.

80 Hay, R.S., Welch, J.R. and Cinibulk, M.K. (1997) TEM specimen preparation and characterization of ceramic coatings on fiber tows. *Thin Solid Films*, **308–309**, 389–92.

81 Hay, R.S., Fair, G., Mogilevsky, P. and Boakye, E.E. (2005) Measurement of fiber coating thickness variation. *Ceramic Engineering and Science Proceedings*, **26**, 11–18.

82 Hay, R.S., Boakye, E.E. and Petry, M.D. (2000) Effect of coating deposition temperature on monazite coated fiber. *Journal of the European Ceramic Society*, **20**, 589–97.

83 Petry, M.D., Mah, T. and Kerans, R.J. (1997) Validity of using average diameter for determination of tensile strength and weibull modulus of ceramic filaments. *Journal of the European Ceramic Society*, **80**, 2741–4.

84 Lam, D.C.C. and Lange, F.F. (1994) Microstructual observations on constrained densification of alumina powder containing a periodic array of sapphire fibers. *Journal of the American Ceramic Society*, **77**, 1976–8.

85 Rahaman, M.N. (1995) *Ceramic Processing and Sintering*, Marcel-Dekker Inc., New York.

86 Davis, J.B., Marshall, D.B. and Morgan, P.E.D. (1999) Oxide composites of LaPO4

and Al_2O_3. *Journal of the European Ceramic Society*, **19**, 2421–6.

87 Fujita, H., Levi, C.G., Zok, F.W. and Jefferson, G. (2005) Controlling mechanical properties of porous mullite/alumina mixtures via precursor-derived alumina. *Journal of the American Ceramic Society*, **88**, 367–75.

88 Levi, C.G., Yang, J.Y., Dalgleish, B.J., Zok, F.W. and Evans, A.G. (1998) Processing and performance of an all-oxide ceramic composite. *Journal of the American Ceramic Society*, **81**, 2077–86.

89 Church, P.K. and Knutson, O.J. (1976) Chromium Oxide Densification, Bonding, Hardening, and Strengthening of Bodies having Interconnected Porosity. US patent 3,956,531.

90 Mogilevsky, P., Kerans, R.J., Lee, H.D., Keller, K.A. and Parthasarathy, T.A. (2007) On densification of porous materials using precursor solutions. *Journal of the American Ceramic Society*, **90** (10), 3073–84.

91 Keller, K.A., Mah, T., Lee, H.D., Boakye, E.E., Mogilevsky, P., Parthasarathy, T.A., (2004) Towards Dense Monazite Coatings in Dense Oxide-Oxide Composites. *28th Annual Cocoa Beach Conference and Exposition on Advanced Ceramics and Composites*, American Ceramic Society, Daytona Beach, FL.

92 Mawdsley, J.R. and Halloran, J.W. (2001) The effect of residual carbon on the phase stability of $LaPO_4$ at high temperatures. *Journal of the European Ceramic Society*, **21**, 751–7.

93 Chawla, K.K. (1993) *Ceramic Matrix Composites*, Chapman & Hall, University Press, Cambridge.

94 Bansal, N.P. and Eldridge, J.I. (1997) Effects of interface modification on mechanical behavior of Hi-nicalon fiber-reinforced celsian matrix composites. *Ceramic Engineering and Science Proceedings*, **18**, 379–89.

95 Morgan, P.E.D. and Marshall, D.B. (1995) Ceramic composites of monazite and alumina. *Journal of the American Ceramic Society*, **78**, 1553–63.

96 Lee, P.-Y., Imai, M. and Yano, T. (2004) Fracture behavior of monazite-coated alumina fiber-reinforced alumina-matrix composites at elevated temperature. *Journal of the Ceramic Society of Japan*, **112**, 628–33.

97 Morgan, P.E.D. and Marshall, D.B. (1993) Functional interfaces for oxide/oxide composites. *Materials Science and Engineering*, **A162**, 15–25.

98 Hikichi, Y. and Nomura, T. (1987) Melting temperatures of monazite and xenotime. *Journal of the American Ceramic Society*, **70**, C252–3.

99 Davis, J.B., Marshall, D.B., Housley, R.M. and Morgan, P.E.D. (1998) Machinable ceramics containing rare-earth phosphates. *Journal of the American Ceramic Society*, **81**, 2169–75.

100 Davis, J.B., Hay, R.S., Marshall, D.B., Morgan, P.E.D. and Sayir, A. (2003) The influence of interfacial roughness on fiber sliding in oxide composites with La-monazite interphases. *Journal of the American Ceramic Society*, **86**, 305–16.

101 Hay, R.S. (2003) (120) and (122) monazite deformation twins. *Acta Materialia*, **51**, 5255–62.

102 Hay, R.S. and Marshall, D.B. (2003) Deformation twinning in monazite. *Acta Materialia*, **51**, 5235–54.

103 Hay, R.S. (2004) Climb-dissociated dislocations in monazite. *Journal of the American Ceramic Society*, **87**, 1149–52.

104 Hay, R.S. (2005) Twin-dislocation interaction in monazite (monoclinic $LaPO_4$). *Philosophical Magazine*, **85**, 373–86.

105 Lee, P. and Yano, T. (2004) The Influence of fiber coating conditions on the mechanical properties of alumina-alumina composites. *Composite Interfaces*, **11**, 1–13.

106 Cazzato, A., Colby, M., Daws, D., Davis, J., Morgan, P., Porter, J., Butner, S. and Jurf, B. (1997) Monazite interface coatings in polymer and sol-gel derived ceramic matrix composites. *Ceramic Engineering and Science Proceedings*, **18**, 269–78.

107 Lewis, M.H. and Tye, A. (1999) Development of interfaces in oxide matrix composites. *Key Engineering Materials*, **164–165**, pp 351–6.

108 Kaya, C., Butler, E.G., Selcuk, A., Boccaccini, A.R. and Lewis, M.H. (2002)

Mullite (Nextel™ 720) fibre-reinforced mullite matrix composites exhibiting favourable thermomechanical properties. *Journal of the European Ceramic Society*, **22**, 2333–42.

109 Davis, J.B., Marshall, D.B. and Morgan, P.E.D. (2000) Monazite containing oxide-oxide composites. *Journal of the European Ceramic Society*, **20**, 583–7.

110 Goettler, R.W., Sambasivan, S. and Dravid, V.P. (1997) Isotropic complex oxides as fiber coatings for oxide-oxide CFCC. *Ceramic Engineering and Science Proceedings*, **18**, 279–86.

111 Boakye, E.E., Hay, R.S., Petry, M.D. and Parthasarathy, T.A. (2004) Zirconia-silica-carbon coatings on ceramic fibers. *Journal of the American Ceramic Society*, **87**, 1967–76.

112 Holmquist, M., Lundberg, R., Sudre, O., Razzell, A.G., Molliex, L., Benoit, J. and Adlerborn, J. (2000) Alumina/alumina composite with a porous zirconia interphase – processing, properties and component testing. *Journal of the European Ceramic Society*, **20**, 599–606.

113 Cinibulk, M.K., Parthasarathy, T.A., Keller, K.A. and Mah, T. (2000) Porous rare-earth aluminate fiber coatings for oxide-oxide. *Ceramic Engineering and Science Proceedings*, **21**, 219–28.

114 Keller, K.A., Mah, T., Cooke, C. and Parthasarathy, T.A. (2000) Fugitive interfacial carbon coatings for oxide/oxide composites. *Journal of the American Ceramic Society*, **83**, 329–36.

115 Saruhan, B., Schmucker, M., Bartsch, M., Schneider, H., Nubian, K. and Wahl, G. (2001) Effect of interphase characteristics on long-term durability of oxide-based fibre-reinforced composites. *Composites A*, **32**, 1095–103.

116 Sudre, O., Razzell, A.G., Molliex, L. and Holmquist, M. (1998) Alumina single-crystal fibre reinforced alumina matrix for combustor tiles. *Ceramic Engineering and Science Proceedings*, **19**, 273–80.

117 Chyung, K. and Dawes, S.B. (1993) Fluoromica coated nicalon fiber reinforced glass-ceramic composites. *Materials Science and Engineering*, **A162**, 27–33.

118 Demazeau, G. (1995) New synthetic mica-like materials for controlling fracture in ceramic matrix composites. *Materials Technology*, **10**, 57–8.

119 Cinibulk, M.K. and Hay, R.S. (1996) Textured magnetoplumbite fiber-matrix interphase derived from sol-gel fiber coatings. *Journal of the American Ceramic Society*, **79**, 1233–46.

120 Fair, G., Shemkunas, M., Petuskey, W.T. and Sambasivan, S. (1999) Layered perovskites as "soft-ceramics." *Journal of the European Ceramic Society*, **19**, 2437–47.

121 Jaskowiak, M.H., Philipp, W.H., Vetch, L.C. and Hurst, J.B. (1992) Platinum interfacial coatings for sapphire/Al_2O_3 composites. *Ceramic Engineering and Science Proceedings*, **13**, 589–98.

122 Wendorff, J., Janssen, R. and Claussen, N. (1998) Platinum as a weak interphase for fiber-reinforced oxide-matrix composites. *Journal of the American Ceramic Society*, **81**, 2738–40.

123 Mamiya, T. (2002) Tensile damage evolution and notch sensitivity of $Al2O_3$ fiber-ZrO_2 matrix minicomposite-reinforced Al_2O_3 matrix composites. *Materials Science and Engineering*, **A325**, 405–13.

124 Chivavibul, P., Enoki, M. and Kagawa, Y. (2002) Size effect on strength of woven fabric Al_2O_3 fiber–Al_2O_3 matrix composites. *Ceramic Engineering and Science Proceedings*, **23**, 685–90.

125 Boakye, E.E., Hay, R.S. and Mogilevsky, P. (2007) Spherical rhabdophane sols, II. Fiber coating. *Journal of the American Ceramic Society*, **90**, 1580–8.

126 Liu, Y., Zhang, Z.-F., Halloran, J. and Laine, R.M. (1998) Yttrium aluminum garnet fibers from metallorganic precursors. *Journal of the American Ceramic Society*, **81**, 629–45.

127 King, B.H. and Halloran, J.W. (1995) Polycrystalline yttrium aluminum garnet fibers from colloidal sols. *Journal of the American Ceramic Society*, **78**, 2141–8.

128 Schmucker, M., Schneider, H., Mauer, T. and Clauss, B. (2005) Temperature-dependent evolution of grain growth in mullite fibres. *Journal of the European Ceramic Society*, **25**, 3249–56.

9
All-Oxide Ceramic Matrix Composites with Porous Matrices

Martin Schmücker and Peter Mechnich

9.1
Introduction

Ceramic materials generally display excellent temperature stability, high hardness, and good corrosion and erosion resistance. However, monolithic ceramics are delicate structural materials due to their inherent brittleness. Many efforts have been made during the last two decades to overcome this problem. To achieve damage tolerant and favorable failure behavior reinforcing components, such as ZrO_2 particles [1], whiskers or chopped fibers [2]), and continuous fibers were employed, the latter showing the most promise [3]).

Ceramics such as nitrides, carbides, borides, or carbon achieve high strength and excellent creep resistance up to elevated temperatures, due to their predominant covalent bonding. However, the fundamental drawback of these materials is their susceptibility to oxidation. Thus, for many years, every endeavor has been made to extend the lifetime of non-oxide materials, in particular non-oxide Ceramic Matrix Composites (CMCs), by protective coatings [4]. Long-term oxidation protection is hard to obtain as a matter of principle, particularly when the coated composites are used under cyclic thermal or mechanical load. As a consequence, there is an increasing interest in oxide fiber-oxide matrix ceramics (oxide/oxide CMCs) as structural materials for high temperature applications in oxidizing atmospheres.

Although both constituents of fiber-reinforced ceramics, that is, ceramic fibers and ceramic matrices, are brittle, the composites display quasi-ductile deformation due to mechanisms such as crack deflection, crack bridging, or fiber pull out [5]. A premise for these mechanisms to work is debonding between fibers and the matrix under mechanical load. To achieve weak fiber-matrix bonding, either suitable fiber coatings have to be employed, for example, cleavable, porous, and low toughness materials ("weak interphase" approach), or alternatively, highly porous matrices ("weak matrix" approach) may be used. Materials making use of the porous matrix concept exhibit only a few local fiber-matrix contacts and hence debonding at the fiber-matrix interface is easily achieved [6].

Ceramic Matrix Composites. Edited by Walter Krenkel
Copyright © 2008 WILEY-VCH Verlag GmbH & Co. KGaA, Weinheim
ISBN: 978-3-527-31361-7

Table 9.1 Requirements of hot-gas conducting structures.

	Monolithic non-oxide ceramics	Monolithic oxide ceramics	Non-oxide CMC	Oxide CMC
Sufficient strength at elevated temperature	+	○	+	○
Oxidation resistance	–	+	–	+
Thermal shock resistance	–	–	+	+
Graceful failure behavior	–	–	+	+

+, good; ○, suitable; –, insufficient.

Current and future applications of oxide/oxide CMCs are focused on gas turbine technology (e.g. combustor liners, nozzles, vanes), space transportation (e.g. heat shields for re-entry space vehicles), chemical engineering and metallurgy (e.g. burner nozzles, catalyst supports, filters, kiln furniture), and energy conversion (e.g. absorbers and heat exchangers of solar power plants). In Table 9.1, the properties required for hot gas conducting structures are compiled, showing how the different groups of materials (oxide, non-oxide; monolithic, CMC) can fulfill the demands. It turns out that oxide/oxide CMCs have the potential to make the best compromise.

9.1.1
Oxide Ceramic Fibers

In order to achieve strength values as high as possible, a fiber microstructure characterized by nano-sized grains is preferred. This is derived from the Hall–Petch-like correlation between strength and crystal size with $\sigma \propto d^{-0.5}$ (σ = strength and d = grain diameter) [7]. Though nano-crystalline fibers have high strength values at room temperature, thermal and mechanical stability at elevated temperatures often decreases because of creep and grain coarsening effects. Thus, the grain size is a key feature for tailoring the fibers' mechanical properties.

It is instructive to compare microstructure and properties of the most used commercial fibers Nextel 610 and 720, both fabricated by 3M, St Paul, MN (USA). Nextel 610 consists of single phase α-alumina with an average grain size of ≈80 nm. However, Nextel 720 consists of a mullite/alumina phase assemblage with a bimodal grain size distribution: ≈80 nm alumina grains co-exist with ≈350 nm mullite crystals. Due to their phase composition, the room temperature strength and Young's modulus of Nextel 610 fibers are significantly higher than those of Nextel 720. However, the creep rate of Nextel 720 is about four orders of magnitude lower with respect to Nextel 610. Temperature-dependent grain coarsening and related strength degradation of Nextel 610, 650 (a less used Al_2O_3/ZrO_2 fiber), and 720 fibers are depicted in Figure 9.1. In general, the onset temperature of grain coarsening is ≈1200 °C. However, relative strength degradation depends

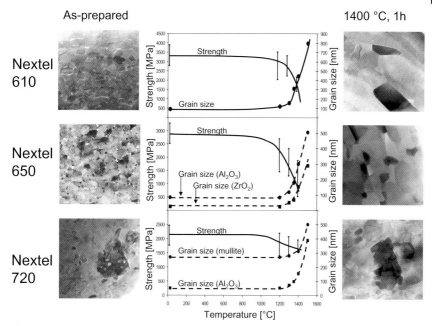

Figure 9.1 Room temperature strength, microstructure and grain size of different Nextel fiber types 610, 650 and 720 after isothermal firing (1 h).

on the starting grain size as the driving force of coarsening. Thus strength deterioration of Nextel 720 is less pronounced than in Nextel 610 and 650.

9.1.2
"Classical" CMC Concepts

The fiber bundles or fabrics have to be introduced homogeneously in suitable oxide matrices (typically alumina, mullite, yttrium aluminum garnet (YAG), and glass ceramics). This is commonly achieved by the liquid phase impregnation of fiber tows or fiber fabrics using dispersed particles ("slurries") polymers, sols, or combinations thereof. Subsequently, the conversion to ceramic bodies is carried out by evaporation of the carrier liquid, pyrolysis, or gelation and a final sintering step.

The preparation of fairly dense and defect-free crystalline matrices is challenging: On the one hand, sintering temperature should not exceed 1300 to 1350 °C, in order to avoid fiber degradation (see above). On the other hand, fibers and fiber fabrics tend to form rigid networks that may inhibit the sintering process. Thus, matrix shrinkage within the rigid fiber network typically leads to a high amount of matrix cracks [5]. To some extent, matrix cracks and porosity caused by shrinkage within the rigid fiber skeleton can be reduced through consolidation via unidirectional hot-pressing or by hot-isostatic pressing (HIP). However, while unidirectional pressing is suitable for structures of simple shape only, HIP is a complicated and costly method. A possible alternative to overcome the shrinkage-related problems is matrix processing via reaction-bonding, using non-oxide, for example, Al metal powders as starting materials, such as (RBAO) *reaction bonded aluminium oxide* and (RBM) *reaction bonded mullite* [8, 9]). The principal idea of this method is that the volume gain due to oxidation should compensate for the sintering-induced shrinkage. Although simple in theory, the RBAO process is complicated, due to the low temperature melting of aluminium and the formation of transient alumina, and also because oxidation and sintering proceed at different temperatures. In the case of RBM, the process requires temperatures of ≈1550 °C [9], which is much too high for CMC processing using polycrystalline oxide fibers. Though mullite formation can be significantly accelerated using suitable additives (CeO_2, Y_2O_3) [10], the transient liquid phase, which is then involved, will react with the fiber surfaces.

It is generally accepted that debonding at the fiber-matrix interface is the precondition for energy dissipating mechanisms, such as crack deflection, crack bridging, and fiber pull-out. Thus, the interface strength must be weak enough to facilitate the above-mentioned mechanisms, yet strong enough to enable a load transfer between fibers and matrix and to provide sufficient cohesion in transverse direction. To tailor the interfacial characteristics of CMCs with dense matrices, suitable fiber coatings have been suggested. These interphases must not react with the fiber and the matrix during processing and under service conditions [5]. In the ideal case, the fiber coating provides optimum interface strength and fracture toughness and protects the fiber against mechanical damage during CMC processing. The *status quo* of interface engineering is compiled in the previous chapter of this book.

9.2
Porous Oxide/Oxide CMCs without Fiber/Matrix Interphase

As implied above, CMC concepts with dense matrices reveal many unsolved problems. Though much effort was made, particularly in the field of fiber coating and composite processing, many questions still remain. In the open literature, there is virtually no example of oxide/oxide CMC with fiber coating and dense matrix produced at a technical or pilot-plant scale. Research activities are mainly restricted to so-called micro- or mini-CMCs consisting of one fiber or one fiber tow, respectively. These CMCs are relatively easy to fabricate and apt for rapid evaluation, though the interpretation of mechanical properties may be misleading [11].

As an alternative to oxide/oxide CMCs with dense matrices and fiber-matrix interphases, composites with porous matrices without interphases have been developed since the mid-1990s [12]. The porous matrices typically consist of alumina or aluminium silicates. The bonding between a porous matrix and the incorporated fibers typically is weak, thus leading to matrix crack deflection along the fiber-matrix interface. Moreover, the matrix strength is too low to allow stress transfer into the fibers high enough to fracture them. The macroscopic deformation and fracture behavior of porous matrix CMCs is similar to "classical" dense matrix-coated fiber CMCs. However, in contrast to these materials, manufacturing of porous matrix composites is more simple and inexpensive, since both costly fiber coatings and elaborate matrix densification techniques are not required. For these reasons, porous matrix CMCs are highly attractive materials. Detrimental effects with respect to CMCs with dense matrices have lower compressive strength, inherent permeability, and lower wear resistance [13, 14].

Figure 9.2 shows the microstructure of a porous oxide/oxide composite compared to a CMC with dense matrix and a fiber-matrix interphase.

To date, major research and development activities on oxide/oxide CMCs with porous matrices are conducted by the University of California (UCSB), the German Aerospace Center (DLR), and the University of Bayreuth (Germany). Meanwhile, porous matrix CMCs are commercially available *inter alia* from ATK-COI ceramics

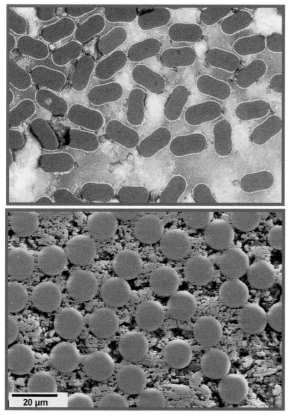

Figure 9.2 Scanning electron micrographs of a conventional CMC with dense mullite matrix obtained by hot-pressing and BN fiber coating (top) compared to a porous matrix CMC (bottom).

Table 9.2 Oxide fiber/mullite matrix composites: fabrication and main characteristics.

Fabricating institution	Infiltration techniques of fiber bundles or fabrics	Shaping of components	Fibers	Matrix	Status
ATK-CO.I. Ceramics, San Diego, CA (USA)	Slurry and polymer	Laminating of woven fabrics, filament winding	Nextel 720, Nextel 610	Aluminosilicate, alumina (porous)	Commercial
University of Bayreuth (Germany)	Slurry	Laminating of woven fabrics	Nextel 720	Aluminosilicate (porous)	Laboratory scale
DLR (Germany)	Slurry	Filament winding	Nextel 720, Nextel 610	Alumina, aluminosilicate (porous)	Pilot scale
University of California, Santa Barbara UCSB (USA)	Slurry and polymer (with fillers)	Laminating of woven fabrics	Nextel 720	Aluminosilicate/ zirconia (porous)	Laboratory scale
Pritzkow (Germany)	Slurry	Laminating of woven fabrics	Nivity, Nextel 440	Aluminosilicate	Commercial
Duotherm Isoliersysteme GmbH (Germany)	Slurry	Filament winding	Nextel 720, Nextel 610	Alumina (porous)	Commercial

(USA), Pritzkow Spezialkeramik, and Duotherm Isoliersysteme GmbH (both Germany), the latter manufacturing CMC materials under license from DLR. Table 9.2 gives an overview of recent activities on all-oxide fiber-reinforced CMCs with porous alumina or aluminosilicate matrices.

9.2.1
Materials and CMC Manufacturing

Similar processing techniques were described for porous matrix oxide/oxide CMCs by UCSB, University of Bayreuth, and ATK-COI [6, 15–17]). First, ceramic textile preforms (typically woven Nextel 610 or 720 fabrics) are infiltrated by a matrix slurry. Vacuum-assisted infiltration may be used to enhance this prepreg fabrication process. Next, the prepregs are stacked to form laminates. The laminates can be molded and subsequently sintered in air at moderate temperatures (1000–1200 °C). The matrix of these CMCs typically consists of alumina or mullite grains bonded either by a siliceous gel, colloidal alumina, or colloidal mullite. In the UCSB processing route, the sintered panels subsequently are subjected to several impregnation cycles with $Al_2(OH)_5Cl$ solution, followed by gelation and pyrolysis, in order to increase the matrix density.

9.2 Porous Oxide/Oxide CMCs without Fiber/Matrix Interphase

In contrast to the infiltration of ceramic textiles, the oxide/oxide composites at DLR (WHIPOX) Wound Highly Porous Oxide Matrix CMC, are manufactured by filament winding. The matrix is derived from a commercial pseudo-boehmite/amorphous silica phase assemblage with overall compositions ranging between 70 and 100 wt.-% Al_2O_3. The processing route is sketched in Figure 9.3. In a first step, the organic sizing of the fibers (Nextel 610 or 720, respectively) is removed by thermal decomposition in a tube furnace. The cleaned fiber tow is continuously infiltrated with a water-based matrix slurry. Then the infiltrated tow is pre-dried to stabilize the matrix and finally wound in one-dimensional-two/dimensional orientation on a mandrel. Green bodies are removed from the mandrel at the moist stage, allowing subsequent stacking, forming, or joining of the prepregs. Once in their final shapes, the green bodies are sintered free-standing in air at ≈1300 °C. High porosity allows easy mechanical finishing of sintered WHIPOX CMCs. Conventional machining methods (drilling, cutting, grinding, milling, etc.) can be applied. Fiber contents of WHIPOX CMCs range between 25 and 50 vol.-%. In contrast to CMC materials prepared by UCSB, COI, and University of Bayreuth, the matrix slurry used for WHIPOX does not consist of a filler and binder phase, resulting in a bimodal particle size distribution. Instead, alumina or aluminosilicate particles of uniform size distribution are employed, having high sinterability due to a special powder pre-treatment.

CMC fabrication via winding process gives a number of benefits with respect to fabric infiltration techniques:

- Complex shapes are possible using a suitable mandrel (Figure 9.4a).
- Tailored reinforcement with adapted fiber orientation can be achieved through variation of the fiber winding angle (Figure 9.4b).
- By variation of the winding pattern continuous fiber layers, open structures (grids), or combination of both can be implemented (Figure 9.4c).
- The fiber fraction can be adjusted by tuning the particle size distribution of the slurry (Figure 9.5) [18].
- Fiber rovings used for the winding process are significantly lower-priced than fiber fabrics (e.g. 750 €/kg vs 1500 €/kg, Nextel 610, 3M, 3000 Den).

Figure 9.6 shows a typical fiber-matrix contact zone of WHIPOX CMCs. The composite consists of Nextel 610 fibers and a matrix with a composition of 95 wt.-% Al_2O_3, 5 wt.-% SiO_2 corresponding to a phase assemblage of α-alumina

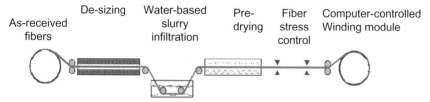

Figure 9.3 Processing route for porous matrix CMCs (WHIPOX) using DLR's filament winding equipment.

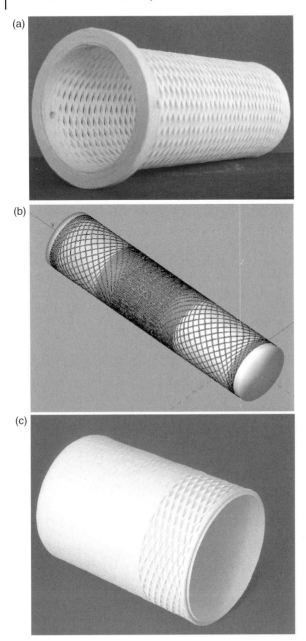

Figure 9.4 (a) Rotation-symmetric WHIPOX component (protection tube); (b) winding pattern with variable fiber orientation (computer simulation); (c) combination of continuous fiber layers and grid-type structures as a result of special winding sequence.

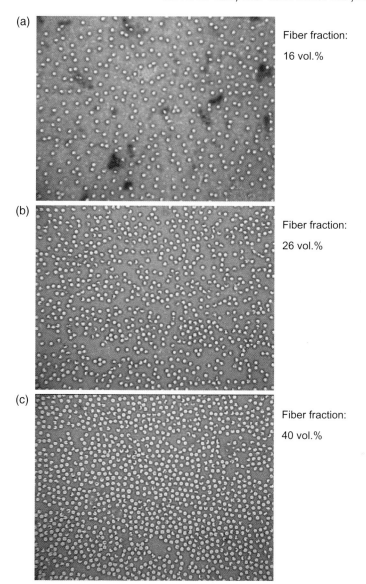

Figure 9.5 Tailoring of WHIPOX CMCs fiber content: (a) 16 vol.-% fiber fraction; (b) 26 vol.-% fiber fraction; (c) 40 vol.-% fiber fraction.

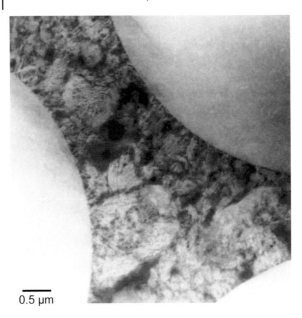

Figure 9.6 Dark field STEM image of fiber/matrix contact area of WHIPOX CMCs.

and minor amounts of mullite. Construction of Dirichlet cells [19] is a suitable tool for the characterization of the composite's mesostructure, providing information on fiber distribution and homogeneity. Fiber cells are areas around individual fiber centers containing all points closer to the respective fiber center than to any other. Figure 9.7 shows the fiber cells corresponding to Figure 9.5a–c with fiber fractions of 16, 26, and 40%, and the corresponding cell size distributions.

9.2.2
Mechanical Properties

Damage tolerance of porous oxide/oxide CMCs is not only due to "classical" fiber-matrix debonding. Matrix pores act as sites of micro-cracking and disintegration, both relating to substantial energy dissipation. These failure mechanisms also reduce the stress concentrations caused by fiber breaks [20]. In porous matrix composites, a crack does not develop a continuous front and matrix failure occurs instead by multiple micro-cracking. Micro-cracking proceeds until the matrix is completely disintegrated, at which point fiber failure and hence total failure of the composite takes place. A sequence of processes is taken into account: matrix micro-cracking → local fiber failure → evolution of bundle damage → coalescence into "bundle cracks" → linkage of bundle cracks, that is, "fiber pull-out" [21].

Early analyses of the mechanical behavior of porous matrix CMCs emanates from models used in conventional CMCs to characterize the debonding behavior of coated fibers. Using a different physical scale, a fiber bundle of a porous matrix

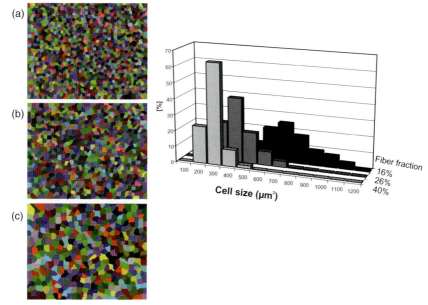

Figure 9.7 Construction of Dirichlet fiber cells corresponding to WHIPOX mesostructures imaged in Figures 9.5a–c. Average cell sizes are inversely proportional to the fiber fraction and cell size distribution is a measure of fiber homogeneity.

CMC was equated to a single filament of a conventional CMC, while the porous matrix was equated to (porous) fiber coatings [22]. According to this consideration, a certain ratio of matrix fracture energy to fiber bundle fracture energy is the precondition for crack deflection in porous matrix CMCs. The critical ratio is of the ratio 1:4, but also depends on elastic mismatch between fibers and matrix and on residual stresses. To fulfill the fracture energy condition, matrix toughness should be low, which is achieved by its high porosity. Furthermore, special demands were made on the architecture of a porous matrix CMC: it was supposed that the distribution of fiber bundles should be heterogeneous in order to create matrix-only pathways for delamination cracks [22].

Systematic investigations on mesostructure and delamination behavior of WHIPOX CMCs suggest that the matrix pathway argument is less important. Although it was shown that thick interlaminar matrix agglomerations may facilitate delamination and hence control the interlaminar shear strength of the material [23], it turned out that damage-tolerant fracture behavior is not affected by the homogeneity of fiber distribution.

Re-examinations of the debonding criteria [24] revealed a significant gap between the limit of debonding and the conditions typically occurring in porous matrix CMCs. As a consequence, the matrix porosity may be reduced to a certain extent to enhance shear strength and off-axis properties, yet without the risk of material embrittlement. Re-infiltration experiments on WHIPOX CMCs showed that the work of fracture is not affected, if the matrix density is increased by up to four infiltration cycles [25]. However, data interpretation is difficult since supplementary infiltration may be more effective in the outer areas of a CMC, thus leading to a matrix density gradient.

Comparison of strength data of porous matrix CMCs published by different authors is difficult due to varying processing and design parameters (fiber bundles or fiber fabric, fiber content, fiber type, matrix composition, matrix porosity, sintering temperature, etc.). The anisotropic design of the oxide/oxide composites leads to different failure mechanisms (Figure 9.8). Depending on the direction of the applied load, matrix-controlled delamination or fiber-controlled fracture can occur. Beyond, intralaminate debonding and miscellaneous failure mechanisms may emerge. To date, highest tensile strength of porous matrix CMCs was published by Simon [17] (Figure 9.9). Ranges of characteristic data obtained for WHIPOX CMCs, using different winding patterns, fiber types, and matrix compositions, are summarized in Table 9.3.

Cyclic fatigue behavior at room temperature was investigated for WHIPOX CMCs, consisting of Nextel 720 fibers and aluminum silicate matrix [26]. For this purpose, Young's moduli were determined as a function of cycle numbers under various mechanical loads. Figure 9.10a shows a shear stress-deformation diagram used to determine the "yield shear strength" τ_{el} of the material. As long as cyclic fatigue experiments were carried out in the elastic range, $\tau_{max}/\tau_{el} \leq 1$, no indication of degradation could be detected, even after 10^4 cycles. On the other hand, if loads greater than the yield strength were applied, a continuous loss in stiffness was observed, indicating fatigue due to an increasing number of defects, such as microcracks and local fiber-matrix debonding (Figure 9.10b). Zawada et al. investigated

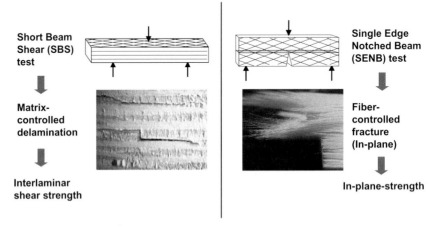

Figure 9.8 Fracture mechanisms in porous matrix CMCs as a response to the anisotropic microstructures; left: matrix-controlled delamination; right: fiber-controlled fracture.

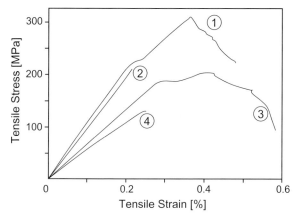

Figure 9.9 Stress strain curves of different porous matrix CMCs: (1) Nextel 610/mullite matrix [17]; (2) Nextel 610/mullite-alumina matrix [6]; (3) Nextel 720/mullite matrix [17]; (4) Nextel 720/mullite-alumina matrix ([6]).

Table 9.3 Ranges of materials properties of WHIPOX CMCs.

Tensile strength	55–120 MPa
In-plane bending strength	80–350 MPa
Young's modulus	40–200 GPa
Interlaminar shear strength	5–30 MPa
Specific weight	1.5–3 g cm^{-3}
Thermal conductivity	1–2 W/mK
Thermal expansion coefficient	4.5–8.5 10^{-6} K^{-1}
Total porosity	25–50 vol.-%

the high temperature fatigue of an oxide/oxide composite manufactured by General Electric (GEN IV) [15]. At each stress level tested in the 1000 °C experiments, a slight decrease in modulus with increased cycle count was observed. It can be assumed that the minor fatigue effects at 1000 °C are due to the free SiO_2 "binder" in the matrix of these materials.

Creep of WHIPOX CMCs under tensile load was recently investigated by Hackemann [27]. In Figure 9.11, creep rates are compiled as a function of stress and temperature. Data in the figure allow estimation of the tolerable load for a given creep rate. For instance, 1 % deformation within 1000 hours can be anticipated at 1050 °C, if a stress of 50 MPa applies. At 1100 °C, the same deformation rate corresponds to only 25 MPa. Further investigations revealed that the processing temperature of the composite has a significant influence on its creep rate. Creep rates are relatively low if high sintering temperatures were used before and vice versa. This can be explained in terms of grain coarsening in the load bearing fibers at

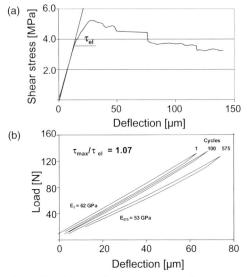

Figure 9.10 (a) Short beam shear test to determine the yield shear strength τ_{el} of WHIPOX CMC; (b) cyclic fatigue test showing degradation if the load exceeds τ_{el}.

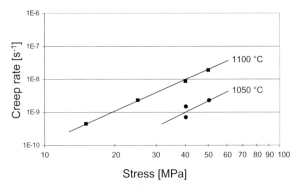

Figure 9.11 Creep behavior of WHIPOX- CMCs at 1050 and 1100 °C.

elevated sintering temperatures. Since creep rates correspond to d^{-3} (d = grain size), in the case of Coble-type diffusional creep, slight changes in average grain size can affect the creep rate distinctly.

9.2.3
Thermal Stability

High temperature applicability of porous matrix oxide/oxide CMCs depends not only on fiber stability (see above) but is limited by the stability of the matrix microstructure against reactions and densification. The influence of thermal aging on the mechanical properties of a porous matrix oxide/oxide CMC manufactured at UCSB was investigated by Carelli et al. [28]. The CMC consists of Nextel 720 woven fabrics with a porous mullite-alumina matrix. The study revealed that strength and failure strain in a 0- to 90-degree orientation remains essentially unchanged after a 1200 °C (1000 h) treatment, despite some strengthening of the matrix and of the fiber-matrix bonding. However, in the 45-degree direction, significant increase in strength and modulus was observed, but at the expense of the damage tolerant fracture behavior. A change in fracture mechanism was observed from matrix cracking, delamination, and fiber scissoring to fiber cracking, when thermally aged samples were tested in the 45-degree direction. The anisotropic degradation behavior was rationalized by the fact that mechanical properties in the 45-degree direction are dominated by the matrix, whereas in the 0- to 90-degree direction, it is largely fiber-dominated. Thus, matrix strengthening is expected to affect the mechanical properties in the diagonal direction rather than in the fiber direction.

Recent investigations of Simon [17] showed that an all-mullite matrix system exhibits the best performance. The strength level of this material remains unchanged up to 1200 °C and retains about 80% of its initial strength, even after exposure at 1300 °C. Sintering effects of the mullite matrix, as monitored by increasing stiffness, occur above 1150 °C, while the stiffness increases above 1000 °C in mullite-alumina matrix composites [13]. CMCs with an alumina-silica matrix undergo rapid degradation, starting at temperatures as low as 1000 °C. In addition to the strength loss, fracture occurs in a brittle manner with virtually no fiber pull-out. Closer inspection of a WHIPOX-type Nextel 720/aluminosilicate matrix composite (fiber composition 85 wt.-% Al_2O_3, 15 wt.-% SiO_2; matrix composition 68 wt.-% Al_2O_3, 32 wt.-% SiO_2) reveals that the silica-rich matrix reacts with the alumina-rich Nextel 720 fibers, leading to a new generation of mullite formed at the fibers' rims (Figure 9.12) [29]. The silica-rich matrices of these heat-treated materials show significant pore agglomeration and densification effects. Thus, due to the poor matrix stability, the fracture behavior of these composites becomes completely brittle. An alumina-rich matrix (95 wt.-% Al_2O_3, 5 wt.-% SiO_2; i.e. α-alumina plus minor amounts of mullite), on the other hand, is thermodynamically stable against the Nextel 720 fiber. Only minor pore agglomeration and particle coarsening were observed upon firing at 1500 °C. Evidence suggests damage-tolerant fracture behavior of WHIPOX CMCs with an alumina-rich matrix, even if temporarily heat treated at 1600 °C (Figure 9.13).

Lifetime predictions of mullite-alumina matrix/Nextel 720 fiber composites were performed by Zok [13]. The propensity of crack deflection was analyzed in terms of toughness and Young's modulus of fibers and matrix considering the development of the matrix properties by thermal treatment. Lifetimes of 60 000 hours were estimated for a mullite matrix composite at 1200 °C, being comparable to service lives of gas turbine components.

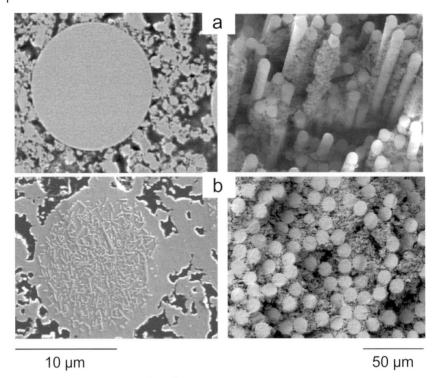

Figure 9.12 WHIPOX CMC with mullite/SiO$_2$ matrix: (a) as-prepared; (b) 1600 °C, 1 h; left: polished section; right: fracture surface. Note that upon firing the matrix shows pore agglomeration, particle coarsening and reaction with the Nextel 720 fibers, thus leading to embrittlement of the composite.

9.2.4
Other Properties

The favorable fracture behavior of porous matrix CMCs corresponds to their excellent thermal shock and thermal fatigue properties. Thermal shock resistance can be explained by local stresses dissipating over short distances and hence thermally induced stresses cause only short-range damages. The excellent thermal shock and thermal fatigue behavior was demonstrated by focally heating of a WHIPOX plate with a solar furnace or an oxyacetylene torch (Figure 9.14). Despite the extreme thermal gradients, no macroscopic deterioration is detectable, even after multiple heating/quenching cycles [26].

Various material properties (e.g. thermal diffusivity, air permeability) have been determined for WHIPOX CMCs. Reliable data are important for potential applications, such as thermal insulators, filters, or burners. Thermal conductivity perpendicular to the fiber orientation is 1 to 2 W/mK at elevated temperatures. Closer

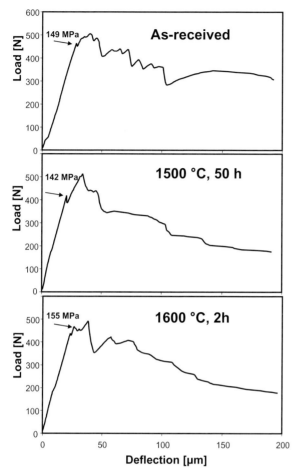

Figure 9.13 Load/deflection curves of WHIPOX CMCs with alumina/mullite matrix. Damage-tolerant fracture behavior upon short-term firing (1 h) at 1500 and 1600 °C is retained.

inspection reveals lower thermal conductivity if a mullite matrix is employed instead of alumina. However, thermal conductivity in fiber direction is somewhat higher than perpendicular to the fiber direction, reflecting the anisotropic architecture of the composite. The air permeability of wound CMCs strongly depends on parameters such as matrix composition and winding pattern. According to Figure 9.15, air penetration rates between 10 and 100 $l\,cm^{-2}h^{-1}$ were determined at overpressures of 250 mbar. It can be anticipated that flow-rate controlling mesopores, typically occurring near fiber bundle intersections, are more ubiquitous, if stiffer 3000 DEN fiber rovings are used instead of more flexible 1500 DEN rovings. Furthermore, these data suggest higher air penetration if a winding angle of ±45

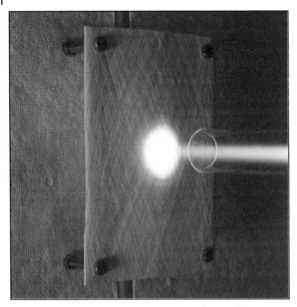

Figure 9.14 Hot spot test using an oxyacetylene torch as a proof of thermal shock and thermal fatigue resistance of WHIPOX CMCs.

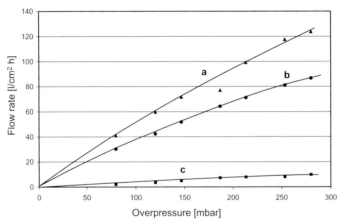

Figure 9.15 Air permeability of different WHIPOX CMCs consisting of Nextel 610 fibers and Al_2O_3/mullite matrix: (a) 3000 DEN fiber roving; fiber orientation ±45 degrees; (b) 3000 DEN fiber roving; fiber orientation ±15 degrees; (c) 1500 DEN fiber roving; fiber orientation ±15 degrees.

degrees (two-dimensional orientation) is used instead of ±15-degree fiber orientation. These findings also can be rationalized by higher amounts of fiber crossover regions occurring in composites of two-dimensional fiber texture. Virtually airtight porous matrix CMCs, on the other hand, may be obtained by subsequent sealing of the surface or by application of external coatings (see below).

9.3
Oxide/Oxide CMCs with Protective Coatings

The chemical stability of mullite and alumina is a serious issue for the long-term application of oxide/oxide CMCs in combustion environments, characterized by the presence of water vapor-rich (exhaust) gases. Basically, mullite and alumina exhibit a relative high chemical stability under quasi-static conditions. However, under highly dynamic flow conditions of powerful industrial burners and combustors, mullite and alumina are prone to decomposition and volatilization. Due to the high water-vapor partial pressure and high gas velocities, volatile Si- and Al-hydroxides are formed and are ablating continuously. In the case of mullite ceramics, a significant recession is observed in flowing water vapor, even at moderate temperatures around 1000 °C. The driving force behind the recession is formation and evaporation of Si-hydroxides, predominantly $Si(OH)_4$ [30], leading to formation of a porous residuum of $\alpha\text{-}Al_2O_3$. However, starting at temperatures of 1300 °C, a significant recession of $\alpha\text{-}Al_2O_3$ is observed. In a similar manner to SiO_2, Al_2O_3 is mobilized via formation of Al-hydroxides, predominantly $Al(OH)_3$ [31]. At temperatures of >1450 °C kinetics of Al_2O_3, recession nearly equals that of mullite [32]. Gas permeability and high specific surface areas associated with the fiber architecture may give rise to even more severe corrosion problems for mullite- and/or alumina-based CMCs.

The application of chemically resistant environmental barrier coatings (EBCs) is considered as a solution for the corrosion problem. EBCs promise higher corrosion resistance and therefore increased service lifetime and temperature capability of components. Suitable EBC materials must exhibit a thermodynamic compatibility to mullite and/or alumina at least to service temperature. In order to minimize mechanical stresses, similar thermal expansion rates of CMC and EBC are beneficial.

In the past, a number of candidate materials were identified and tested to be used as EBC, but with focus on protection of silicon-based, non-oxide structural ceramics and composites [33]. Among the first investigated promising EBC materials were yttrium-containing silicates Y_2SiO_5 and $Y_2Si_2O_7$, exhibiting water vapor recession rates of two to three orders of magnitude lower than mullite and alumina [34]. In particular, the monosilicate Y_2SiO_5 displayed an excellent corrosion resistance under simulated gas turbine environments. Analogous, highly corrosion resistant compounds were found among the rare earth silicates $Yb_2Si_2O_7/Yb_2SiO_5$ and $Lu_2Si_2O_7/Lu_2SiO_5$ [35, 36]. Due to their relative low thermal expansion, zirconium silicate (zircon $ZrSiO_4$) and hafnium silicate (hafnon, $HfSiO_4$) were considered as potential EBC materials. However, the former displayed only minor corrosion resistance in water-rich environments [37, 38].

Regarding the chemical nature of common oxide/oxide CMCs, reactions between EBC and CMC become a key issue. In state-of-the-art CMC, alumina is a dominating component. Hence, critical reactions between Al_2O_3 and silicate EBCs can be anticipated. A eutectic composition exists in the ternary system $Al_2O_3\text{-}SiO_2\text{-}Y_2O_3$ at only around 1360 °C. Therefore, service temperatures of Y-silicate EBCs are restricted below this point, in order to avoid any co-melting and disintegration of the EBC/CMC system. In reality, ubiquitous impurities are considered, even to lower the maximum service temperatures.

Silica-free compounds are attractive EBC materials to avoid the reaction problem. In alumina/alumina CMCs, Y-aluminates such as YAG (yttrium-aluminum-garnet $Y_3Al_5O_{12}$) are promising since there is no low-melting eutectic and thermal expansion coefficients of coating and CMC are similar [CTE: Al_2O_3: $9.6 \times 10^{-6} K^{-1}$; $Y_3Al_5O_{12}$: $9.1 \times 10^{-6} K^{-1}$). YAG-based EBCs with a coating thickness of about 100 μm were fabricated on DLR's WHIPOX-type CMCs via thermally activated metal-organic chemical vapor deposition (MO-CVD) and subsequent heat treatment [39]. The main benefits of this technique are high deposition rates (tens of microns/h) and low (self) shadowing, which is favorable in cases of complex shaped substrates. However, under simulated combustor environments, a superficial decomposition of YAG to the Y-enriched phase $Y_4Al_2O_9$, and perhaps even to Y_2O_3, was observed at temperatures above 1350 °C. This behavior is obviously due to Al_2O_3 hydroxylation and subsequent mobilization and may give rise to a continuous outbound diffusion of Al_2O_3 and hence an instability of the entire EBC/CMC system [40].

Due to its thermodynamic compatibility and low recession rate up to high temperatures (>1400 °C), yttria stabilized zirconia (Y-ZrO_2, YSZ) is considered one of the most attractive EBC materials for alumina- and mullite-based CMCs. Due to their low thermal conductivity, ZrO_2 coatings may also provide additional thermal protection. Since oxide/oxide CMCs typically show highly irregular surface structure, smoothing of the CMC surface prior to coating deposition was found to be helpful. This is achieved by pre-grinding the CMC surface and subsequent deposition of a thin ceramic layer. A thin, reaction-bonded alumina (RBAO) coating, deriving from a dip-coated powder mixture of Al and Al_2O_3 was found to be suitable [41]. At DLR, different types of ZrO_2-based coatings were developed for WHIPOX-type oxide/oxide CMCs [42, 43]. Thin (~20 μm) and relative dense YSZ-EBCs were successfully deposited on top of the RBAO-interphase by means of vacuum plasma spraying (Figure 9.16, left) and magnetron-sputtering (Figure 9.16, right). A less sophisticated but effective ZrO_2 deposition technique for WHIPOX-type CMCs was realized by dispersion dip-coating and subsequent reaction-bonding (Figure 9.17). In this case, ZrO_2 and zirconium nitride ZrN, as non-oxide starting materials, were employed. Thick EBCs with pronounced thermal barrier functionality require a different morphology displaying improved strain tolerance. A favorable columnar microstructure along with high deposition rated is realized by electron-beam physical vapor deposition (EB-PVD). Figure 9.18 shows a 200 μm-thick YSZ-EBC on a WHIPOX-type CMC, which is equipped with tilted cooling channels.

A combined thermal-environmental protection system for oxide/oxide CMCs has been introduced by Siemens-Westinghouse Power Corporation Inc. (USA). So-called FGI (friable graded insulation) coatings consist of ceramic hollow spheres and ceramic filler bonded by a ceramic, preferably aluminium phosphate binder [44]. The green FGI material is applied to the CMC surface in a paste-like state, sintered and finally machined to the desired coating thickness.

9.3 Oxide/Oxide CMCs with Protective Coatings

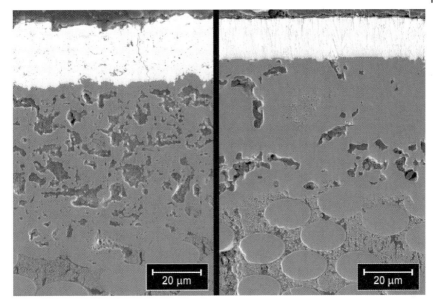

Figure 9.16 WHIPOX-type CMC with ~20 μm thick YSZ environmental barrier top coatings fabricated by vacuum plasma spraying (VPS) left and RF-magnetron sputtering (right).

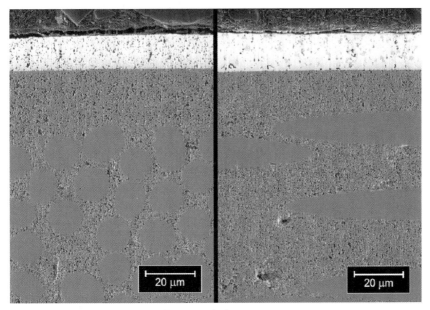

Figure 9.17 WHIPOX-type CMC with ~20 μm thick YSZ environmental barrier top coatings fabricated slurry dip-coating and subsequent reaction-bonding.

Figure 9.18 WHIPOX-type CMC with tilted cooling channel as an example of the combination of the oxide/oxide CMC approach with a film-cooling concept. The ~200 μm thick YSZ environmental barrier top coating was fabricated by EB-PVD.

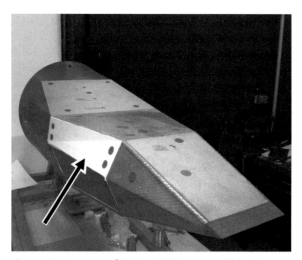

Figure 9.19 Nose cap of DLR's SHEFEX space vehicle. The segmented heat shield consists of different ceramic panels, one of them (arrow) being a WHIPOX-type porous oxide/oxide CMC with RBAO coating (courtesy H. Weihs).

9.4
Applications of Porous Oxide/Oxide CMCs

The technical and commercial importance of oxide fiber-reinforced oxide matrix composites significantly increased in recent years. Oxide fiber/oxide matrix composites have become candidate materials for many turbine engines, spacecraft, and other industrial applications [45]. Key applications are oxide/oxide components for combustion chambers, allowing higher wall temperatures and a significant contribution to increased system efficiency. Recently, a combustor liner consisting of structural filament wound Nextel 720 fiber/alumina matrix composite with an FGI-type oxide insulation layer developed by ATK-COI and Siemens Power Generation, performed successfully in a Solar Turbines engine for more than 20 000 hours [46, 47]. Due to their low specific weight and low thermal conductivity, porous oxide/oxide CMCs gained attraction as heat shields for reusable spacecrafts. In the framework of the DLR-mission SHEFEX (sharp edge flight experiment), a WHIPOX- type CMC panel with a RBAO coating was tested successfully as part of the segmented nose cap of the SHEFEX space vehicle (Figure 9.19). For a short time, components of industrial furnaces, such as firing nozzles were fabricated by Duotherm Isoliersysteme (Germany) under DLR license [48]. A wide variety of customer-specific thin-walled CMC products is fabricated by Pritzkow Spezialkeramik (Germany), ranging from flame tubes, hot-gas guidings, troughs for aluminium casting, and so on, to prototype pore burners and solar receivers [49, 50].

References

1 Koyama, T., Hayashi, S., Yasumori, A., Okada, K., Schmücker, M. and Schneider, H. (1996) Microstructure and mechanical properties of mullite/zirconia composites prepared from alumina and zircon under various firing conditions. *Journal of the European Ceramic Society,* **16**, 231–37.

2 Hirata, Y., Matsushita, S., Ishihara, Y. and Katsuki, H. (1991) Colloidal processing and mechanical properties of whisker-reinforced mullite matrix composites. *Journal of the American Ceramic Society,* **74**, 2438–42.

3 Chawla, K.K. (1998) *Composite Materials, Science and Engineering,* 2nd edn, Springer, New York, pp. 212–51.

4 Strife, J.R. and Sheehan, J.E. (1988) Ceramic coatings for carbon-carbon composites. *Ceramic Bulletin,* **67**, 369–74.

5 Chawla, K.K. (2003) *Ceramic matrix composites,* 2nd edn, Kluwer Academic Publishers, Boston, pp. 291–354.

6 Levi, C.G., Yang, J.Y., Dalgleish, B.J., Zok, F.W. and Evans, A.G. (1998) Processing and performance of an all-oxide ceramic composite. *Journal of the American Ceramic Society,* **81**, 2077–86.

7 Kingery, W.D., Bowen, H.K. and Uhlmann, D.R. (1975) *Introduction to Ceramics,* 2nd edn, John Wiley & Sons, New York, p. 794.

8 Wu, S., Holz, D. and Claussen, N. (1993) Mechanisms and kinetics of reaction bonded aluminium oxide ceramics. *Journal of the American Ceramic Society,* **76**, 970–80.

9 Holz, D., Pagel, S., Bowen, C., Wu, S. and Claussen, N. (1996) Fabrication of low-to-zero shrinkage reaction bonded mullite. *Journal of the European Ceramic Society,* **16**, 255–60.

10 Mechnich, P., Schmücker, M. and Schneider, H. (1999) Reaction sequence and microstructural development of CeO_2-doped reaction-bonded mullite. *Journal of the American Ceramic Society,* **82**, 2517–22.

11 Kerans, R.J., Hay, R.S., Parthasarathy, T.A. and Cinibulk, M.K. (2002) Interface design for oxidation-resistant ceramic composites. *Journal of the American Ceramic Society,* **85**, 2599–632.

12 Lange, F.F., Tu, W.C. and Evans, A.G. (1995) Processing of damage-tolerant, oxidation resistant ceramic matrix composites by precursor infiltration and

pyrolysis method. *Materials Science and Engineering*, **A195**, 145–50.
13 Zok, F.W. (2006) Developments in oxide fiber composites. *Journal of the American Ceramic Society*, **89**, 3309–24.
14 Staehler, J.M. and Zawada, L.P. (2000) Performance of four ceramic-matrix composite divergent flap inserts following ground testing on an F110 turbofan engine. *Journal of the American Ceramic Society*, **83**, 1727–38.
15 Zawada, L.P., Hay, R.S., Lee, S.S. and Staehler, J. (2003) Characterization and high-temperature mechanical behavior of an oxide/oxide composite. *Journal of the American Ceramic Society*, **86**, 581–90.
16 Tandon, G.P., Buchanan, D.J., Pagano, N.J. and John, R. (2001) Analytical and experimental characterization of thermomechanical properties of a damaged woven oxide-oxide composite, in *25th Annual Conference on Composites, Advanced Ceramics, Materials, and Structures* (eds M. Singh and T. Jessen), American Ceramic Society, Westerville, OH, pp. 687–94.
17 Simon, R.A. (2005) Progress in processing and performance of porous-matrix oxide/oxide composites. *International Journal of Applied Ceramic Technology*, **2** (2), 141–9.
18 Kanka, B. Einstellung des Faservolumengehaltes in oxidkeramischen Faser-Verbundwerkstoffen. Patent pending EU 061144924.1-2111.
19 Pyrz, R. (1994) Quantitative description of the microstructure of composites. *Composites Science and Technology*, **50**, 197–208.
20 Heathcote, J.A., Gang, X.Y., Yang, J.Y., Ramamurtz, U. and Zok, F.W. (1999) In-plane mechanical properties of an all-oxide ceramic composite. *Journal of the American Ceramic Society*, **82**, 2721–30.
21 Levi, C.G., Zok, F.W., Yang, J.-Y., Mattoni, M. and Löfvander, J.P.A. (1999) Microstructural design of stable porous matrices for all-oxide ceramic composites. *Zeitschrift für Metallkunde*, **90**, 1037–47.
22 Tu, W.C., Lange, F.F. and Evans, A.G. (1996) Concept for a damage-tolerant ceramic composite with "strong" interfaces. *Journal of the American Ceramic Society*, **79**, 417–24.
23 Schmücker, M., Grafmüller, A. and Schneider, H. (2003) Mesostructure of Whipox all oxide CMCs. *Composites A*, **34**, 613–22.
24 Zok, F.W. and Levi, C.G. (2001) Mechanical properties of porous-matrix ceramic composites. *Advanced Engineering Materials*, **3**, 15–23.
25 She, J., Mechnich, P., Schneider, H., Schmücker, M. and Kanka, B. (2002) Effect of cyclic infiltrations on microstructure and mechanical behavior of porous mullite/mullite composites. *Materials Science and Engineering*, **A325**, 19–24.
26 Göring, J., Hackemann, S. and Schneider, H. (2003) Oxid/Oxid-Verbundwerkstoffe: Herstellung, Eigenschaften und Anwendungen(in German), in *Keramische Verbundwerkstoffe* (ed. W. Krenkel), Wiley-VCH Verlag GmbH, Weinheim, pp. 123–48.
27 Hackemann, S. unpublished.
28 Carelli, E.A.V., Fujita, H., Yang, J.Y. and Zok, F.W. (2002) Effects of thermal aging on the mechanical properties of a porous-matrix ceramic composite. *Journal of the American Ceramic Society*, **85**, 595–602.
29 Schmücker, M., Kanka, B. and Schneider, H. (2000) Temperature-induced fiber/matrix interactions in porous alumino silicate ceramic matrix composites. *Journal of the European Ceramic Society*, **20**, 2491–7.
30 Opila, E.J., Fox, D.S. and Jacobson, N.S. (1997) Mass spectometric identification of Si-O-H(g) species from the reaction of silica with water vapor at atmospheric pressure. *Journal of the American Ceramic Society*, **804**, 1009–12.
31 Opila, E.J. and Myers, D.L. (2004) Alumina volatility in water vapour at elevated temperatures. *Journal of the American Ceramic Society*, **899**, 1701–5.
32 Yuri, I., Hisamatsu, T. and Rate, R. (2003) Prediction for ceramic materials in combustion gas flow, Proceedings of ASME TURBO EXPO 2002, 1–10, paper GT 2003-38886.
33 Lee, K.N. (2000) Current status of environmental barrier coatings for

Si-based ceramics. *Surface and Coating Technology*, **133–134**, 1–7.
34. Klemm, H., Frisch, M. and Schenk, B. (2004) Corrosion of ceramic materials in hot gas environment. *Ceramic Engineering and Science Proceeding*, **25**, 463–8.
35. Lee, K.N. (2005) *Recent progress in the development of advanced environmental barrier coatings*, 29th International Conference on Advanced Ceramics and Composites (ICACC), Daytona Beach, FLA, USA.
36. Ueno, S., Jayaseelan, D.D., Ohji, T. and Lin, H.T. (2006) Recession mechanism of $Lu_2Si_2O_7$ phase in high-speed steam jet environment at high temperatures. *Ceramics International*, **32**, 775–8.
37. Ueno, S., Jayaseelan, D.D., Ohji, T. and Lin, H.T. (2005) Corrosion and oxidation behaviors of $ASiO_4$ (A = Ti, Zr and Hf) and silicon nitride with an $HfSiO_4$ environmental barrier coating. *Journal of Ceramic Processing and Research*, **6**, 81–4.
38. Ueno, S., Ohji, T. and Lin, H.T. (2007) Corrosion and recession behavior of zircon in water vapor environment at high temperature. *Corrosion Science*, **49**, 1162–71.
39. Tröster, I., Samoilenkov, S.V., Wahl, G., Braue, W., Mechnich, P. and Schneider, H. (2005) Metal-organic chemical vapor deposition of environmental barrier coatings for all-oxide ceramic matrix composites. *Ceramic Engineering and Science Proceeding*, **26**, 173–9.
40. Fritsch, M. and Klemm, H. (2006) *Hot gas corrosion of oxide materials in the Al2O3-Y2O3 system and of YAG coated alumina in combustion environments*, 30th International Conference on Advanced Ceramics and Composites (ICACC), Cocoa Beach, FL, USA.
41. Mechnich, P., Braue, W. and Schneider, H. (2004) Multifunctional reaction-bonded alumina coatings for porous continuous fiber-reinforced oxide composites. *International Journal of Applied Ceramic Technology*, **1**, 343–50.
42. Mechnich, P. and Braue, W. (2005) Schutzschichtkonzepte für oxidkeramische Faserverbundwerkstoffe, *CFI/Ber. Deutsche Keramische Gesellschaft*, **82**, 61–4.
43. Mechnich, P. and Kerkamm, I. (2007) Reaction-bonded zirconia environmental barrier boatings for all-oxide ceramic matrix composites, 31th International Conference on Advanced Ceramics and Composites (ICACC), Daytona Beach, FLA, USA.
44. Merrill, G.B., Morrison, J.A. and Temperature, H. (2000) Insulation for Ceramic Matrix Composites, US Patent 6,013,592.
45. Newman, B. and Schäfer, W. (2001) Processing and properties of oxide/oxide composites for industrial applications, in *High temperature ceramic matrix composites* (eds W. Krenkel, R. Naslain and H. Schneider), Wiley-VCH Verlag GmbH, Weinheim, pp. 600–9.
46. Szweda, A. (2004) *Unique Hybrid Oxide/Oxide Ceramic Technology Successfully Demonstrated, Advanced Ceramics Report Autumn 2004*, United States Advanced Ceramics Association.
47. Lane, J., Morrison, J., Mazzola, B. and Marini, S. (2007) Oxide-based CMCs for combustion turbines, 31th International Conference on Advanced Ceramics and Composites (ICACC), Daytona Beach, FLA, USA.
48. www.duotherm-stark.de (accessed 7.12.2007).
49. Pritzkow, W. (2005) Keramikblech, ein Werkstoff für höchste Ansprüche, *cfi/Ber. Deutsche Keramische Gesellschaft*, **82**, 40–2.
50. www.keramikblech.com (accessed 7.12.2007).

10
Microstructural Modeling and Thermomechanical Properties

Dietmar Koch

10.1
Introduction

Ceramic matrix composites (CMC) show, in comparison to monolithic ceramics, high fracture toughness and damage tolerance, which only can be achieved if both brittle components, the ceramic matrix and the ceramic fibers, interact with each other in an efficient way. When CMCs are loaded, cracks are generally initiated in the matrix. For enhanced failure tolerance, the fibers must remain intact even when the cracks in the matrix propagate toward the fiber-matrix interface. This is realized if strength and fracture toughness of the fibers are well adapted to crack resistance of the other components, such as the matrix and fiber-matrix interface. Two general adjustments of the CMC microstructure allow this behavior:

- The crack resistance of the fiber matrix interface is low enough to ensure debonding between fiber and matrix. This type of CMC is called Weak Interface Composite WIC.

- The matrix strength is low enough to induce multiple cracking while the fibers remain intact and are responsible for sufficient strength and damage tolerance. This type of CMC is called Weak Matrix Composite WMC.

Compared to natural composites such as nacre, where a very thin weak organic layer between stiff and brittle aragonite platelets assures that propagating cracks deviate or stop, the damage tolerance behavior of CMCs are achieved by the WIC and WMC concepts. In both cases, the CMCs are characterized by the combination of stiff and strong fibers with at least one weak micro-structural component, which prevents premature failure of the fibers. These basic approaches will be described and discussed in this chapter.

CMCs benefit from their low density and excellent specific properties compared to metallic materials and are therefore mainly developed for high temperature applications under short-term and long-term conditions in oxidative as well as in inert atmospheres. The second part of this chapter discusses the mechanical behavior of CMCs under relevant application conditions. The mechanical potential

Ceramic Matrix Composites. Edited by Walter Krenkel
Copyright © 2008 WILEY-VCH Verlag GmbH & Co. KGaA, Weinheim
ISBN: 978-3-527-31361-7

and limits of different types of CMC will be presented and evaluated by considering the complex micro-structural properties necessary for damage-tolerant behavior with high fracture toughness and non-brittle failure. The various aspects of mechanical behavior of CMCs and the correlation with microstructure are also found in some recently publications [1–3].

10.2
General Concepts of CMC Design, Resulting Properties, and Modeling

For enhanced mechanical behavior and high fracture toughness, the interaction between the CMC components fiber, interface/interphase, and matrix must be adopted in a sufficient way. The fibers as reinforcing elements offer higher strength and stiffness values compared to conventional monolithic ceramics. Their strength and stiffness define the achievable mechanical properties of CMCs. The production and the latest developments of ceramic fibers are described in Chapter 1. In general, the type of the fiber influences the production process to be selected and the composite design concerning definition of interface/interphase and matrix. For production of complex geometries, continuous endless as well as short fibers can be used. The latter are used, for example, for manufacturing fiber reinforced ceramic brakes [4, 5], but are not discussed any further here.

The matrices typically show inferior properties, due to the occurrence of manufacturing process cracks, pores, or other microstructural defects. During mechanical loading, first cracks are initiated in the matrix and propagate through the composite. The general demand is that the fibers remain intact during crack propagation. Therefore, the crack resistance of the fiber-matrix interface is lowered in order to allow debonding between fiber and matrix (WIC). Alternatively, the matrix must be weak enough to allow crack deviation and branching close to the fibers, to prevent fiber failure (WMC).

In the following, the micro-structural interaction of the components fiber, matrix, and interface/interphase will be described. The two different technical approaches, WIC and WMC, leading to the flaw tolerance of CMCs, are also described below.

10.2.1
Weak Interface Composites WIC

The primary developments of CMC were focused on composites with dense matrices, characterized by high stiffness and strength. Therefore, weak fiber matrix interfaces had to be incorporated in order to achieve crack deviation around the fibers. With well established micro-structural models, the mechanical behavior of these WICs can be described. Typical WIC are CVI derived SiC matrix, DiMOx derived Al_2O_3 matrix, and glass matrix composites. Also, some SiC-matrix composites produced via liquid silicon infiltration are associated with this material class.

The idealized mechanical behavior of WIC may be described by tensile loading of a uni-directionally reinforced composite (Figure 10.1). In the linear-elastic range, the stiffness can be calculated by the rule of mixture where the Young's modulus of the composite results from stiffness of fibers and matrix and their individual volume content [6]. Matrix cracks and other voids induced during manufacturing do not grow and contribute to the initial matrix properties. Beyond a critical stress level, defined as matrix cracking stress (1), debonding and crack initiation, crack propagation, and opening occur. The initial cracks propagate first in the matrix as the fibers are stronger and show higher strain to failure. When the matrix crack approaches the fiber matrix interface, the stress concentration at the interface does not induce fiber failure but initial fiber matrix debonding. Then the fibers bridge the matrix crack. If the interfacial strength is not adopted correctly, the matrix crack may induce fiber failure and brittle premature failure of the composite is observed.

With further loading, the debonding proceeds along the fiber-matrix interface, called progressive debonding. In this debonding area, the stresses are transferred back from the fibers to the matrix by friction, until the matrix cracking strength is reached again. This process leads to an increasing matrix crack density. If it is assumed that matrix strength, fiber matrix debonding, and internal fiber matrix friction are defined values without statistical scatter, then matrix cracks arise all over the loaded volume at this stress level in an equidistant manner. The crack distances are then calculated, depending on the interfacial properties.

Even, if in real composites, this stress level (1) is often not noticeable, the overall stiffness E will be reduced continuously with further loading. In this range (2),

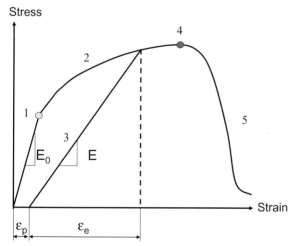

Figure 10.1 Idealized tensile stress-strain-behavior of CMC with stiff and load carrying matrix and a sufficiently weak interface (WIC). (1) matrix cracking stress; (2) progressive fiber failure and stress transfer from fibers to matrix; (3) actual stiffness; (4) composite strength, additional strain due to fiber pull-out). E_0 and E represent the initial stiffness and the actual stiffness, with ε_p and ε_e as residual strain and corresponding elastic strain, respectively.

the load is increasingly transferred from the matrix to the fibers and fiber failure is initiated sequentially, leading to the increase of the fraction of failed fibers due to the statistic scattering of the fiber strength (following a Weibull distribution). The decrease of Young's modulus of the composite E may be evaluated by unloading/reloading cycles (3) and the damage D is calculated from $D = 1 - E/E_0$. The maximum applicable stress, that is the strength of the composite (4), is reached when about 15 to 30% of the fibers have failed. If the fibers do not fail within the matrix crack level but within the intact matrix, a reduced load can still be carried by the composite (5), as then the fibers may be pulled out of the matrix. This additional work of fracture is rarely observed in real tensile tests.

The importance of tailoring the interfacial properties of WIC is shown in Figure 10.2, where the conditions of brittle and non-brittle failure are shown according to the calculation presented by He and Hutchinson [7]. If the ratio of fracture energy of the interface Γ_I and fracture energy (surface energy) of the fiber Γ_F is below the borderline shown in Figure 10.2, then non-brittle failure occurring as initial debonding can then be realized. If fiber and matrix have similar Young's moduli, E_F and E_M, respectively, then the boundary condition $\Gamma_I/\Gamma_F \leq 0.25$ must be fulfilled in order to prevent premature fiber failure. This case is practically reached in, for example, CMC with CVI-derived SiC matrices. These CMCs are typical WIC materials where fiber coating for reduction of interfacial fracture energy is necessary. Today, mainly pyrocarbon, SiC, BN, and BCN layers, or combinations of layers are applied on the fibers in order to allow individual filament

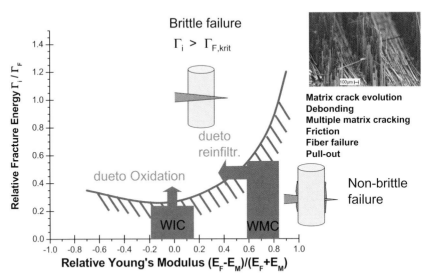

Figure 10.2 Non-brittle and brittle behavior with respect to relative fracture energy of interface and fiber dependent on stiffness ratio of fiber and matrix [7]. Qualitatively, the influence of interfacial embrittlement, due to oxidation in case of WIC, and the enhancement of stiffness of matrix, due to matrix reinfiltration steps in case of WMC, are displayed.

bridging of matrix cracks [8–15]. For adopted interfacial properties in oxide/oxide CMCs fiber coatings, for example, carbon/ZrO_2 and Al_2O_3, rare-earth phosphates, and other mixed-oxide compounds such as La-monazite ($LaPO_4$), are used [16, 17].

Other concepts to adopt fiber matrix interface are the introduction of controlled porosity into the interphase. This is realized by fugitive coating where CVI-derived carbon or organic precursors are applied to the fiber surface and then removed by oxidation at moderate temperatures [18]. The concept of porous interphases can also be extended and combined with the WMC approach, where weak matrices are adjusted by controlling the porosity of the CMC matrix and the stiffness of the matrix is much lower than that of the fibers. In the He–Hutchinson-diagram (Figure 10.2), the relative Young's modulus of the composite then deviates from zero and the boundary curve allows an increase of the critical ratio Γ_I/Γ_F. This will be further discussed for WMC in the following section.

If the composites are used at high temperatures in an oxidative atmosphere, the applied fiber coatings may be attacked by the environmental conditions and the mechanical properties may change and lead to increased relative fracture energy of fiber and interface. Figure 10.2 shows this effect qualitatively in the case of WIC. If the change of the interfacial properties leads to values of relative fracture energy, which are above the borderline, then brittle failure will occur. Therefore, the applied fiber coating has to fulfill not only mechanical functions in order to provide debonding, but must also be stable enough under thermal and environmental loading. Figure 10.3 shows the effect of oxidation of the fiber-matrix interface in a SiC/DiMOx composite measured with a single fiber push-in test, an excellent method for the characterization of the interfacial properties. While at room temperature initial debonding occurs, no debonding can be measured after oxidation

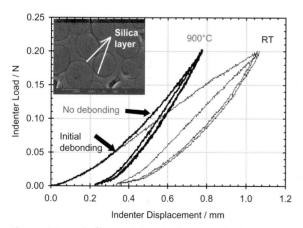

Figure 10.3 Single fiber push-in test of SiC/DiMOx shows initial debonding at room temperature and formation of silica resulting in embrittlement of the interface after oxidation at 900 °C.

at 900 °C. The micrograph in Figure 10.3 shows that the former carbon interphase has been replaced by a much stiffer and stronger silica layer, which enhances fiber matrix bonding. This results in brittle failure of the composite in macromechanical tests.

The effect of interface properties on the macroscopic behavior of CMC is shown in Figure 10.4, where the interfacial properties measured with a single fiber push-in technique [19] and the corresponding stress-strain curves from tensile tests of large specimens are shown for CVI derived SiC/SiC. The push-in curves allow the measurement of interfacial strength, as well as friction between fiber and matrix after initial debonding. In high bonding strength, a first load drop is only observed at a high load of about 0.3 N, which can be calculated to interfacial fracture energy of about $10\,J\,m^{-2}$. The succeeding hysteresis area, during unloading and reloading, is dependent on the reciprocal value of friction between fiber and matrix. As the area is very low, the interfacial friction is very high. This result can be well correlated to the related tensile test, where brittle behavior with failure at a low stress level of 120 MPa is observed. According to Figure 10.2, the ratio of the fracture energies of interface and fiber is too high, resulting in premature brittle failure. The other two push-in curves show slight deviations from the first curve at about 0.05 N. Initial debonding starts here and is calculated to fracture energies of about $0.2\,J\,m^{-2}$ (low friction state) and $0.4\,J\,m^{-2}$ (optimized state). The interfacial friction stresses calculated from the hysteresis loops, are 36 MPa and 93 MPa, respectively. The corresponding tensile tests are explained by these changes of interfacial microstructure.

While low interfacial fracture energy is necessary to prevent brittle failure when matrix cracks occur (Figure 10.2), the internal friction is a measure of load transfer between fiber and matrix within the debonded fiber matrix section. In case of low friction, the load transfer from fiber to matrix is not as efficient, as it is in the optimized case of higher friction. Under low friction the fibers are loaded on higher stress levels along larger fiber intervals. According to the Weibull strength distribution, the failure probability of the fibers is increased and leads to reduced overall strength, and strain to failure is strongly enhanced due to pronounced

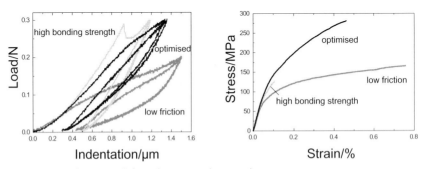

Figure 10.4 Push-in tests (left) and corresponding tensile tests (right) on CVI-derived SiC/SiC with different fiber matrix interface (strong bonding, low friction and optimized).

sliding and fiber pull-out. In the optimized case, the interfacial properties fiber matrix bonding and friction are well adjusted, resulting in high strength as well as in high strain to failure. The interpretation of the push-in test and the tensile test allow improved design of the interfacial properties of CMC [20].

10.2.2
Weak Matrix Composites WMC

During the last few years, mainly liquid infiltration processes such as liquid polymer infiltration, liquid silicon infiltration, or ceramic slurry infiltration have been developed to produce CMCs. In general, these manufacturing routes provide matrices with a relatively high content of porosity and resulting low mechanical properties. Therefore these composites are called Weak Matrix Composites (WMC) [21–23]. Compared to composites with stiff matrices, cracks easily propagate within the weak matrix and deviate close to the fiber within the matrix, which results in damage tolerant behavior, even in the case of a strong fiber-matrix interface. The reduced importance of the interfacial fracture energy can be explained in Figure 10.2. As the Young's modulus of fiber and matrix differ strongly in WMC, the ratio of fracture energy of fiber and interface may increase significantly without inducing brittle failure.

During loading of WMC, the matrix fails at low stresses but the composite can be loaded well above the matrix cracking stress as long as the overall load is still carried by the fibers. The stress redistribution from fiber to matrix does not occur further in a significant amount. The fibers do not break on a local scale but in a large volume throughout the component when final failure occurs (Figure 10.5). The mechanical behavior of WMC is strongly dominated by the properties of the fibers and therefore the mechanical response of WMC depends on fiber orientation and corresponding loading direction. Especially in off-axis loading mode and in compression load, the strength of WMC is low.

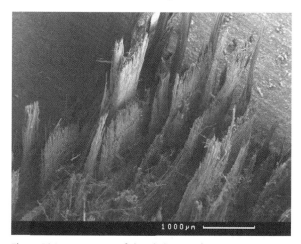

Figure 10.5 Macroscopic failure behavior of WMC (C/C).

It has to be emphasized that the general concepts described as WIC and WMC are typical boundary examples, while most of the real composites are to be characterized somewhere in between. A typical example is the manufacturing route of liquid polymer infiltration. By several re-infiltration cycles, the porosity of the matrix can be reduced significantly. The resulting enhanced stiffness of the matrix directly influences the overall mechanical behavior (Figure 10.2). If the matrix stiffness increases due to re-infiltration, the interfacial properties become more important again and therefore the fibers have to be coated with an interphase. Otherwise the borderline to brittle behavior is passed and the composite fails at reduced stresses [24].

10.2.3
Assessment of Properties of WIC and WMC

The difference between typical WIC and WMC materials can easily be demonstrated in Figure 10.6. Due to the stiff matrix in WIC (Figure 10.6a [25]), the mechanical properties of the composites are almost independent of the loading orientation. The initial Young's modulus is about 200 GPa and identical in both loading orientations, while the strength in 0/90 degrees is slightly higher than in +45/−45-degree orientation. However, WMC show a strong dependence on loading orientation and reinforcement direction, as mentioned earlier. While in 0/90 degree loading, high stiffness of about 100 GPa and high strength of above 350 MPa is measured in +45/−45 degrees, orientation of the composite behavior is characterized by strong nonlinearity and low stiffness of below 25 GPa and strength of just above 50 MPa. It becomes obvious that WIC and WMC not only can be distinguished by interface and matrix properties, but also by their significantly different mechanical behavior. In Figure 10.7, the ratios of Young's modulus are shown for different CMC measured in on-axis (0/90 degrees) and off-axis (+45/−45 degrees) loading, respectively. A high value of $E_{0°}/E_{45°}$ indicates WMC (e.g. C/C), while a ratio of 1 is typical for WIC (e.g. CVI derived SiC/SiC). Composites, such as oxide/oxide CMC or CMC derived from liquid polymer infiltration, are listed between these boundary values.

10.2.4
Modeling of the Mechanical Behavior of WMC

As mentioned above, the damage and failure mechanisms of WMC cannot be described sufficiently by the micromechanical approach, which is useful for WIC. As generally large volume failure occurs, a macro-mechanical approach is more suitable to adequately describe the mechanical behavior, that is, stress-strain response and strength. The experimentally derived macroscopic properties of WMC under tensile, shear, and compressive loading are used to establish a finite element model, which allows the prediction of mechanical behavior of these composites (Figure 10.8). The model is based on continuum damage mechanics and allows the separate calculation of inelastic deformation and stiffness degradation,

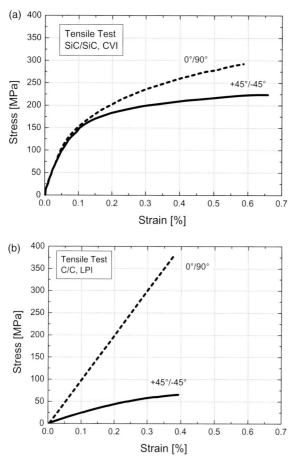

Figure 10.6 Representative tensile stress strain curves of: (a) WIC (CVI SiC/SiC) [25]; (b) WMC (LPI C/C) with on-axis (0/90°) and off-axis (±45°) loading orientation.

inducing yield surfaces and damage surfaces with the assumptions of isotropic hardening and associated flaw rules. The hardening functions, which calculate inelastic deformation and stiffness reduction, are coupled by using the same equivalent stress. The implementation of the model into a finite element code in MARC allows the calculation of the stress-strain behavior of composites loaded with different angles between fiber orientation and loading direction [33–36].

The established FE model is also applicable for prediction of failure behavior of complex-shaped samples. An example is the stress-strain behavior of DEN (double end notch) specimens with various ligament widths, which are tested under 0/90-degree and +45/−45-degree loading conditions, which is experimentally measured and theoretically calculated. In Figure 10.9, a sketch of a DEN specimen, as well as representative fracture surfaces, is shown. The 0/90-degree loaded specimen

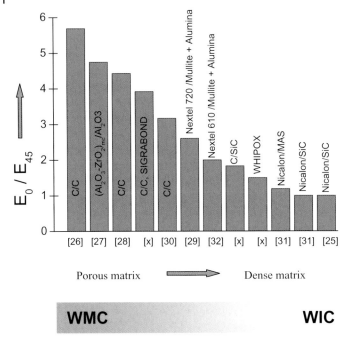

Figure 10.7 Classification of various composites according to the stiffness ratio $E_{0°}/E_{45°}$ when loaded in on-axis mode and off-axis mode, respectively. High ratio corresponds with WMC, $E_{0°}/E_{45°} = 1$ corresponds to WMC. Values from literature [25–32] and own measurements (marked with [x]).

(Figure 10.9, centre) shows failure within the ligament zone and extensive damage above and below the ligament, which corresponds to the FE-modeling (see inlet in Figure 10.11a), where locally induced stress concentrations at the notch tip and multi-axial loading conditions lead to damage and succeeding stress redistributions above and below the ligament. The FE calculations of strength values are in good agreement with the measured values (Figure 10.10a).

If DEN specimens are loaded in off-axis orientation (+45/−45 degrees), the strength values decrease strongly with increasing ligament width and decreasing notch length and this can be verified by calculating the strength values using ligament width as the reference cross-section (Figure 10.10a). The results suggest a notch-induced improvement of material properties, but if the fracture micrograph is evaluated (Figure 10.9, right) it becomes obvious that the load carrying cross-section has to be defined as ligament width plus the length of one notch, and not as ligament width alone. With this calculation, the DEN results are almost identical with respect to the tensile tests performed in +45/−45-degree orientation, and notch insensitivity is proofed (Figure 10.10b).

In Figure 10.11, the measured and calculated stress-strain curves of DEN tests are shown. The distribution of the maximum equivalent stresses calculated with

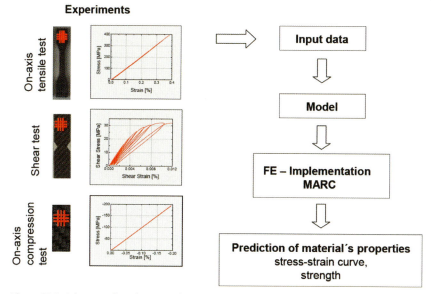

Figure 10.8 Schematic flow diagram of modeling the inelastic deformation and damage of WMC.

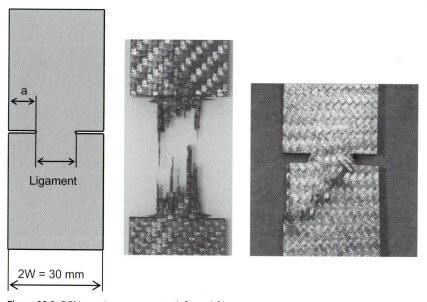

Figure 10.9 DEN specimen geometry (left) and fracture surfaces of 0/90 degree and +45/−45 degree specimens (center and right).

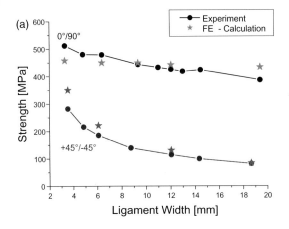

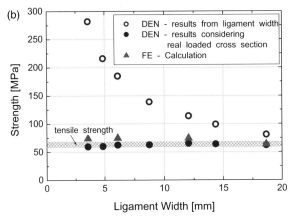

Figure 10.10 DEN tests and respective modeling of strength dependent on ligament width and fiber orientation.

FE is displayed in the inlets of Figure 10.11. Elongation was measured using a laser extensometer with a gauge length of 25 mm and ligament in the center. This gives an integral measurement of inhomogeneous strain distribution in the damaged area above and below the load carrying cross-section. As for the 0/90-degree orientation (Figure 10.11a), damage is concentrated within the gauge length, the calculation and the experimental values corresponding accurately with the FE-model. In the case of testing DEN specimens with +45/−45-degree orientation (Figure 10.11b), the measured strain values do not fit the calculated curves in the same perfect manner as the damage zone, and the gauge length does not.

In conclusion, these calculations show that the established model is able to calculate the behavior of complex-shaped parts under multi-axial loading conditions. The design of CMC components and their failure behavior can be predicted by this FE tool.

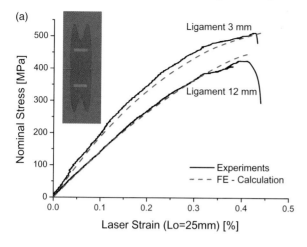

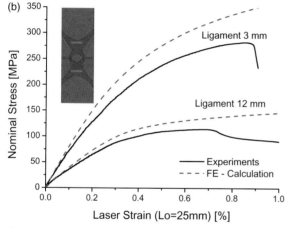

Figure 10.11 Stress-strain curves of DEN tests and corresponding calculated curves dependent on ligament width and fiber orientation (a) 0/90 degrees; (b) +45/−45 degrees.

10.2.5
Concluding Remarks

It can be concluded that WIC with relatively strong and stiff matrices and obligatory fiber coating provide high fracture toughness values. Following the micromechanical mechanisms of debonding and the effect of the micro-cracking pattern, these CMCs are relatively notch insensitive with the highest achievable stress being fairly independent of the fiber and applied stress orientation. Size effects are also not significant and low scattering of strength is observed. Micromechanical modeling is a well-established method to evaluate the mechanical properties of WIC.

WMCs are composites with compliant and porous matrix and show a mechanical behavior, which strongly depends on fiber orientation and loading direction. The design of components must consider this strong anisotropic behavior. A newly developed finite element based model allows the prediction of the mechanical behavior even if complex loading is applied. However, this model is no longer based on microstructure but takes into consideration the mechanical behavior of WMC under basic loading conditions (tensile, compression, and shear loading).

WIC and WMC are conceptual approaches in order to classify CMC. The real composites produced by various processing methods can be arranged in order between WIC and WMC and represent equivalent mechanical behavior.

10.3
Mechanical Properties of CMC

Fiber reinforced ceramic composites are designed to perform enhanced toughness and damage tolerant behavior. It has been shown that there are two concepts to achieve the desired properties, that is, WICs and WMCs. Compared to monolithic ceramics, CMCs show a quasi-ductile behavior resulting in enhanced strain to failure, inelastic deformation, and reduction of stiffness with increasing load. However, the high strength values of monolithic ceramics cannot be reached. On the other hand, CMCs show a reduced scattering of properties, increasing the reliability for the design of components.

10.3.1
General Mechanical Behavior

As mentioned earlier, the fundamental mechanisms for damage tolerant behavior are related to matrix crack deviation and branching, fiber-matrix interface debonding, and fiber failure. This leads to local reduction of stress concentrations and therefore CMCs do not show significant notch sensitivity [31, 37–39]. The resulting energy dissipation during crack propagation due to applied load is favorably measured in single edge notch beam tests (SENB), allowing controlled fracture experiments with pre-notched specimens and stable crack growth. The response curve in the SENB-tests describes a rising nonlinear slope with a more or less pronounced maximum, followed by a reduction of the applicable load, which reflects the stable character of final crack growth up to final fracture of the composite. Depending on microstructure crack branching and deviation, multiple microcracking and shear deformation are observed. Therefore, the exact crack propagation is hardly to be measured and a simplified procedure is applied to determine a nominal stress intensity factor $K_{I,nom}$, being calculated from the applied load F and the initial ratio of the notch depth and specimen thickness $\alpha = a_0/W$, according to the following [40]:

$$K_{I,\text{nom}} = \frac{F}{B\sqrt{W}} Y$$

$$Y = \frac{L \cdot 3 \sqrt{\alpha} \cdot [1.99 - \alpha(1-\alpha)(2.15 - 3.93\alpha + 2.7\alpha^2)]}{W \quad 2(1+2\alpha)(1-\alpha)^{3/2}}$$

The SENB-specimens are usually 50 mm long with a height $W = 10$ mm and a width B, corresponding to the thickness of the CMC plates of 3 to 10 mm. The relative notch depth α corresponds to a third to half of the specimen height. L is the distance between the two load supports in the 3-point bending device ($L = 40$ mm). For a wide variety of CMC materials, this nominal fracture toughness procedure has been shown to deliver characteristic K values for a given material, independent of the initial notch depth, the specimen size, and testing mode (3- or 4-point bending). $K_{I,\text{nom}}$ is used as a first and simple measure to describe the characteristic differences in the toughening effect of different CMC materials. This nominal value provides a conservative evaluation of the actual fracture toughness, which would be obtained if the actual crack length could be determined and used instead of a_0 [40].

A comparison of fracture behavior of C/C and C/SiC composites is shown in Figure 10.12. The maximum of the K-curves is measured for newly developed C/C-SiC CVI/LSI. This shows that the combination of different matrix processing methods (CVI and LSI) is a successful technique to provide high toughness-CMCs.

In tensile loading mode, different types of stress-strain curves are observed, depending on the properties of fiber, matrix, and interface (Figure 10.13). WIC

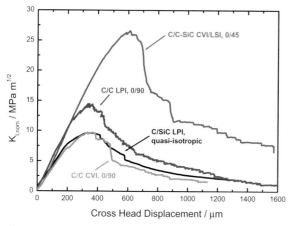

Figure 10.12 Stress intensity curves of new types of CMC and C/C materials. $K_{I,\text{nom}}$ is calculated from the results of SENB-tests ($L = 40$ mm) with the initial notch depth as crack length a_0 [40].

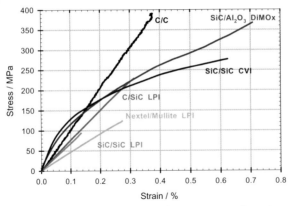

Figure 10.13 Stress-strain diagram of various WIC and WMC materials.

materials (SiC/SiC CVI, SiC/Al$_2$O$_3$ DiMOx) show a strong nonlinear behavior beyond the matrix cracking stress and fail at high stress levels and corresponding high strain to failure with more than 0.6%. In contrast, CMC with weak matrices (C/SiC LPI, Nextel/Mullite LPI, C/C, SiC/SiC LPI) represent an almost linear elastic behavior up to final failure. In C/C, high strength of about 400 MPa is measured while the matrix fails at very low stress. In contrast, the SiC/SiC LPI material presented here is characterized by low overall strength. This composite was developed in the mid-1990s and clearly shows that brittle failure occurs if neither the interface nor the matrix is weak enough to induce crack deviation at the fiber-matrix interface. Therefore, the composite fails around the matrix cracking stress. Today's LPI-derived C/SiC or SiC/SiC composites, with 0/90-degree fiber reinforcement, reach strength values of well over 300 MPa.

In both WIC and WMC, matrix cracks occur and final failure is dominated by the fibers. With a stiff matrix, redistribution of the load to be carried is possible with load transfer from fibers to matrix, resulting in nonlinear behavior. A weak matrix is unable to carry significant load and therefore the stress-strain behavior remains linear-elastic. Nevertheless, micro-structural damage evolution occurs in the matrix and can be observed in WMC by evaluating acoustic emission signals, which are generated during loading in the tensile mode [41]

Recent progress in the development of oxide/oxide composites is shown in Figure 10.14. Due to the use of sol–gel-derived oxide matrix with fine porosity in combination with almost defect free fine-grained microstructure, an enhanced strength and strain to failure could be achieved in comparison to previously developed composites [42, 43].

10.3.2
High Temperature Properties

As CMCs are mainly developed for high temperature applications, testing rigs are necessary to provide mechanical testing of specimens in application relevant con-

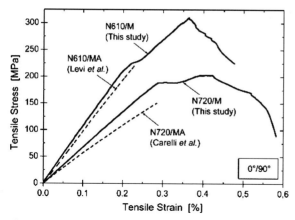

Figure 10.14 Tensile stress-strain behavior of the N610/M and N720/M composites compared with data of N610/MA [22] and N720/MA [29] (from [43] [marked as "This study"]).

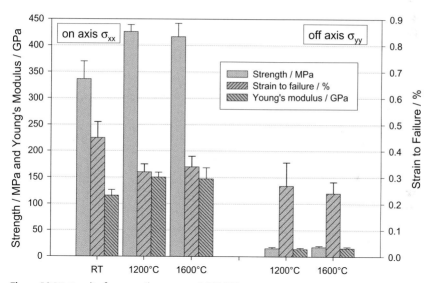

Figure 10.15 Results from tensile tests on C/SiC LPI at various temperatures under on-axis and off-axis loading conditions.

ditions with high heating and cooling rates, and the adjustment of atmosphere in order to determine the evolution of mechanical properties depending on temperature, time, and atmosphere [44].

Carbon fiber reinforced composites generally show an increase of strength with increasing temperature, as the fiber properties are improved with temperature (Figure 10.15). Young's modulus and strength of C/SiC LPI rise under on-axis loading conditions from 120 GPa and 330 MPa at room temperature up to 150 GPa and over 400 MPa at 1200 and 1600 °C in an inert atmosphere, respectively. This C/SiC LPI is classified as a typical WMC, as the mechanical properties strongly depend on the loading direction and fiber orientation. Stiffness and strength are significantly reduced under off-axis loading conditions, while the strain to failure is almost comparable to on-axis loading. As carbon fibers are creep resistant up to temperatures above 1600 °C and as the properties of WMC under on-axis loading are dominated by the fibers, the applicable maximum load in the tensile creep mode is in the range of 90% of the measured tensile strength, with time to failure up to 30 hours. Under off-axis loading, shorter lifetimes are observed as the matrix has to carry the load.

In bending tests, the matrix dominates the mechanical response. While at 1200 °C the matrix is still creep resistant, the composite reacts with a high resistance against bending (Figure 10.16a). In contrast, the creep rupture test at 1500 °C shows strong creep of the composite leading to high deformation (Figure 10.16b). The fibers are still creep resistant, but the matrix creeps strongly, allowing reorientation of the fibers without inducing failure. It can be summarized that the fibers control the overall creep behavior of C/SiC LPI. In general, the matrix is less creep resistant than the fibers, inducing a redistribution of stresses from the matrix to the fibers. While at 1200 °C the matrix remains creep resistant as no creep strain of the composite is measured, enhanced bonding between fiber and matrix induces failure at reduced stresses. If the load transfer remains unaffected by temperature and atmosphere, high creep resistance is observed and high loads corresponding to 90% of the strength can be carried. This is observed at 1500 °C, where the matrix creeps strongly while the fibers are able to carry the transferred load without failure. In general, it has to be considered that the loading direction always plays an important role in the overall behavior of WMC. As the matrix is weak and not able to carry high loads, strength and creep resistance are reduced significantly under off-axis conditions.

The creep tests have shown that oxidation processes during creep exposure may influence the mechanical behavior of CMCs at high temperatures. While all oxide composites are inherently oxidation resistant if used, interfacial layers are not susceptible to oxidation, and non-oxide composites have to be protected against oxidation at elevated temperatures. Especially CMCs with carbon fibers or carbon fiber coating are endangered, as carbon can be oxidized at intermediate temperatures around 600 °C. Progressive oxidation can be prevented, for example, by passivating oxidation layers, which emerge as dense silica layers on top of the SiC matrix or within the microstructure. These layers provide a minor self-healing effect if cracks occur and are normally acceptable only for short-term applications. For higher reliability and long-term applications, additional oxidation protection systems (OPS) are developed, which cover the CMC component completely. These OPS consist of several sublayers (Figure 10.17). A first bonding layer prevents failure due to internal stresses induced by thermal mismatch between the layer system and the CMC substrate. Succeeding sublayers prevent oxygen diffusion from the surface to the bulk material, even if cracks occur, as they provide a chemically reactive self-healing mechanism. On top of the OPS, an erosion resistant

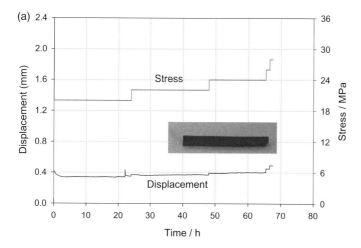

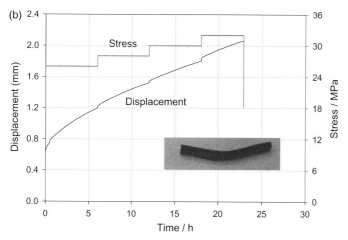

Figure 10.16 Short beam bending tests on C/SiC LPI performing: (a) no creep at 1200 °C; (b) strong creep at 1500 °C.

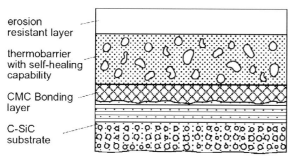

Figure 10.17 Schematic OPS applied on non-oxide CMC.

layer has to be applied in order to prevent mechanical damage during servicing of the composite [45, 46].

Thermo-mechanical cyclic tests, with thermal transients and gradients and superimposed mechanical loading, are performed in order to evaluate the effectiveness of such complex OPS simulating re-entry conditions of space vehicles (Figure 10.18). In order to qualify the erosion resistant layer, the surface can be pretreated by initial impact loading. The damaged OPS surface leads to reduced strength after five thermo-mechanical cycles, compared to specimens without predamage or with a repaired coating [44].

Thermal degradation data of oxide-fiber/oxide-matrix composites are available to a much lower extent; the experiments are restricted to intermediate temperatures, for example, 1100 to 1300 °C and are generally performed with CMCs containing Nextel fibers from recent generations, that is, Nextel 610 and Nextel 720 [16, 47–50].

Beside re-entry applications, CMCs are also promising candidates for use in gas turbines and aero engines. Typical loading conditions are thermal cycling up to 1200 °C with temperature differences of 500 K in different atmospheres [51]. After thermal cycling, the residual strength in tensile mode is measured and compared to the properties of virgin material (Figure 10.19). The residual strength of CVI-derived C/SiC drops to 89% after 50 thermal cycles in a wet oxygen atmosphere, while it remains almost unaffected in argon atmosphere. Additionally, the Young's modulus is reduced after thermal cycling, indicating induced damage of the matrix microstructure.

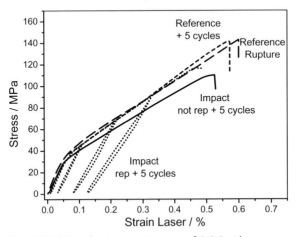

Figure 10.18 Tensile stress-strain curves of C/SiC with oxidation protection system as coating after pretreatment by impact and five thermal cycles simulating reentry conditions.

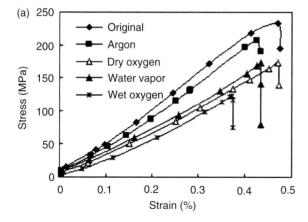

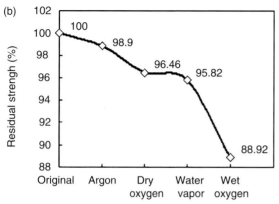

Figure 10.19 (a) Typical stress-strain curves; (b) statistical results of residual strength; in argon, dry oxygen, water vapor and wet oxygen atmospheres (compared with original strength, which was unquenched) after 50 thermal cycles [51].

10.3.3
Fatigue

Most applications of CMCs require long-term stability at high temperatures and under air. Beside the development of OPS, self-healing matrices are developed based on Si-B-C systems, which are reinforced by SiC fibers [52]. The composites are investigated using static fatigue tests. Beside measurement of deformation, acoustic emission was applied as a versatile tool, in order to characterize and correlate damage of the composite with internal fracture processes such as matrix cracking and internal friction between matrix debris, and between fiber and matrix. These effects are also dominant under cyclic mechanical loading. Wear of the fiber coating leads to a reduction of the interface sliding stress.

The effects of internal friction can be demonstrated when CVI-derived SiC/SiC is cyclically loaded at 100 Hz with stepwise increased maximum loads (Figure 10.20). If the applied stress is lower than the matrix cracking stress of 80 MPa, no hysteresis appears in the cyclic test. However, at higher stresses the stiffness is strongly reduced and the internal friction induces heat release, which can be measured as overall warming up of the specimen of about 30 K (Figure 10.21). With further increasing stresses, the temperature continues to rise and hysteresis

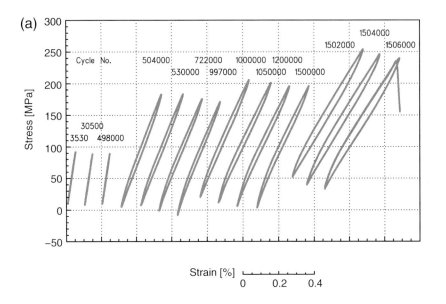

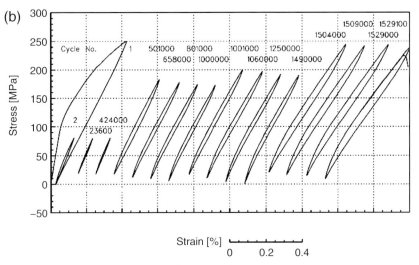

Figure 10.20 Fatigue testing at 100 Hz of CVI-derived SiC/SiC without (a) and with (b) preloading before the first cycle.

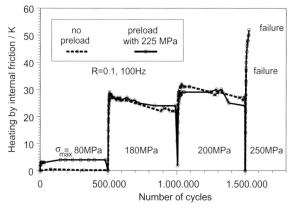

Figure 10.21 Change of intrinsic heating of CVI-derived SiC/SiC due to cyclic loading and resulting internal friction, as typical fatigue effects without and with preload.

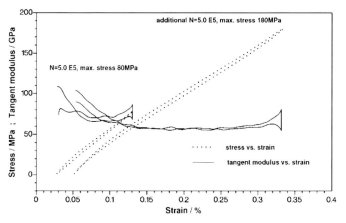

Figure 10.22 Evaluation of single hysteresis loops showing typical S-curve with increase of tangent modulus close to the stress reversal points.

loops open until finally the specimens fail at high stresses just around the initial strength. If SiC/SiC is preloaded by a tensile test up to 225 MPa (Figure 10.20b), then matrix cracks are already induced at the very beginning of the fatigue test. Even at low stresses below 80 MPa, internal friction then leads to heating of the specimen and reduction of stiffness compared to the undamaged specimen. When the initial preload is reached in the stepwise enlarged maximum cyclic stress, no significant difference of the mechanical behavior of the specimens with and without preloading can be observed [53, 54].

On evaluating the stress-strain loops during cycling, it becomes evident that the hysteresis curves show a significant S-shape. This means that close to the minimum and maximum applied stresses, an increase of tangent modulus (i.e. actual stiffness) is observed (Figure 10.22). This S-curve can be explained by use of a micromechanical approach, taking into consideration internal friction along the fiber-matrix interface, according to [55]. If cyclic loading is performed in a region where a fiber bridges a matrix crack, then sliding of fiber within the matrix will occur. With the assumption of constant friction, the debonding length and the sliding length during loading and unloading is calculated. The internal friction, which also induces an increase in specimen temperature, is a measure of energy dissipation and additionally a hint that internal surfaces are smoothed due to internal sliding. In Figure 10.23, a model is proposed where the frictional stress τ between fiber and matrix is reduced in the area where fiber-matrix sliding occurs. With this assumption, a macroscopic hysteresis loop is calculated, which shows an S-curve as experimentally observed.

It has been shown that the surface temperature of the specimen increases due to internal heat generation from sliding friction processes between constituents of the composite under cyclic loading. This increase can be directly related to the frequency and applied cyclic stress level. If LPI-derived C/SiC is cycled at 375 Hz, localized oxidation at the fiber surface is observed, resulting in reduced fatigue life at this frequency [56, 57]. The oxidation of carbon fibers becomes even more relevant in cases of high temperature testing, causing a reduction of cycles to failure with lower frequencies at elevated temperature in comparison to tests at room

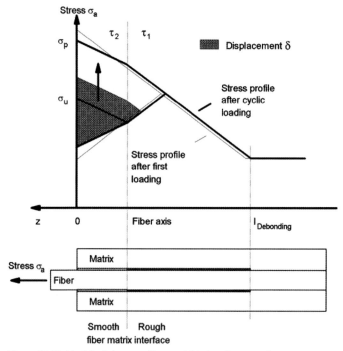

Figure 10.23 Model of change of internal friction due to cyclic loading as base for calculation macroscopic S-curved hysteresis loops.

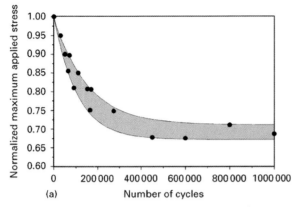

Figure 10.24 Normalized maximum applied stress vs fatigue life (Woehler curve) of uncoated three-dimensional SiC/SiC composite [64].

temperature [58]. The fatigue life of CMC with weak fiber-matrix interfaces typically decreases as the loading frequency increases [59–62]. This decrease is again attributed to frictional heating and frequency dependent damage of matrix, interface, and fiber [63].

A typical display of fatigue, mainly relating to metallic materials, are the so-called Woehler curves, where the maximum applied stress is plotted against the number of cycles to failure. It gives distinct information on acceptable maximum stress for fatigue tested specimens without rupture. It is shown that in cases of three-dimensional-reinforced SiC/SiC composites at stress level of 70% of ultimate tensile strength, a lifetime of 1 million cycles can be achieved [64]. At higher stresses the matrix will be destroyed significantly, leading to reduced interlaminar shear strength (Figure 10.24).

10.3.4
Concluding Remarks

The mechanical response of CMCs strongly depends on the composition, the manufacturing process, and the resulting microstructure. It has been shown that WMC and WIC are suitable boundary concepts, used to classify CMCs and to explain their overall mechanical behavior. For design of CMCs, not only must the loading regime, temperature, and environmental conditions be considered, but also the specific anisotropic behavior of CMCs and oxidative susceptibility have to be kept in mind. In combination with enhanced testing techniques, such as use of acoustic emission to detect and to assign the occurring damage to internal micro-structural processes, new non-destructive evaluation techniques, finite element modeling, and highly qualified experiments, a closed understanding of mechanical behavior of CMC can be achieved.

Acknowledgment

The author would like to thank Georg Grathwohl, Lotta Gaab, Jürgen Horvath, Ralf Knoche, Meinhard Kuntz, and Kamen Tushtev for their valuable support.

References

1 Bansal, N.P. (ed.) (2005) *Handbook of Ceramic Composites*, Kluwer Academic Publisher.
2 Chawla, K.K. (1993) *Ceramic Matrix Composites*, Chapman & Hall, London.
3 Warren, R. (ed.) (1992) *Ceramic-Matrix Composites*, Chapman & hall, New York.
4 Krenkel, W. (ed.) (2003) *Keramische Verbundwerkstoffe*, Wiley-VCH Verlag GmbH, Weinheim.
5 Krenkel, W., Heidenreich, B. and Renz, R. (2002) C/C-SiC composites for advanced friction systems. *Advanced Engineering Materials*, **4** (7), 427–36.
6 Hashin, Z. (1979) Analysis of properties of fiber composites with anisotropic constituents. *Journal of Applied Mechanics*, **46**, 543–50.
7 He, M.Y. and Hutchinson, J.W. (1989) Kinking of a crack out of an interface. *Journal of Applied Mechanics*, **56**, 270–8.
8 Stumm, T., Fitzer, E. and Wahl, G. (1992) Chemical vapor-deposition of very thin coatings on carbon-fiber bundles. *Journal De Physique III*, **2** (8), 1413–20.
9 Dietrich, D., Roll, U., Stockel, S., Weise, K. and Marx, G. (2002) Structure and composition studies of chemical vapour-deposited BCN fibre coatings. *Analytical and Bioanalytical Chemistry*, **374** (4), 712–14.
10 Nubian, K., Wahl, G., Saruhan, B. and Schneider, H. (2001) Fiber-coatings for fiber-reinforced mullite/mullite composites. *Journal De Physique IV*, **11** (PR3), 877–84.
11 Morscher, G.N., Yun, H.M., DiCarlo, J.A. and Thomas-Ogbuji, L. (2004) Effect of a boron nitride interphase that debonds between the interphase and the matrix in SiC/SiC composites. *Journal of the American Ceramic Society*, **87** (1), 104–12.
12 Keller, K.A., Mah, T.-I., Parthasarathy, T.A., Boakye, E.E., Mogilevsky, P. and Cinibulk, M.K. (2003) Effectiveness of monazite coating in oxide/oxide composites after long-term exposure at high temperature. *Journal of the American Ceramic Society*, **86**, 325–32.
13 Kern, F. and Gadow, R. (2004) Deposition of ceramic layers on carbon fibers by continuous liquid phase coating. *Surface and Coatings Technology*, **180–181**, 533–7.
14 Hopfe, V., Weiss, R., Meistring, R., Brennfleck, K., Jackel, R., Schonfeld, K., Dresler, B. and Goller, R. (1997) Laser based coating of carbon fibres for manufacturing CMC. *Key Engineering Materials*, **127**, 559–66.
15 Grathwohl, G., Kuntz, M., Pippel, E. and Woltersdorf, J. (1994) The real structure of the interlayer between fibre and matrix and its influence on the properties of ceramic composites. *Physica Status Solidi A* **146**, 393–414.
16 Zok, F.W. (2006) Developments in oxide fiber composites. *Journal of the American Ceramic Society*, **89** (11), 3309–24.
17 Hay, R.S., Boakye, E. and Petry, M.D. (2000) Effect of coating deposition temperature on monazite coated fiber. *Journal of the European Ceramic Society*, **20** (5), 589–97.
18 Evans, A.G., Marshall, D.B., Zok, F. and Levi, C. (1999) Recent advances in oxide-oxide composite technology. *Advanced Composite Materials*, **8** (1), 17–23.
19 Kuntz, M. and Grathwohl, G. (2001) Advanced evaluation of push-in data for the assessment of fiber reinforced ceramic matrix composites. *Advanced Engineering Materials*, 371–9.
20 Wendorff, J., Janssen, R. and Claussen, N. (1998) Model experiments on pure oxide

composites. *Materials Science and Engineering A*, **250** (2), 186–93.
21. Tu, W.C., Lange, F.F. and Evans, A.G. (1996) Concept for a damage-tolerant ceramic composite with "strong" interfaces. *Journal of the American Ceramic Society*, **79** (2), 417–24.
22. Levi, C.G., Yang, J.Y., Dalgleish, B.J., Zok, F.W. and Evans, A.G. (1998) Processing and performance of an all-oxide ceramic composite. *Journal of the American Ceramic Society*, **81**, 2077–86.
23. Zok, F.W. and Levi, C.G. (2001) Mechanical properties of porous-matrix ceramic composites. *Advanced Engineering Materials*, **3** (1–2), 15–23.
24. Mattoni, M.A., Yang, J.Y., Levi, C.G. and Zok, F.W. (2001) Effects of matrix porosity on the mechanical properties of a porous-matrix oxide ceramic composite. *Journal of the American Ceramic Society*, **84** (11), 2594–602.
25. Camus, G. (2000) Modelling of the mechanical behavior and damage processes of fibrous ceramic matrix composites: application to a 2-D SiC/SiC. *International Journal of Solids and Structures*, **37**, 919–42.
26. Neumeister, J., Jansson, S. and Leckie, F. (1996) The effect of fiber architecture on the mechanical properties of carbon/carbon fiber composites. *Acta Materialia*, **44** (2), 573–85.
27. Mamiya, T., Kakisawa, H., Liu, W.H., Zhu, S.J. and Kagawa, Y. (2002) Tensile damage evolution and notch sensitivity of Al_2O_3 fiber-ZrO_2 matrix minicomposite-reinforced Al_2O_3 matrix composites. *Materials Science and Engineering A*, **325** (1–2), 405–13.
28. Hatta, H., Denk, L., Watanabe, T., Shiota, I. and Aly-Hassan, M.S. (2004) Fracture behavior of carbon-carbon composites with cross-ply lamination. *Journal of Composite Materials*, **38** (17), 1479–94.
29. Carelli, E.A.V., Fujita, H., Yang, J.Y. and Zok, F.W. (2002) Effects of thermal aging on the mechanical properties of a porous-matrix ceramic composite. *Journal of the American Ceramic Society*, **85** (3), 595–602.
30. Goto, K., Hatta, H., Takahashi, H. and Kawada, H. (2001) Effect of shear damage on the fracture behavior of carbon-carbon composites. *Journal of the American Ceramic Society*, **84** (6), 1327–33.
31. McNulty, J.C., Zok, F.W., Genin, G.M. and Evans, A.G. (1999) Notch-sensitivity of fiber-reinforced ceramic-matrix composites: effects of inelastic straining and volume-dependent strength. *Journal of the American Ceramic Society*, **5**, 1217–28.
32. Heathcote, J.A., Gong, X.-Y., Yang, J.Y., Ramamurty, U. and Zok, F.W. (1999) In-plane mechanical properties of an all-oxide ceramic composite. *Journal of the American Ceramic Society*, **82** (10), 2721–30.
33. Tushtev, K., Horvath, J., Koch, D. and Grathwohl, G. (2004) Deformation and failure modeling of fiber reinforced ceramics with porous matrix. *Advanced Engineering Materials*, **6** (8), 664–9.
34. Tushtev K. and Koch D. (2005) Finite-Element-Simulation der nichtlinearen Verformung von Carbon/Carbon-Verbundwerkstoffen *Forschung im Ingenieurwesen* **69**, 216–222.
35. Koch, D., Tushtev, K., Kuntz, M., Knoche, R., Horvath, J. and Grathwohl, G. (2005) Modelling of deformation and damage evolution of CMC with strongly anisotropic properties. International Conference on Advanced Ceramics and Composites, Cocoa Beach Proceedings, Vol. 26, Issue 2 (ed. E. Lara-Curzio), pp. 107–14.
36. Tushtev, K., Koch, D., Horvath, J. and Grathwohl, G. (2006) Mechanismen und Modellierung der Verformung und Schädigung keramischer Faserverbundwerkstoffe. *International Journal Materials Research*, **97** (10), 1460–9.
37. Haque, A., Ahmed, L. and Ramasetty, A. (2005) Stress concentrations and notch sensitivity in woven ceramic matrix composites containing a circular hole; an experimental, analytical, and finite element study. *Journal of the American Ceramic Society*, **88** (8), 2195–201.
38. Mackin, T.J., Purcell, T.E., He, M.Y. and Evans, A.G. (1995) notch sensivity and stress redistribution in three ceramic-matrix composites. *Journal of the American Ceramic Society*, **78**, 1719–28.
39. Mattoni, M.A. and Zok, F.W. (2005) Strength and notch sensitivity of porous-matrix oxide composites. *Journal of the*

American Ceramic Society, **88** (6), 1504–13.

40 Kuntz, M. (2001) Keramische Faserverbundwerkstoffe, in *cfi-Yearbook*, Göller Verlag Baden–Baden, pp. 54–71.

41 Koch, D., Tushtev, K., Horvath, J., Knoche, R. and Grathwohl, G. (2006) Evaluation of mechanical properties and comprehensive modeling of CMC with stiff and weak matrices. *Advances in Science and Technology*, **45**, 1435–43.

42 Simon, R.A. and Danzer, R. (2006) Oxide fiber composites with promising properties for high-temperature structural applications. *Advanced Engineering Materials*, **8** (11), 1129–34.

43 Simon, R.A. (2005) Progress in processing and performance of porous-matrix oxide/oxide composites. *International Journal of Applied Ceramic Technology*, **2** (2), 141–9.

44 Knoche, R., Koch, D., Grathwohl, G. and Trabandt, U. (2005) C/SiC Faserverbundkeramiken im simulierten Wiedereintrittstest von Raumtransportern. *cfi/Ber. Deutsche Keramische Gesellschaft*, **82**, 69–74.

45 Wilshire, B. (2002) Creep property comparisons for ceramic-fibre-reinforced ceramic-matrix composites. *Journal of the European Ceramic Society*, **22** (8), 1329–37.

46 Trabandt, U., Esser, B., Knoche, R., Koch, D. and Tumino, G. (2005) Ceramic matrix composites life cycle testing under reusable launcher environmental conditions. *International Journal of Applied Ceramic Technology*, **2** (2), 150–61.

47 Antti, M.L., Lara-Curzio, E. and Warren, R. (2004) Thermal degradation of an oxide fibre (Nextel 720)/aluminosilicate composite. *Journal of the European Ceramic Society*, **24** (3), 565–78.

48 Peters, P.W.M., Daniels, B., Clemens, F. and Vogel, W.D. (2000) Mechanical characterisation of mullite-based ceramic matrix composites at test temperatures up to 1200 °C. *Journal of the European Ceramic Society*, **20**, 531–5.

49 Radsick, T., Saruhan, B. and Schneider, H. (2000) Damage tolerant oxide/oxide fiber laminate composites. *Journal of the European Ceramic Society*, 545–50.

50 Belmonte, M. (2006) Advanced ceramic materials for high temperature applications. *Advanced Engineering Materials*, **8** (8), 693–703.

51 Mei, H., Cheng, L.F., Zhang, L.T., Luan, X.G. and Zhang, J. (2006) Behavior of two-dimensional C/SiC composites subjected to thermal cycling in controlled environments. *Carbon*, **44** (1), 121–7.

52 Moevus, M., Reynaud, P., R'Mili, M., Godin, N., Rouby, D. and Fantozzi, G. (2006) Static fatigue of a 2.5D SiC/[Si-B-C] composite at intermediate temperature under air. *Advances in Science and Technology*, **50**, 141–6.

53 Koch, D. and Grathwohl, G. (1995) S-curve behavior and temperature increase of ceramic matrix composites during fatigue testing. *Ceramic Transactions*, **57**, 419–24.

54 Koch, D. and Grathwohl, G. (1996) An analysis of cyclic fatigue effects in ceramic matrix composites, in *Fracture Mechanics of Ceramics* (eds R.C. Bradt, D.P. Hasselman and D. Munz), Plenum Press, New York, pp. 121–34.

55 Marshall, D.B. (1992) analysis of fiber debonding and sliding experiments in brittle matrix composites. *Acta Metallurgica et Materialia*, **40**, 427–41.

56 Staehler, J.M., Mall, S. and Zawada, L.P. (2003) Frequency depence of high-cycle fatigue behavior of CVI C/SiC at room temperature. *Composites Science and Technology*, **63**, 2121–31.

57 Mall, S. and Engesser, J.M. (2006) Effects of frequency on fatigue behavior of CVIC/SiC at elevated temperature. *Composites Science and Technology*, **66** (7–8), 863–74.

58 Liu, X.Y., Zhang, J., Zhang, L.T., Xu, Y.D. and Luan, X.G. (2006) Failure mechanism of C/SiC composites under stress in oxidizing environments. *Journal of Inorganic Materials*, **21** (5), 1191–6.

59 Shuler, S.F., Holmes, J.W., Wu, X. and Roach, D.H. (1993) Influence of loading frequency on the room temperature fatigue of a carbon-fiber SiC-matrix composite. *Journal of the American Ceramic Society*, **76**, 2327–36.

60 Reynaud, P., Dalmaz, A., Tallaron, C., Rouby, D. and Fantozzi, G. (1998)

Apparent stiffening of ceramic-matrix composites induced by cyclic fatigue. *Journal of the European Ceramic Society*, **18** (13), 1827–33.

61 Vanswijgenhoven, E., Holmes, J., Wevers, M. and Szweda, A. (1998) The influence of loading frequency on the high-temperature fatigue behavior of a Nicalon-fabric-reinforced polymer-derived ceramic-matrix composite. *Scripta Materialia*, **38** (12), 1781–8.

62 Sorensen, B.F., Holmes, J.W. and Vanswijgenhoven, E.L. (2002) Does a true fatigue limit exist for continuous fiber-reinforced ceramic matrix composites? *Journal of the American Ceramic Society*, **85** (2), 359–65.

63 Pailler, F. and Lamon, J. (2005) Micromechanics based model of fatigue/oxidation for ceramic matrix composites. *Composites Science and Technology*, **65** (3–4), 369–74.

64 Kostopoulos, V., Vellios, L. and Pappas, Y.Z. (1997) Fatigue behaviour of 3-D SiC/SiC composites. *Journal of Materials Science*, **32**, 215–20.

11
Non-destructive Testing Techniques for CMC Materials
Jan Marcel Hausherr and Walter Krenkel

11.1
Introduction

Non-destructive testing (NDT), also called non-destructive evaluation (NDE) or non-destructive inspection (NDI), are testing methods that analyze an object without damaging or destroying the structure. They are the method of choice for applications that require the test object to remain intact, such as tests regarding lifetime analysis, safety relevant structures, or expensive manufacturing processes that justify NDT in regard to costs.

Other uses of NDT methods are the optimization of processing routes and improvement of the manufacturing process itself by analyzing the materials and the quality of the components manufactured. Commonly used destructive testing methods usually provide more reliable data, but the destruction of the tested specimen usually makes this type of test more costly than NDT and furthermore renders the object useless.

These aspects are true especially in the field of ceramic matrix composites (CMC). The manufacturing of CMC structures involves high manufacturing costs, justifying the use of NDT. Most CMC components are employed in highly critical environments, requiring a 100% assurance of their structural integrity. This makes the NDT of the structures a necessary requirement.

In addition, many damage mechanisms that occur in CMCs are not yet sufficiently described. Extensive research is needed to gain an understanding of how damage develops and how to avoid critical damage from occurring [1].

This chapter will introduce five current NDT methods that are suited to evaluate oxide and non-oxide CMCs. The components analyzed are either made of carbon fiber reinforced silicon carbide (C/SiC) manufactured via the liquid silicon infiltration(LSI) process or intermediate states that occur during LSI processing. The methods are capable of detecting typical defects occurring, such as cracks, delaminations, inhomogenities, pores, or inclusions. The methods described are:

- Optical/haptic
- Ultrasonic analysis

Ceramic Matrix Composites. Edited by Walter Krenkel
Copyright © 2008 WILEY-VCH Verlag GmbH & Co. KGaA, Weinheim
ISBN: 978-3-527-31361-7

- Thermographic/infrared analysis
- Radiography
- X-ray computed tomography.

11.2
Optical and Haptic Inspection Analysis

The most basic but often underestimated NDT methods are those based on the fundamental human senses. These include the visual, haptic, and abstract geometric analyses. Modern NDT methods enhance the human senses by using mechanical and computer aids, permitting a faster and reliable method of testing. For example, specifically designed computer algorithms can rapidly identify defects by comparison of undamaged samples and predefined damage criterions, faster and more reliably than a human being could.

The visual inspection remains a key NDT method and is automatically present in all manufacturing processes that include human interaction. Certain quality control systems can be performed by fully automated computer aided visual inspection methods. With precisely formulated rules, computers can perform a quality control. Automated systems are used in a large variety of manufacturing processes [2].

The basic use of computer aided visual NDT consists of a digital camera that analyzes specimens and compares the acquired figures with a reference figure. An example of this technique is the testing of ceramic tiles for geometric imperfections such as warped surfaces, cracks, or missing pieces. Detection of a damaged part results in rejection.

Similar to the visual inspection, the detection of surface deformations by touch can be used to detect variations in surface structure. Automated systems are based on direct touch systems such as geometric analyzers or laser-based interference systems that scan the surface topology of a specimen. The computer aided methods again compare the acquired data to existing sample a data and decide on a good-bad decision if the data are within defined specifications.

Examples of computer aided haptic NDT are laser topographic systems that determine the external geometric dimensions.

11.3
Ultrasonic Analysis

Ultrasonic NDT was developed in the early 1950s. After successful application in medical research and diagnostics, the procedure was adapted for materials research in the 1960s. Similar to the reflection of electromagnetic radiation from aircrafts and ships, the ultrasonic NDT analyzes the acoustic energy reflected within a body. It has been developed into one of the most reliable non-destructive testing methods available and is in widespread use. The method can reliably detect pores, inserts, cracks and delaminations.

11.3.1
Physical Principle and Technical Implementation

Ultrasonic analysis is based on the partial reflection of high frequency ultrasonic signals when encountering changes in densities. Typical frequencies are between 50 kHz and 200 MHz [3–5].

The acoustic energy that is reflected at a density gradient (R) and the amount of energy that crosses a density gradient (D) is based on two material constants: the speed of sound within the two materials (c_1, c_2) and the density of the materials (ρ_1, ρ_2). The dependency is described in Equations 11.1 and 11.2:

$$R = \left(\frac{(\rho_2 c_2 - \rho_1 c_1)^2}{(\rho_2 c_2 + \rho_1 c_1)^2} \right) \tag{11.1}$$

and

$$D = [1 - R] \Rightarrow D = \left[1 - \left(\frac{(\rho_2 c_2 - \rho_1 c_1)^2}{(\rho_2 c_2 + \rho_1 c_1)^2} \right) \right] \tag{11.2}$$

The ultrasonic analysis therefore detects changes in densities within a specimen. This permits detection of all types of defects and damages resulting in a change in densities, such as delaminations, cracks, pores, or inserts [6].

The amount of ultrasonic energy reflected during the transfer from several common materials is shown in Table 11.1. The reflectivity of ultrasonic energy from air into other materials is very high compared to the reflectivity from water into other materials. Ultrasonic analysis is therefore mostly performed under water.

Table 11.1 Examples of total reflected and total pass-through acoustic energy in sample material gradients.

	Reflection	Pass-through
Air → Steel	0.9999	2.6e-5
Air → CFRP	0.999	2.1e-4
Air → SiC	0.999	2.1e-4
Water → Steel	0.81	0.19
Water → CFRP	0.15	0.85
Water → SiC	0.76	0.24

Defects, such as delaminations or pores, are usually air filled. Due to the very high reflectivity between air and other materials, these defects can be easily detected.

The ultrasonic principle can be used to analyze specimens in two distinct methods: the transmission analysis and the pulse-echo analysis.

11.3.2
Transmission Analysis

Transmission style ultrasonic analysis is based on an analysis of the acoustic energy that passes through a specimen. Acoustic energy is directed toward the specimen from one side. A detector placed on the other side of the specimen detects the ultrasonic energy that passes through the material [6, 7].

Figure 11.1 shows the principle function of the transmission style ultrasonic analysis. Without any specimen, 100% of the ultrasonic energy that is emitted can be detected (Figure 11.1, left).

Analysis of an intact specimen (Figure 11.1, middle) shows the acoustic signal weakened due to reflection occurring at the top of the specimen and a second reflection when emerging from inside the specimen. Assuming a 50% reflection of the acoustic energy at each density gradient, the acoustic energy that passes through an intact specimen is 25% of the total energy initially emitted.

In the case of a damaged area within the specimen (Figure 11.1, right), the signal strength is further weakened. The ultrasonic signal has to pass through an additional set of two density gradients, reducing signal strength by an additional factor. Assuming again a partial reflection of 50% at each density gradient, the final signal strength received is reduced to 6% of the initial signal.

Due to the simple set-up, the transmission method reliably detects defects. A disadvantage is the necessity of having detector and receiver on opposing sides of the specimen, requiring access to both sides of the specimen.

Figure 11.2 shows a transmission style ultrasonic figure. The specimen analyzed is a 6 mm-thick CFRP plate with simulated delaminations in different depths [7].

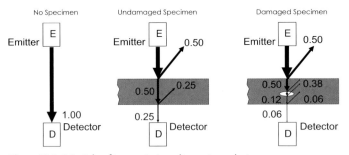

Figure 11.1 Principle of transmission ultrasonic analysis showing the absolute signal strength that reaches the detector.

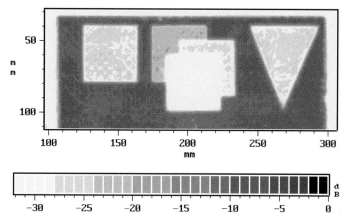

Figure 11.2 Transmission image of a CFRP plate with simulated triangular and square defects in different depths.

All delamination defects can be clearly detected. Resolution of the method is approximately 0.2 mm. Overlaying defects are visible as one single defect.

11.3.3
Echo-Pulse Analysis

The echo-pulse method, similar to the transmission method, is based on reflections at density gradients. Instead of measuring the signal that passes through the specimen, the echo-pulse analysis analyzes the acoustic energy that is reflected within the specimen and returns to the surface of the sample.

The advantage is that both emitter and receiver are located on one side of the sample, enabling the scan of complex and closed structures such as pipes or structures with an irregular topology. Another advantage is the possibility to detect the depth of the damaged areas. Knowing the speed of sound, it is possible to calculate the location of the damage within the specimen by measuring the time that passes between emitting and the return of the signal. Figure 11.3 shows the principle function of the echo-pulse method.

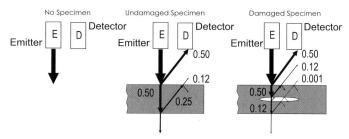

Figure 11.3 Principle of the echo-pulse analysis.

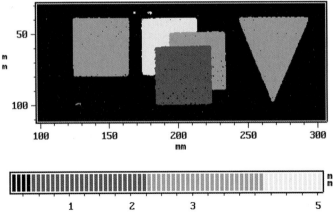

Figure 11.4 Echo pulse figure of the CFRP plate with defects. Similar distances from the surface are color coded.

The advantage of this method is seen in Figure 11.4, where the damage depth of the specimen measured with transmission style is shown as variations in return time. The individual depth is shown as changes in grayscale. Three inserts are positioned at an identical depth, while the remaining two inserts are positioned lower and above [7].

11.3.4
Methods and Technical Implementation

A typical area scanning ultrasonic analyzer is depicted in Figure 11.5. The analyzer consists of a two-axis *x-y*-manipulator, a basin, an extension that holds the ultra-

Figure 11.5 Water-coupled ultrasonic analyzer with water coupling basin.

sonic transducers in the water, a sample holder, and a computer workstation. For an ultrasonic analysis, the manipulator moves the transducers over the specimen surface, measuring the ultrasonic signal at each point. A typical measurement of a 300 by 300 mm plate requires approximately 10 minutes.

Most ultrasonic analyzers require the specimen to be immersed in water. This is due to the fact that ultrasonic energy cannot easily overcome the density gradients imposed by any occurring air gaps. A transmission style ultrasonic analysis with air gaps would require the ultrasonic energy to pass:

Emitter/Air Gap → Air Gap/Specimen → Specimen/Air Gap → Air Gap/Detector

Due to the high reflection at the air/material density gradient, the received signal strength is extremely low, making air-coupled ultrasonic analysis difficult [6, 8]. The solution was to substitute the air gaps with water by immersing the object to be tested in water and thereby increasing the signal strength. However, this poses a problem for materials that are sensitive to water, such as highly porous structures or objects composed of water soluble materials.

This problem was solved in the late 1990s by the development of new air-coupled ultrasonic analyzers [5, 9]. The technical challenges only permit air-coupled analysis using transmission style analysis and require a long measurement time, increasing the duration of a measurement by a factor of approximately five.

11.3.5
Ultrasonic Analysis of CMC

Ultrasonic analysis can be used to detect defects that result in a change of density, specifically delaminations and variations in porosity. Figure 11.6 shows delaminations within a sample carbon-carbon plate with a porosity greater than 20% [9].

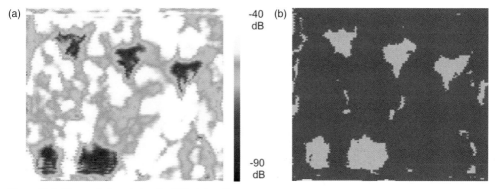

Figure 11.6 Air coupled analysis of a C/C plate with delaminations (a) and enhanced visualization of delaminations (b).

Figure 11.7 Process control of a C/SiC plate using air-coupled ultrasonic analysis. From left to right: CFRP state, after pyrolysis, after post-pyrolysis, and the final C/SiC state (DLR, Germany).

Figure 11.7 shows the ultrasonic analysis used during process control of fiber reinforced C/SiC plates. The four figures are subsequent scans of an identical plate taken after each processing step. The signal strength that was detected after passing the plate varies from a strong signal strength (light gray) to weak signal strength (dark gray). The CFRP state shows a constant strong signal strength with a minor inhomogeneity due to variations in porosity in the middle left-hand side.

The following pyrolysis and post-pyrolysis step result in a highly porous C/C-state. The C/C material contains air-filled micro-cracks that weaken the signal significantly. The following post-pyrolysis introduces an almost constant micro-crack system in the whole plate, yet again reducing signal strength.

The final infiltration of silicon results in an almost homogeneous C/SiC material with only minimal absorption. The dark gray areas in the lower right corner are the result of residual silicon remaining on the surface.

The ultrasonic analysis indicates that the theory of micro-cracking is essentially correct. The initial CFRP plate initiates micro-cracks in the first pyrolysis that are fully pronounced after the high-temperature post-pyrolysis. After infiltration, all cracks are filled with silicon and converted to SiC [7].

11.4
Thermography

Compared to ultrasonic analysis, thermogaphic analysis is a passive system that detects damages by analyzing changes in temperature. Thermographic analysis was initially developed in the 1950s for military purposes, resulting in the first infrared detectors and film systems [10]. The application of new infrared sensitive semi-conducting materials, such as InSb or InGaAs in the late 1980s, led to the development of highly sensitive digital detection arrays, allowing for a new class of detectors capable of detecting temperature differences within a range of mK.

Thermographic analysis is based on the detection of infrared energy that radiates from an object, mostly in the range of 3 to 5 μm by using specially adapted infrared detectors. Usually, the heat radiating from the surface of a structure is constant and does not change over time. Damages within a structure modify the heat flux and either increase or decrease the local surface temperature. Thermographic NDT methods utilize this effect. This chapter will introduce three methods capable of determining defects and their corresponding locations:

- Thermal Imagery (Infrared Photography)
- Lockin Thermography
- Ultrasonic Thermography

11.4.1
Thermal Imaging (Infrared Photography)

Infrared photography is the oldest and most widespread thermal based NDT method. Developed for military reconnaissance in the 1950s, the concept of thermal imaging is based on cameras adapted for capturing infrared images. The widespread use of thermographic imagery was achieved with the development of infrared sensitive photodiode arrays in the 1980s, permitting direct digital imagery and computer aided analysis [10].

Thermal imaging is a passive NDT method that measures the infrared radiation emitted by an object. This requires the object to have a temperature difference to the outside temperature and the necessity of an internal heat flux to the surface of the object [11].

The amount of infrared energy radiating away from the surface of the object varies on the geometric and material properties, resulting in a unique radiation signature that is detectable by an infrared camera.

A typical thermal image signature of a rotary furnace can be seen in the middle of Figure 11.8. The variations in gray symbolize varying surface temperature due to different thicknesses of insulation and other local geometric variations. Small circumferential lines show areas where the ceramic insulators are bolted together. In Figure 11.8 (a) a local increase in temperature can be seen as a white area. The higher local temperature (b) clearly indicates a breach of the insulation, resulting in the increase of surface temperature.

The advantage of thermal imagery analysis is its capability of in-situ monitoring, even during operation. Passive objects without an internal heat flux cannot be analyzed with this procedure, requiring an object to possess its own heat source.

270 | *11 Non-destructive Testing Techniques for CMC Materials*

(a)

(b)
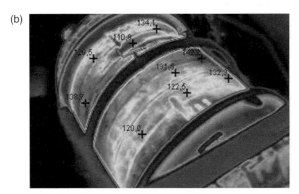

(c)
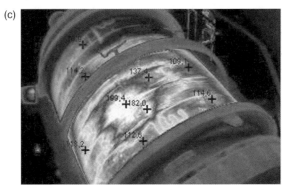

Figure 11.8 Thermographic imagery of a rotary oven (a) with intact (b) and defective (c) insulation (Industriethermografie Harald Schweiger).

11.4.2
Lockin Thermography

The non-destructive Lockin thermographic analysis was developed at the University of Stuttgart in the late 1990s [12–14]. The Lockin analysis does not depend on an object having its own internal heat source. Instead, an external heat flux is applied and a time-dependent response is analyzed [15]. The heat flux within the structure is induced by an external heat source, such as high power halogen lamps onto the surface of the object.

The infrared detector observes the change in surface temperature as the heat is absorbed by the structure. Any defect, such as a crack, delamination, or material inhomogenity, creates a thermal barrier that hinders the heat flux into the object. The thermal wave is reflected back toward the surface, creating a local rise in temperature (Figure 11.9). The size of the local temperature change enables the determination of the defect size [13].

By using Lockin thermography, it is possible to determine the depth of the defect. Similar to the echo-pulse ultrasonic analysis, analysing the time elapsed until the heat wave returns to the surface can give details about the location of the defect [14]. To enhance the time-dependent data, the Lockin analysis modulates the external heat applied to the object, permitting a phase-shift analysis. Low modulation frequencies permit deep penetration, detecting any damages within the whole volume scanned. An increase in modulation frequency reduces the scanned volume, only detecting damages between the surface and a certain distance from the surface of the object.

A variation of the Lockin thermography analysis is flash thermography. Instead of applying a modulated heat source, the surface is subjected to a brief, intense flash of heat, having a duration of several microseconds. Typical sources are high powered lasers or xenon-type flash lamps. The brief increase in temperature at the surface generates a thermal wave that moves into the object.

Lockin and flash thermography have physical limits that need to be considered. As the surface-induced thermal wave travels into the object, it propagates in all directions. It is difficult to detect small defects, such as individual pores or cracks, as small defects deep within the structure do not provide an efficient thermal

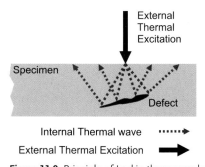

Figure 11.9 Principle of Lockin thermography.

barrier – the heat wave propagates around the defect without being reflected. The correlation between detectable defect size vs. distance from the surface is approximately 1 to 1. This means that a defect of 2 mm diameter can only be detected within 2 mm from the surface. The detectable defect size therefore varies, depending on how far the defect is located from the surface.

11.4.3
Ultrasonic Induced Thermography

A different approach is done by ultrasonic induced thermography (US-thermography). Similar to other thermal methods, US-thermography analyzes the increased temperature on the surface of a structure. However, the change in surface temperature is induced by friction heat that emanates from defects inside the object [16, 17]. To initiate friction within defects, the object is excited by a strong ultrasonic vibration applied from the outside. The contact surfaces of a defect are each subjected to a slightly asynchronous vibration, resulting in frictional heat [18, 19].

The main advantage of the US-thermography is the possibility to detect small cracks not detectable using Lockin thermography. Instead of blocking the externally-induced thermal waves, the ultrasonic excitation lets the defect itself become the heat source [20].

11.4.4
Damage Detection Using Thermography

Figure 11.10 shows a CFRP plate with artificial delaminations analyzed using Lockin thermography. The modulation frequency was selected to investigate a volume up to 3 mm from the surface. The figure shows the delaminated areas. The triangular delamination and topmost square delamination 1 mm from the

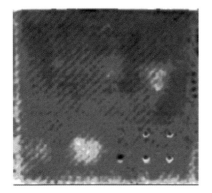
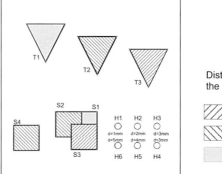
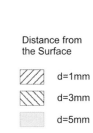

Figure 11.10 Lockin analysis of a sample CFRP plate with 3 mm penetration depth.

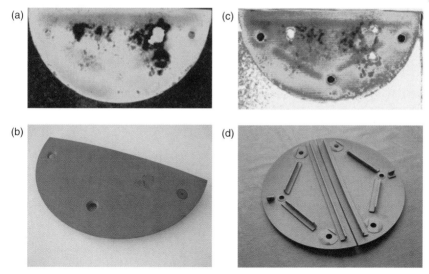

Figure 11.11 Lockin analysis of a defective reentry shield (DLR–Germany). (a), (b) Surface-near debonded areas, (c), (d) in-depth analysis with visible rear attachments.

surface, as well as the square areas 3 mm from the surface are clearly visible. The triangular area 3 mm from the surface is also discernible.

A major strength of the Lockin thermography is its capability to detect the debonding of surface coatings. Figure 11.11 shows a defective re-entry heat tile prepared for the FOTON mission. The thermal barrier coating had debonded in several places, visible as blue areas in the surface near analysis (Figure 11.11, left). An in-depth analysis additionally detected the rear-mounted stabilizing ribs and inserts (Figure 11.11, right) [1, 22].

11.5
Radiography (X-Ray Analysis)

Radiography is one of the most common methods used for NDT. Based on the absorption of X-rays within materials, radiographic methods have been used for industrial quality control since the late 1920s.

X-Rays consist of high energy photons with a wavelength between 0.01 nm and 10 nm, permitting the penetration of materials that are opaque to visible light. They are generated by accelerating electrons in an electric field and bombarding a target consisting of a heavy metal, usually tungsten. The kinetic energy of the accelerated electrons is transferred to the tungsten atoms, discharging X-ray photons. Modifying the strength of the electric field can vary the intensity of the X-rays emitted.

The principle of X-ray analysis is based on the attenuation of X-rays within the specimen and depends on both the density of the material as well as the thickness of the material that is penetrated. The radiographic image depicts a silhouette figure representing the X-ray radiation that was not absorbed within the specimen observed. Local variations inside the specimen, such as differences in thickness or density, pores, cracks, or inserts with different density cause a change in absorption of the X-rays [10, 22]. The intensity of radiation I that is recorded after passing through a sample is defined as an exponential function:

$$I = I_0 \cdot e^{-\mu \cdot L} \tag{11.3}$$

I_0 being the total radiation emitted from the X-ray source, being a material dependant absorption coefficient and L corresponding to the thickness of the material penetrated [23].

11.5.1
Detection of X-Rays

As X-rays cannot be detected by the human eye, it is necessary to use an intermediate medium or measurement to make the radiation visible. This can be achieved by different methods, utilizing chemical or physical effects. Due to the difference of the underlying principle, each method has certain advantages and disadvantages.

11.5.1.1 X-Ray Film (Photographic Plates)
Photographic film is the classic method for detecting X-rays. Consisting of an X-ray sensitive material, usually silver bromide or silver iodide with an ordered crystalline silver structure, the film is placed behind the object and exposed to radiation. The silver bromide absorbs the X-rays, releasing the silver atoms in the process. A chemical development fixates the remaining silver bromide and produces an image that shows the radiation intensity as variations in grayscale.

The advantages of X-ray film are a high flexibility of the film (enabling the bending onto curved surfaces) as well as the high resolutions (up to 10 µm) that are attainable. Once developed and fixated, the image is stable and can be stored for archiving purposes.

A major disadvantage is the fact that the image cannot be seen "live." To view the results, the film needs to be developed via a chemical process that requires several minutes. Incorrect X-ray parameters, such as too high/low acceleration voltage and filament intensities can render the figure unusable, requiring experienced personnel.

11.5.1.2 X-Ray Image Intensifier
Consisting of a phosphorous screen with an attached photomultiplier, the image intensifier was the first all-electronic method for detecting X-rays. The X-rays detected are absorbed in the phosphorous screen and converted into visible light.

The photomultiplier detects the light emitted from the screen, producing a scanning signal that can be viewed on a monitor.

The main advantage is the possibility of seeing a live image during measurement and subsequently being able to adapt the X-ray intensity to view the specimen at optimal contrast. Disadvantages are the low resolution, limited to mostly 500 by 500 pixels with a typical resolution of about 0.5 mm and a relatively bulky system due to the size of the image intensifier.

11.5.1.3 Solid State Arrays

Derived from the image intensifier, solid state arrays are based on large arrays of photodiodes. Similar to a CCD sensor in a digital camera, the detector consists of an array of highly sensitive photodiodes that detect the light emitted from a phosphorous screen. Typical arrays consist of typically 512 by 512, 1000 by 1000, or 2000 by 2000 photodiodes. The sensitive area is typically 20 or 40 cm square.

The major disadvantage of the solid state arrays is the gradual degradation of the photodiodes due to the X-ray radiation. Depending on the X-ray intensity, the semi-conducting structure can be damaged, degrading the photodiodes and requiring replacement after several years of operation.

11.5.1.4 Gas Ionization Detectors (Geiger Counter)

Gas ionization detectors utilize the ionization capabilities of X-rays. The detector consists of a metallic tube filled with a noble gas and a thin wire running through the center of the tube. The X-rays enter the cylinder at one end through a small polymer window and ionizes the gas inside, creating an electric charge that corresponds to the intensity of the X-ray radiation.

The system is maintenance free and has no degradable components. The major drawback is the very low detector resolution of 1 mm to 2 mm due to the minimal dimensions of the cylinder.

11.5.2
Application of Radiography for C/SiC Composites

Radiographic images are usually grayscale images with high attenuation depicted as white areas and low attenuation as dark areas. Changes in material composition, varying thicknesses, or defects change the attenuation and result in a change of color.

C/SiC ceramics consist solely on the elements carbon and silicon, and consist of three different material states: carbon, silicon carbide, and pure silicon. The attenuation of each material state varies (Table 11.2). As the attenuation in carbon is about 40% of the attenuation in SiC, it is possible to discern between areas with a high amount of carbon, and areas with large amounts of silicon and silicon carbide. Air-filled defects such as large pores or cracks are visible as areas with even less attenuation.

A typical radiographic image of a short-fiber reinforced C/SiC ceramic brake disk is shown in Figure 11.12 (a). The brake disk consists of short carbon-fibers

Table 11.2 Density and X-ray attenuation of different materials at 100 kV [25].

Material	Density (g cm^{-3})	Attenuation at 100 kV (cm^{-1})
Carbon	1.5	0.2271
Silicon Carbide	3.2	0.5360
Silicon	2.3	0.4220
Air	0.001	0.0001

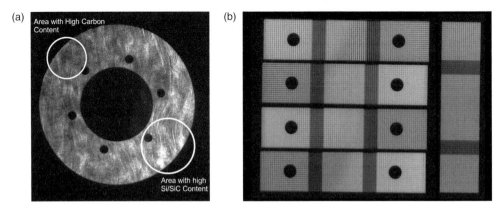

Figure 11.12 X-ray images of C/SiC composites with varying attenuation. (a) a short fibre reinforced brake disk; (b) incompletely infiltrated fabric C/SiC specimens (DLR, Germany).

embedded in a SiC matrix with a constant thickness. Any visible differences in attenuation should therefore be the result of density changes within the material or defects, such as cavities/pores within the sample [25].

The top left side of the brake disk shows an area with reduced attenuation, indicating an area with a higher content of carbon. The right part of the disk is dominated by irregular areas of high attenuation, visible as wavy light gray lines and light gray areas. These are cracks filled with Si/SiC oriented to carbon fiber bundles and areas completely filled with SiC or Si. The lack of reference specimens makes it impossible to discern between the absolute amount of Si and SiC present.

Figure 11.12 (b) shows a radiograph of an irregular batch of rectangular C/SiC specimens made up of 12 layers of carbon fiber fabric, including an intact reference plate. Due to an error during the infiltration process, the plates were subjected to more silicon than intended, resulting in a local increase of SiC. The carbon fiber orientation of the woven fabric is visible as a square criss-cross pattern of dark gray. Compared to the reference specimen in the figure, the areas of excess SiC are visible as light gray areas.

11.5.3
Limitations and Disadvantages of Radiography

Radiography is a useful tool for determining material inhomogenities such as pores, imbedded defects, and variations in densities. However, there are several types of defects that cannot be recognized.

Being an integrating procedure, radiographic analysis can only measure the total attenuation of the X-rays passing from the source to the detector. It is not possible to distinguish the attenuation occurring in several thin layers of a material or in a single layer of material as long as the total thickness is identical (Figure 11.13, middle). This makes the detection of delaminations impossible. Similarly, it is not possible to detect cracks in a material unless the crack is exactly in line with the X-ray source and the detector, allowing radiation to pass through the low attenuated region filled with air (Figure 11.13, right).

11.6
X-Ray Computed Tomography

The deficiencies occurring when trying to analyze complex three-dimensional structures using radiography was a major drawback, creating a desire for a true three-dimensional visualization procedure. Mathematical theories that permitted the description of a three-dimensional volume through a series of two-dimensional radiographic figures existed, but the algorithms were complex and required computing power that only became available in the 1980s.

The complete three-dimensional visualization of a specimen is the major advantage of X-ray computed tomography. All the structures scanned can be visualized and viewed from any angle or position, resulting in a new quality of NDT. Sample preparation is unnecessary and depending on the sample it is possible to discern defects smaller than 1 micron in size.

11.6.1
Functional Principle of CT

There are two basic technical designs of computed tomography systems: the helical scan and rotational CT [26].

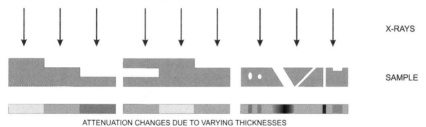

Figure 11.13 Variations in attenuation due to variations in thickness and defects.

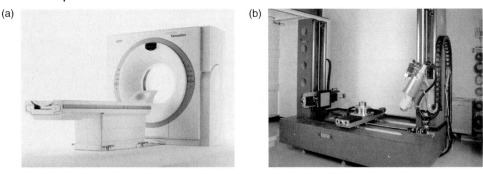

Figure 11.14 Different types of CT systems. (a) a medical CT system (Siemens AG); (b) A NDE-CT system (Phoenix X-Ray).

The helical scanning systems are used primarily in medical diagnostics. The set-up consists of a X-ray source and a detector matrix that are mounted in a fixed frame. During analysis, both source and detector are rotated around the specimen in a helical motion, moving constantly down the length of the specimen (Figure 11.14, (a)). Helical computer tomography systems are capable of measuring very long objects such as helicopter blades or tubing [27].

In a rotational computer tomography system, the X-ray source and detector are stationary. Instead, the specimen is placed on a turntable and rotated in small increments, permitting a more stable and vibration-free set-up (Figure 11.14, (b)).

The mathematical reconstruction algorithms that are the basis of computed tomography dictate that a sufficiently large number of two-dimensional radiographic figures can describe the three-dimensional structure of any body. In a three-step process, the individual radioscopic images are first converted into two-dimensional parallel slices and then converted into a three-dimensional model of the object (Figure 11.15). Using special visualization software, the object can then be virtually dissected and viewed at any position.

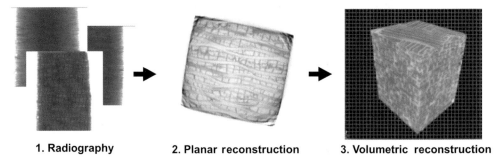

1. Radiography   2. Planar reconstruction   3. Volumetric reconstruction

Figure 11.15 The three steps required during reconstruction.

11.6 X-Ray Computed Tomography

As the procedure is based on the attenuation of X-rays, it can discern any density dependent variations within the specimen. Apart from detecting structural defects, CT is capable of analyzing density dependent morphological attributes.

11.6.2
Computed Tomography for Defect Detection

Defects and damaged areas within a structure, such as open cracks or delaminations, typically consist of an air-filled space that reduces X-ray attenuation. The defect determination is based on detecting these local changes in attenuation and determining their dimensions. Damages such as cracks or pores are visible as areas of a darker gray/black (Figure 11.16). The images show a cylindrical domed rod with a diameter of 35 mm made of monolithic silicon nitride. The interior is severely damaged due to shrinkage cracking that is clearly visible in the computed tomography figures.

Although CMCs typically are more heterogeneous, consisting of a more complex morphology with fibers and matrix material, the reduced attenuation of air-filled cracks and pores can clearly be discerned from the CMC. Changes in the CMC itself, such as inhomogeneous carbon-rich areas in C/SiC, are also discernible.

Figure 11.17 demonstrates the capabilities to detect delaminations in CFRP and C/C materials. The CFRP sample on the left contains a 100 µm thick triangular inserted plastic foil of low density that was inserted between two fabric layers as an artificial delamination. The insert can be clearly discerned. Similarly, an artificial air-filled delamination in a highly porous C/C is clearly visible in the right-hand figure. The reduced attenuation within the air-filled gap is clearly visible. The chosen resolution of 150 µm also shows individual fiber bundles, visible as a regular criss-cross pattern.

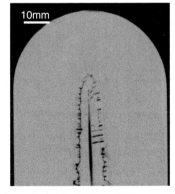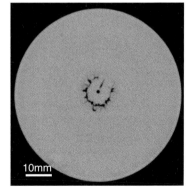

Figure 11.16 CT analysis of a seemingly intact 35 mm Si_3N_4 component (right). Perpendicular and parallel layer view of shrinkage damage inside the specimen.

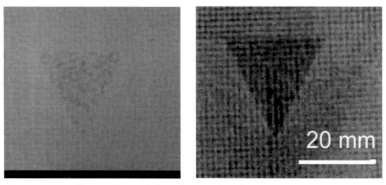

Figure 11.17 CT figure of CFRP (left) with triangular plastic foil insert and a pyrolysed high-porous C/C plate with delamination (right).

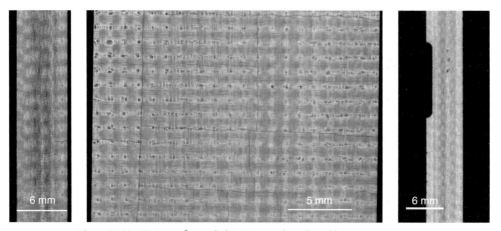

Figure 11.18 CT views of a graded C/SiC sample with visible cracks, fiber bundles and pores. Middle: thermal cracks in a depth of 1.5 mm.

Figure 11.18 shows the analysis of a graded C/SiC plate. The surface-near layers have an increased density (visible in the lighter gray) due to a higher silicon carbide content. The increase in carbon toward the middle of the specimen is clearly visible as a darker gray (Figure 11.18, left). The middle image shows air-filled relaxation cracks, visible in a layer approximately 1 mm from the surface. Also visible are individual translaminar cracks filled with silicon carbide and several pores with a diameter of approximately 400 μm (Figure 11.18, right).

11.6.3
Micro-structural CT-Analysis

The high resolution capabilities of computed tomography are not only suited for defect detection, but can also be implemented for micro-structural investigations.

The three-dimensional acquisition of the complete volume permits morphological investigations, such as crack propagation, material distribution, and the determination of phase distribution and occurring porosity.

To investigate the morphological changes occurring within the specimen during the LSI process, several micro-computed tomography scans were performed. Having a resolution of 3 μm [28], the micro-CT was capable of discerning individual fiber bundles and fibers. Beginning in the CFRP state, the micro-structural analysis showed the orientation of individual carbon fibers within a fiber bundle, the distribution of the phenolic resin and initial pores and micro-cracks that occurred within a single fiber bundle (Figure 11.19). The slightly increased attenuation of the denser carbon fibers (light gray) permitted the differentiation between fibers and less dense phenolic matrix (darker gray).

In a following CT analysis of the intermediate carbon-carbon state, the three-dimensional orientation of the trans-laminar cracking is revealed (Figure 11.20). The three-dimensional volume permits the detailed description of the spatial crack orientation and the precise determination of the crack width, height, and length.

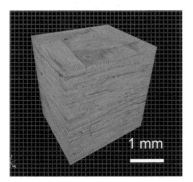
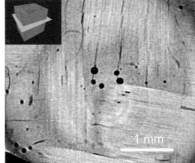

Figure 11.19 View of CFRP with visible pores, cracks and individual fibers.

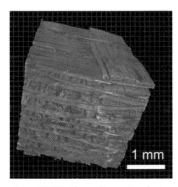

Figure 11.20 Three-dimensional view of a C/C specimen with visible crack structure (resolution 3 μm), vertical center layer of the specimen.

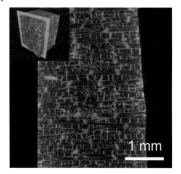 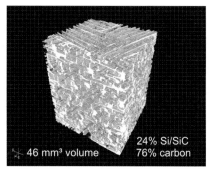

Figure 11.21 Left: High resolution CT-micrograph slice of C/SiC specimen. Dark gray represents carbon, light gray represents Si/SiC. Right: Distribution of Si/SiC (total volumetric content).

The detection of all air-filled regions, including pores, delaminations, and trans-laminar cracks, permits the non-destructive determination of the volumetric porosity, including closed pores that would be undetectable using conventional methods, such as mercury porosimetry.

Analysis of the final C/SiC state shows the heterogeneous volumetric distribution of the SiC and carbon fibers within the material. The trans-laminar crack system present in the carbon-carbon state is filled with Si/SiC. By correlating the attenuation data with the density of the materials present, it is possible to extract the volumetric dispersion of the occurring silicon carbide (Figure 11.21, right) determining the volumetric content of SiC and carbon.

11.6.4
Process Accompanying CT-Analysis

To evaluate the morphological changes occurring during the LSI process, a specimen was accompanied through the complete LSI process. Beginning in the CFRP state, the specimen was scanned with a maximum magnification before the next process step. During every CT, the identical volume was investigated.

The area chosen for comparison contained a resin pocket, visible as a light gray area in the left side of Figure 11.22. The initial CFRP material clearly shows the individual fibers within the fiber bundle. It is possible to clearly discern between fibers and resin.

After pyrolysis, the CFRP material was converted into the intermediate carbon-carbon state (Figure 11.22, middle). The segmentation cracks within the carbon fiber bundles are clearly visible. Additional pronounced and random cracking is visible in the area of the former resin pocket. The foreign inclusion is still visible and has not changed.

The final C/SiC of the specimen shows the segmentation cracks completely filled with SiC. Due to a slight warping of the specimen, the resin pocket is moved

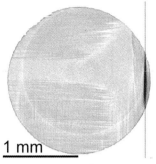 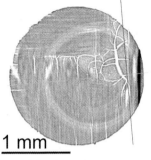 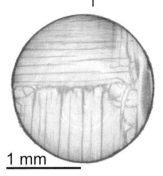

Figure 11.22 Sagittal CT-micrograph slices of the identical area accompanied through the LSI-process. From left to right: CFRP state with visible resin pocket, visible microcracked C/C state, cracks filled with SiC in the final C/SiC state.

out of the plane and is only partly visible in the figure. A large crack in the area of the former resin pocket has not been completely filled, resulting in an air-filled, closed pore.

11.7 Conclusions

NDT is an important aspect in the manufacture of CMC structures. The damage mechanisms that occur in materials during failure are not yet fully understood, requiring extensive analysis to ensure the manufacturing of intact structures. Furthermore, the high costs of CMC structures and their use in safety relevant applications require the integration of NDT methods, not only to ensure the integrity of the manufactured CMC structures, but also as a useful tool to analyze the manufacturing process for optimization and improvements.

Each of the introduced methods has its advantages and disadvantage. The choice of the appropriate method or methods necessitates a careful evaluation of the needed requirements.

The most robust and common method used for damage assessment is ultrasonic analysis. Current available systems are compact and capable of detecting delaminations, large pores, and debonded areas occurring in CMC materials. New generations of air-coupled ultrasonic analyzers provide detection capabilities in air, eliminating the need of water or a coupling medium, thereby providing true non-contact NDE.

The use of thermography, specifically the Lockin thermography system, is suited for mobile use. The available thermographic systems function under almost any operating circumstances and are very tolerant. However, the limited damage detection capabilities need to be considered.

Of all methods, computed tomography is the most sophisticated and detailed non-destructive method available. The capability to analyze complex structures without any sample preparation makes this the most effective testing method available, providing unparalleled information. However, the method is technically limited to small structures with dimensions less than 50 cm. High costs, lack of mobility, and the use of X-ray radiation make computed tomography a NDT method that can only be used for specific analysis.

Instead of computed tomography, radiography is a cheap and fast method that can detect most occurring defects and damages. Due to the fast analysis, X-ray radiography can be implemented in mass production.

References

1 Aoki, R., Eberle, K., Maile, K., Udoh, A. and Maier, H.-P. (2006) NDE assessment of ceramic matrix composite (CMC) structure, in *Damage and Its Evolution in Fiber Composite Materials: Simulation and Non-destructive Evaluation*, ISD Verlag.

2 Ye, S. (1998) Automated optical inspection for industry: theory, technology, and applications II, in *Society of Photo-Optical Instrumentation Proceedings Series*, Vol. 3558, Engi.

3 Hillger, W. (2000) Ultrasonic Testing of Composites – From Laboratory Research to Infield Inspections. WCNDT Rome 2000, 15–21 October 2000, published on CD-ROM.

4 Krautkramer, J. and Krautkramer, H. (1990) *Ultrasonic Testing of Materials*, 4th/revised edn, Springer Verlag.

5 Solodov, I. and Döring, D. (2006) Ultrasonics for NDE of fiber-composite materials, in *Damage and its Evolution in Fiber Composite Materials: Simulation and Non-destructive Evaluation*, ISD Verlag.

6 Stoessel, R., Krohn, N., Pfleiderer, K. and Busse, G. (2002) Air-coupled ultrasound inspection of various materials. *Ultrasonics*, 40, 159–63.

7 Hausherr, J.M. and Krenkel, W. (2005) Zerstörungsfreie Prüfung von Keramiken: Vergleich Computer Tomografie und Ultraschall. Poster DKG/DGM Symposium Hochleistungskeramik, Selb, 12–13 October 2005.

8 Hillger, W. (2001) HFUS 2400 AirTech – ein bildgebendes Ultraschallprüfsystem für Luft- und konventionelle Ankopplung. DGZfP-Jahrestagung 2001, ZfP in der Forschung und Entwicklung, Berlin, 21–23 May 2001, Vol. 75 on CD-ROM.

9 Hausherr, J.M. (2004) *New NDT technologies for analyzing highly porous C/C materials*, Fifth International Conference on High Temperature Ceramic Matrix Composites, 12–16 September 2004.

10 Shull, P.J. (2002) *Nondestructive Evaluation: Theory, Techniques, and Applications*, Marcel Dekker Ltd.

11 Fischer, G. (1987) Einsatz der Thermographie bei der Prüfung von Kunststoffteilen. *Qualitätstechnik, QZ*, 32, 425–9.

12 Busse, G., Wu, D. and Karpen, W. (1992) Thermal wave imaging with phase sensitive modulated thermography. *Journal of Applied Physics*, 71 (8), 3962–5.

13 Wu, D., Steegmüller, R., Karpen, W. and Busse, G. (1995) Characterization of CFRP with Lockin thermography. *Review of Progress in Quantitative NDE*, 14, 439–46.

14 Wu, D. (1996) Lockin-Thermografie für die zerstörungsfreie Werkstoffprüfung und Werkstoffcharakterisierung, Dissertation. Universität Stuttgart.

15 Zweschper, T. (1998) Lockin-Thermographie in der Fügetechnik, PhD Thesis. Fakultät 13, University of Stuttgart.

16 LaRiviere, S. (2007) Introduction to NDT of Composites, Boeing Commercial Aircraft, Manufacturing Research & Development. http://otrc.tamu.edu/Pages/Established%20NDE%20Technology.pdf (accessed January, 10, 2008).

17 Dillenz, A., Zweschper, T. and Busse, G. (2000) Elastic wave burst thermography for

NDE of subsurface features. *Insight*, **42** (12), 815–17.

18 Zweschper, T., Dillenz, A. and Busse, G. (2001) Ultraschall Burst-Phasen-Thermografie. DGZfP-Jahrestagung 2001, Berichtsband 75.

19 Zweschper, T., Dillenz, A., Riegert, G., Scherling, D. and Busse, G. (2003) Ultrasound excited thermography using frequency modutated elastic waves. *Insight*, **45** (3), 178–82.

20 Zweschper, T. (2000) Zerstörungsfreie und berührungslose Charakterisierung von Fügeverbindungen mittels Lockin Thermografie, *ZfP-zeitung*, **71**, 43–6.

21 Krenkel, W., Hausherr, J.M., Reimer, T. and Frieß, M. (2004) Design, Manufacture and Quality Assurance of C/C-SiC composites for Space Transportation Systems. Cocoa Beach.

22 Friedrich, C. (1995) 100 Jahre Röntgenstrahlen. Erster Nobelpreis für Physik. *Materialwissenschaft und Werkstofftechnik*, **26** (11–12), 598–607.

23 Gerward, L. (1993) X-Ray attenuation coefficients: current state of knowledge and availability. *Radiation Physics and Chemistry*, **41**, 783–9.

24 Hubbell, J. and Seltzer, S. (2004) Tables of X-Ray Mass Attenuation Coefficients and Mass Energy-Absorption Coefficients (version 1.4), National Institute of Standards and Technology, Gaithersburg. Available at: http://physics.nist.gov/xaamdi (20 June 2007).

25 Krenkel, W. (2000) Entwicklung eines kostengünstigen Verfahrens zur Herstellung von Bauteilen aus keramischen Verbundwerkstoffen, PhD-Thesis. University of Stuttgart.

26 Grangeat, P. and Amans, J.-L. (1996) *Computational Imaging and Vision Volume 4: Three Dimensional* Figure Reconstruction in Radiology and Nuclear Medicine, Kluwer Academic Publishers, London.

27 Bösiger, P. and Teubner, B.G. (1985) Kernspin-Tomografie für die medizinische Diagnostik, Stuttgart.

28 Hausherr, J.M., Fischer, F., Krenkel, W. and Altstädt, V. (2006) *Material characterisation of C/SiC: comparison of computed-tomography and scanning electron microscopy*, Conference on Damage in Composite Materials, Stuttgart.

12
Machining Aspects for the Drilling of C/C-SiC Materials
Klaus Weinert and Tim Jansen

12.1
Introduction

When we look at economic and technological developments of the past few years, we see that the efficient reduction of main and non-productive time in production processes is becoming increasingly important. In addition, a continuous further development of materials and the development of new materials mean that new demands are placed on machining processes. This also relates to the group of Ceramic Matrix Composites (CMC). This class of materials is typically characterized by heterogeneous and porous microstructures and high fiber volume content. CMCs form a group of materials with superior high temperature resistance and especially good strength/weight relationships. These are excellent properties for manifold applications.

Generally, the cost for producing mass market items should be kept low. However, the high hardness, as well as the inhomogeneous structure of the CMCs, causes unstable processes and high tool wear when machining. Therefore, it is important to use an appropriate process configuration. This chapter focuses on one essential machining aspect: a method for producing drill holes in carbon fiber-reinforced silicon carbide composites (C/C-SiC). This method is being investigated within the current research project at the Institute of Machining Technology (Institut für Spanende Fertigung, ISF) at the Dortmund University, Germany, supported by the German Research Foundation (DFG). The material selected is already being used for various applications including mass produced components [1], for example, ceramic brake discs in the automotive industry [1, 2]. However, these brake discs are used only in premium class vehicles because of high costs. The current situation shows that the machining procedure should be optimized so that costs can be lowered considerably. Weck and Kasperowski [3] describe how hard processing represents an essential cost factor that can account for up to 80% of the total cost for all of the components.

To optimize the drilling process in C/C-SiC, the procedure was analyzed systematically (Figure 12.1). This procedure can be subdivided into three steps:

Ceramic Matrix Composites. Edited by Walter Krenkel
Copyright © 2008 WILEY-VCH Verlag GmbH & Co. KGaA, Weinheim
ISBN: 978-3-527-31361-7

12 Machining Aspects for the Drilling of C/C-SiC Materials

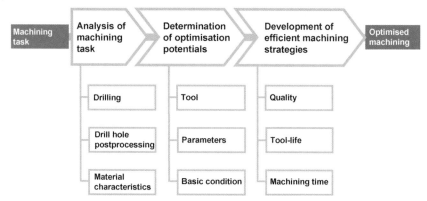

Figure 12.1 Procedure for optimized machining.

analysis of the machining task, the determination of the optimization potential, and the development of efficient machining strategies. The first thing that has to be done when analysing the machining task is to differentiate between "drilling," such as center drilling, and "drill hole post-processing," such as drill out. Center drilling means that a hole is originated in the solid workpiece, whereas center drilling drill out means that a previously produced hole is enlarged. Depending on the drilling type, the tool is stressed differently and also wears out differently. Compared to center drilling, where the tool wear is basically axial, drill out causes radial tool wear. Furthermore, knowledge of the "material characteristics" is fundamental to an analysis of the machining task.

In the second step, depending on the machining task, the possibilities for optimization should be clarified. This can be achieved by modifying the tools and by adapting the process parameters as well as the basic conditions. In the third step, efficient machining strategies should be developed to achieve higher hole quality, lower tool wear, and reduced manufacturing times.

12.2
Analysis of Machining Task

Ceramic materials have a special place in machining technology and their properties can clearly be distinguished from other materials. Their brittleness contrasts with their high hardness and their temperature resistance.

The fiber-reinforced ceramic C/C-SiC material Sigrasic 6010 GNJ [4], discussed in this section, is specified in Figure 12.2. As already mentioned, the characteristics are typical of CMC composites: A quasi-ductile behavior is the result of embedding carbon fiber bundles in a brittle SiC matrix (Figure 12.2, right).

In processing high-performance ceramics, the material is removed by machining with a geometrically undefined cutting edge [5]. Figure 12.3 illustrates the removal of material by chip formation during the grinding of ductile and brittle

12.2 Analysis of Machining Task

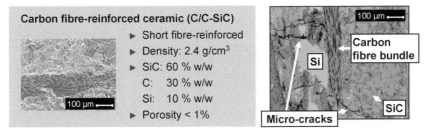

Figure 12.2 Characteristics of the work piece material.

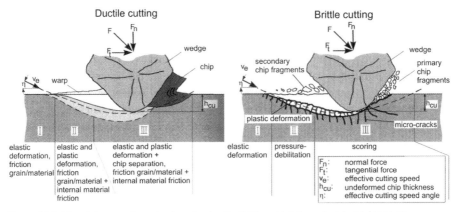

Figure 12.3 Material removal mechanism for ductile (left) and brittle materials (right) [6, 7].

materials. The grinding grains move on cycloidal paths relative to the work piece. The effective cutting grains are arranged within the grinding wheel. Their position varies within certain ranges. The chip formation model of ductile materials is based on König's analysis [6] (Figure 12.3, left). The cutting grain passes through three different stages. In the first stage, elastic deformation occurs, changing into material flow within the work piece.

On the other hand, Saljé and Möhlen describe in their model the chip formation of brittle materials [7] (Figure 12.3, right). The behavior of the brittle material leads to the formation of micro-cracks, which finally cause the breaking off of microscopic fragments.

The C/C-SiC material examined explicitly shows quasi-ductile behavior. This means that this ceramic behaves under specific loads more ductile than other ceramics, but when machining, certainly when grinding, the same material is brittle. Furthermore, as a complicating factor, this ceramic shows an overall inhomogeneous structure because of the embedded short fibers, so that the controlled breaking off of microscopic fragments from the fiber-reinforced ceramic cannot be ensured.

C/C-SiC consists of silicon and carbon, as well as the resulting reaction product silicon carbide. With 60 % weight by weight, silicon carbide makes up the largest

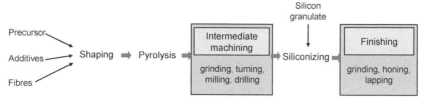

Figure 12.4 Production of C/C-SiC by the LSI method [8].

part of the composite material and is also the hardest component. The inhomogeneity of this material, as mentioned above, influences the drilling process.

Depending on the machining task, a choice must be made between drilling or drill hole post-processing. The composite material is usually produced by the Liquid Silicon Infiltration (LSI) method (Figure 12.4) [8]. Before the C/C preform is infiltrated with liquid silicon and has attained its nearly final contour, final hardness, and form, the work piece can be machined relatively easily. After siliconizing, the work piece requires post-processing. This is because of material expansion, leading to flaws so that the work pieces are characterized by insufficient surface quality. Drill holes produced during intermediate machining must be post-processed. One of the aims in drilling is to reduce the main and the non-productive times. Therefore, the silicon infiltrated work piece should be directly drilled into so that post-processing of the drilling holes is not necessary.

The ISF used as a test machine, the conventional machining center of type Chiron FZ 12 S. With this machine it is possible to reach a rotation speed of $n = 15\,000\,\text{min}^{-1}$. This machine also makes possible an internal or external supply of coolants. The hollow taper shank used attains a high axial precision as well as stiffness.

When drilling, it is necessary to differentiate between blind holes and through holes. To create drilling diameters, which approximately correspond to the tool diameters, axially force-loaded tools are primarily used. To create drilling diameters larger than the tool diameters, drilling is done by circular tool movement with a superimposed axial infeed [9]. This circular processing causes an additional radial stress on the tools used. That is why these tools must be explicitly designed to handle this additional stress. The investigations presented in this section refer exclusively to the axially stressed tools.

12.3
Determination of Optimization Potentials

12.3.1
Tool

Conventional tools with geometrically defined cutting edges, such as cemented carbide drilling tools, cannot be used to drill C/C-SiC, because of their low hardness. Also, tools with geometrically defined, very hard polycrystalline diamond (PCD) cutting edges are unsuitable for this process [9]. This is because of rapid tool wear, which prevents a consistent hole quality.

In line with research at the ISF, drilling by cross-side grinding with mounted points in the form of so-called hollow drilling tools has been and is being carried out. The cutting speed of a drilling tool is not constant over the diameter of the tool. Based on the formula for cutting speed $v_c = d \cdot \pi \cdot n$, the maximum cutting speed exists on the outer side of the tool, whereas in the center (diameter $d = 0$ mm) the cutting speed is $v_c = 0$ m/s, so that only bruise and friction processes are present. These processes cause significant tool wear. By using hollow drilling tools, corresponding bruise and friction processes can be eliminated and, moreover, a high-capacity supply of coolant can be realized. The mounted points consist of a hollow shaft with a soldered grinding head (Figure 12.5). The coolant supply serves as a process coolant and also as a means of transporting chips from the drill hole. The grinding head, consisting of metallic bonding and diamonds as cutting grains, has flutes on the end face and convex surface. The purpose of the flutes is to remove the abrasive coolant/chip composite out of the drill hole, preferably with minor effects on the convex surface and the drilling wall [10, 11]. The advantage of a tool with metallic bonding is its higher heat conductance compared to resin or ceramic grinding wheels. Furthermore, a tool with a metallic bonding has a higher friction resistance compared to one with resinoid bonding [6].

Because of high tool stress – particularly in the region of the flutes – which can lead to total tool failure, a sufficient stability of the grinding head has to be assured, even for thinner wall thickness. A tool failure expresses itself, for example, by formation of cracks in the flutes, which spread erratically from grain to grain and lead finally to tearing of the tool (Figure 12.6).

A characteristic of trepanning with mounted points is the primary axial wear on the tool's end face [9]. This depends on the quantity and allocation of the diamond grains. The wear is characterized by flattening, slivering, or breaking out of the cutting grains. The end face topography varies, depending on the wear of the grains involved in the process. Due to this, a bevel is formed on the inner or outer region of the end face [10]. The axial wear of the tools can, nevertheless, be seen as linear over the whole drill length (Figure 12.7a).

When the tools are first used, or are used with different process parameters, they show a grind-in behavior. This expresses itself as an increase in the axial

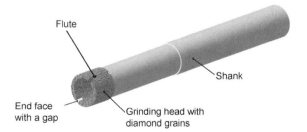

Figure 12.5 Mounted point for drilling in C/C-SiC.

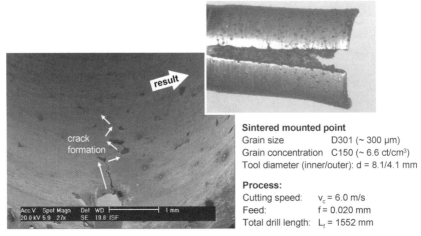

Figure 12.6 Crack formation at the inner surface of the mounted point groove through tool overload.

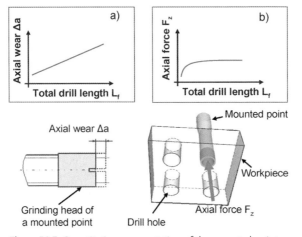

Figure 12.7 Quantitative representation of the mounted points used.

process forces (Figure 12.7b), combined with a reset of the bond and flattening of the diamond grains due to wear. When the grains are evenly distributed and the process parameters are constant, a resulting constant axial force level appears.

Grind-in behavior can also be ascertained in part by the drill hole quality. The reasons for this are radial deviations of the tools, leading to an increase in local tool wear. This expresses itself in a flattening of the diamond grains jutting out of the bond and the accompanying bond wear. The integrated profiling of the mounted points has proved itself to be effective in reducing grind-in behavior [12].

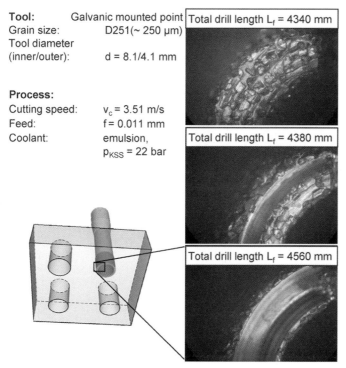

Tool: Galvanic mounted point
Grain size: D251 (~ 250 μm)
Tool diameter (inner/outer): d = 8.1/4.1 mm

Process:
Cutting speed: v_c = 3.51 m/s
Feed: f = 0.011 mm
Coolant: emulsion, p_{KSS} = 22 bar

Figure 12.8 Wear on a galvanic mounted point.

In the investigations already carried out, tools with different bond concepts were employed. A galvanic mounted point, a mounted point electroplated with only one layer of diamond grains, reached its tool life after a significantly shorter drill length. Figure 12.8 shows, in sequence, the end face of the tool employed and the corresponding drill length. As can be seen, diamond grains in the medial radius of the end face are flattened after a total drill length of L_f = 4340 mm. Because of wear, gouging follows at L_f = 4380 mm, which increases with continuing tool application, because the ceramic that has to be machined is harder than the tool base plate, which was to be machined originally. The tool's life was reached after approximately L_f = 4500 mm, using the parameters shown.

Sintered mounted points with bronze bonds have the advantage, compared to galvanic mounted points, that the wear of diamond grains is associated with a setting back of the soft bond in which more grains are bonded. Consequently, new cutting grains, which were hidden in the bond up to that point, appear at the end face. Because of this, a constant grinding behavior is made possible until the end of the tool's life. Finally, the tool's lifetime is determined by the length of the grinding head. Comparative tests with the galvanic tool were broken off after a total drill length of L_f = 4000 mm, because changes were not expected in the further course of the process [12].

12.3.2
Parameters

The specification of the tool with respect to grain size and grain concentration is a significant variable in the drilling process. As already shown, process forces, as well as wear behavior, depend on this specification. A high grain concentration can lead to insufficient bonding of the grains, if the grains lay directly next to each other and are not completely locked in the bond material. Grains which are too large can cause instabilities in the tool. Because of the normally thin wall thickness of hollow drilling tools, large grains can lead to an insufficient cohesion of the bond, particularly in the area of the flutes (compare Figure 12.6). But in general, larger grains wear out more slowly. In experiments, tools with an outer radius of d = 8.1 mm and grain sizes of 300 μm (D301), as well as a grain concentration of 6.6 ct/cm^3 (C150), have proved themselves suitable.

Further variables in the drilling process are the parameters: feed f and cutting speed v_c. So the choice of suitable process parameters has an influence on drill hole quality. In a statistic design for the experiments, a comprehensive parameter field was used. Depending on the choice of feed f and tool rotation speed n, the diameter deviations, evaluated by the fundamental tolerance IT, could be reduced (Figure 12.9). When economic considerations are taken into account, a feed of f = 0.007 mm and a rotation speed of n = 13 400 min^{-1} are the optimal process parameters. With tools that have not been further optimized, diameter deviations of at least IT 8 occur with a feed velocity of v_f = 93.6 mm/min [10].

12.3.3
Basic Conditions

The application of a coolant is fundamental to the process. Cooling of the tool is thus made possible and also the chips can be transported away quickly without interrupting the drilling process. For the process, a coolant supply was achieved that allowed a pump pressure of p_{KSS} = 22 bar at a volume flow of 15 l/min. The

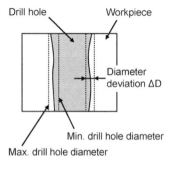

Figure 12.9 ISO system of limits and fits: fundamental tolerance IT [13].

high pressure supports a flushing out of the drilling bottom and decreases the risk of the bore core seizing up in the hollow drilling tool.

12.4
Process Strategies

On the basis of the knowledge acquired regarding drilling with mounted points, advanced optimization measures can be derived. The following investigations concentrate on the application of adapted mounted point concepts.

Flutes on the convex surface of the tools support the transportation of coolant chip composite. For tools with geometrically defined cutting edges, it is known that the chip transportation of twist drills is superior to that of straight fluted drills and so tools with twist drills have a longer lifetime. Corresponding to these findings, mounted points have been used, whose flutes are twisted in a defined helix angle of $\delta = 20$ degrees and $\delta = 45$ degrees around the grinding head. The process results were compared to so-far straight fluted ($\delta = 0$ degrees) mounted points. Axial process forces, measured when using these tools, are shown in Figure 12.10. Before the experiments started, the tools were grounded in. Tools with twisted flutes (Type B and C) show greater axial process forces compared to tools with straight flutes (Type A). This leads to the conclusion that there is an accumulation of coolant chip composite at the bottom of the drill hole. The coolant, applied with a pressure of $p_{KSS} = 22$ bar, cannot be conducted away quickly enough due to the twisted flutes. After drilling, the drill holes produced were measured and evaluated with respect to core roughness depth on the drilling wall.

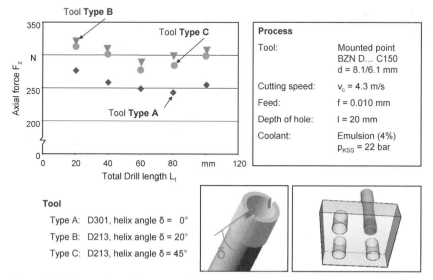

Figure 12.10 Process forces for mounted points with different helix angles.

Figure 12.11 shows that the core roughness depth Rk increases with an incremental helix angle. So the hypothesis that a coolant/chip composite is conducted over the convex surface of the grinding head is confirmed. Also, as can be seen in the photographs taken with the scanning electron microscope (SEM) (Figure 12.12), axial scratches are a sign that coolant/chip removal over the convex surface has taken place. Because of this, an inferior quality on the drill hole wall results, which is reflected in higher values for the core roughness depth.

Another criterion for evaluating the quality of a drilling process is the diameter deviation as a function of the depth of the hole. The fundamental tolerance IT corresponds to the above-mentioned deviation. Results for the mounted points with different helix angles are shown in Figure 12.13. The median deviation of the

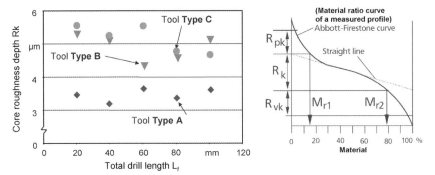

Figure 12.11 Core roughness depth of drillings made by mounted points with different helix angles.

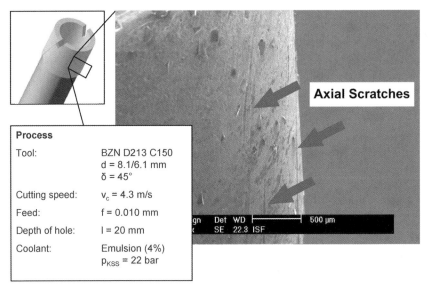

Figure 12.12 Convex surface of a mounted point with flutes with helix angle $\delta = 45$ degrees after application.

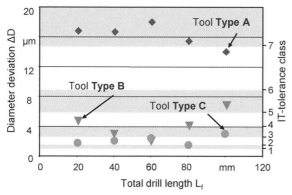

Figure 12.13 Diameter deviations of drillings, depending on the helix angle of the flutes.

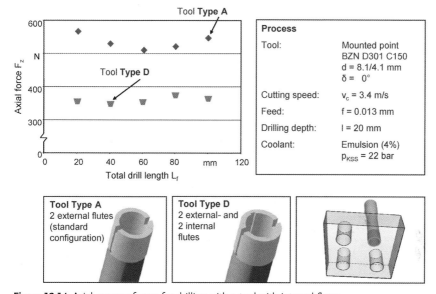

Figure 12.14 Axial process forces for drilling with a tool with internal flutes.

hole diameter, compared to multiple drillings with grounded tools, amounts to IT 6 with a helix angle of $\delta = 20$ degrees and IT 4 with $\delta = 45$ degrees. However, the straight fluted tool produced definitely inferior results; drillings using such a tool have a tolerance of IT 8.

The core produced from trepanning can seize up inside the tool when there is heavy friction contact. Therefore, as a further modification of the grinding head, in addition to external flutes, internal flutes are mounted on the tool to solve these problems. Furthermore, more coolant should reach the active region because of the internal flutes. The application of a corresponding tool, in comparison to tools used before, shows that two additional internal flutes can decrease the process forces (Figure 12.14), but then the tool operates erratically. The diameter deviations of the drill holes worsen from IT 7 to IT 8. In addition, the core roughness depth Rk on the drilling wall increases from Rk = 6.11 µm up to Rk = 10.07 µm.

When blind hole drillings are produced by trepanning, care should be taken to make sure that a drilling core remains. This core has to be removed manually. Normally, this leads to a rough drill hole bottom at the line of breakage. Furthermore, the grinding head does not move completely through the drill hole. Thus, an uneven tool wear can be ascertained. On further application of this tool, the grinding head takes on a conical form [9]. During through borings, this effect can be avoided by running the whole grinding head completely through the drill hole. Between the tool and drilling wall, there exists a corresponding contact, which has an influence on the quality of the drill hole.

To decrease the inferior quality of the drilling and to minimize tool wear, alternative tool concepts with regard to the geometry of the convex surface were examined. A tool with a blasted surface and an electroplated one were used and compared to the tools used before (Figure 12.15). All tools had the same specifications, with bronze as the compound material and diamond cutting grains of approximately 300 µm (D301) with a concentration of 6.6 ct/cm^3 (C150). According to the manufacturer, the standard tool Type A was reworked by turning. On the convex surface, single diamond grains are jutting out from the bonding. Tool Type E was blasted on the convex surface, so that the bonding material and the badly bonded grains that were jutting out were removed. As the third tool concept (Type F), a sintered tool was applied, which had been electroplated with diamond grains of approximately 50 µm (D54). Because the nickel layer is harder than bronze, the diamond grains distributed on the convex surface have a superior bonding to the compound and can jut out farther from it. Furthermore, small diamond grains on the tool perimeter allow lower roughness depths on the drill hole wall. In addition, the diamond grains and the nickel layer form a wear protection layer; the axial material removal takes place on the end face, because of the larger and more resistant diamond grains.

The core roughness depth on the drill hole wall can be reduced by application of Type F tools, due to high concentration of small grains (Figure 12.16). However, the tool with the blasted surface points has no influence on the core roughness depth. Previous research has shown that because of profiling of the convex surface and the related flattening of the diamond grains, core roughness depths can be decreased slightly [12]. In this case, the flattened diamonds are conducive to an improved roughness.

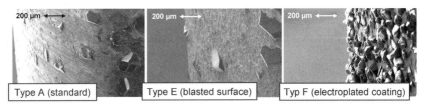

Figure 12.15 Photos of different convex surfaces made by a SEM.

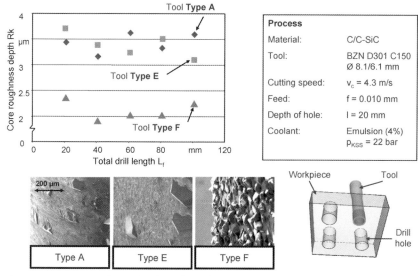

Figure 12.16 Core roughness depth resulting from different tools.

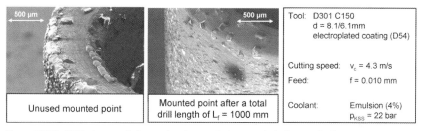

Figure 12.17 SEM photos of electroplated coated sinter tools before and after use.

As expected, the radial wear of the coated tool was marginal. The electroplated coat shows signs of wear in the form of flattening for the parameters used – not so much in the form of diamond grains breaking out – after a total drill length of approximately $L_f = 1000$ mm (Figure 12.17). A debonding of the electroplated coat cannot be detected.

The quality of the drill hole, as determined by the diameter deviations measured, could be improved by using both tool Type E and tool Type F (Figure 12.18). Drill holes produced with the blasted tool show diameter deviations within the tolerance IT 6; the electroplated tool allows drill holes within the tolerance IT 4. With the standard tool Type A, only tolerance IT 8 is achievable. Deviations in the fundamental tolerance IT show that the geometry of the convex surface has a significant influence on drill hole quality. The bonding on the surface of the blasted tool is rough, compared to that on the surface of the standard tool. On the convex surface of the tool with electroplated coating, troughs are formed by the diamond grains,

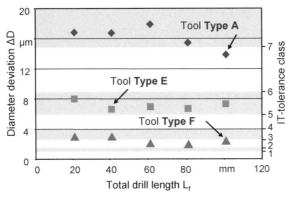

Figure 12.18 Diameter deviation of the drilling resulting from different tools.

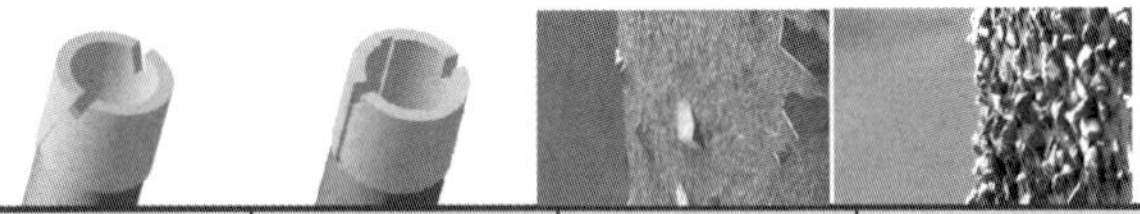

	Helical flutes	Additional internal flutes	Blasted surface	Electroplated coating
Axial force F_z	↑	↓	↑	↑
Diameter deviation ΔD of drilling	↓	↑	↓	↓
Core roughness depth R_k of drilling	↑	↑	–	↓

Figure 12.19 Influence of different mounted point concepts on the result when drilling C/C-SiC.

so that the surface roughness of this tool is greater. Furthermore, this surface property remains longer because of the hardness of the diamond grain.

12.5
Conclusions

Brittle-rigid materials are often machined by grinding. This process is also shown to be practicable when drilling fiber-reinforced ceramics C/C-SiC. The application of trepanning by grinding makes high feed velocities possible, so that the primary machining time can be favorably reduced. Furthermore, hard machining reduces non-productive times and avoids clamping failures. The present chapter has given an overview of the performance characteristics of mounted points, which are in the form of hollow drilling tools. Experiments with different concepts of mounted points show the different ways of optimizing the drilling process. Figure 12.19 summarizes the effects of the tool concepts used in drilling the fiber-reinforced ceramic C/C-SiC.

The experiments have provided evidence of the positive effects of using a tool with electroplated coatings. This type of tool enables a better drill hole quality in comparison to tools used up to now, as to tolerance and roughness depth. At the same time, radial tool wear is also decreased.

References

1 Krenkel, W., Renz, R. and Henke, T. (2000) Ultralight and wear resistant ceramic brakes, in *Anecdotal Report: Euromat99, Materials for Transportation Technology*, Munich, 27.-30.09.1999, Vol. 1, Wiley-VCH Verlag GmbH, pp. 89–94.

2 Wüllner, A. (2003) Carbon-Keramik-Bremsen – Revolution im Hochleistungsbereich! Statusbericht zwei Jahre nach Serieneinführung, in *Symposium Keramik im Fahrzeugbau, Mercedes-Forum Stuttgart*, 06.-07.05.2003 (eds J. Heinrich and H. Gasthuber), Deutsche Keramische Gesellschaft (DKG), Köln, pp. 29–33.

3 Weck, M. and Kasperowski, S. (1997) Integration of lasers in machine tools for hot machining. *Production Engineering*, IV (1), 35–8.

4 N.N. (2005) SIGRASIC 6010 GNJ, Data Specification, SGL Technologies GmbH, Meitingen.

5 Warnecke, G. (2000) Zuverlässige Hochleistungskeramik. Tagungsband zum Abschlusskolloquium des Verbundprojektes "Prozesssicherheit und Reproduzierbarkeit in der Prozesskette keramischer Bauteile", Rengsdorf, 5–6 April 2000.

6 Klocke, F. and König, W. (2005) *Fertigungsverfahren. Schleifen, Honen, Läppen. 4. Auflage*. Springer Verlag, Berlin Heidelberg.

7 Saljé, E. and Möhlen, H. (1987) Prozessoptimierung beim Schleifen keramischer Werkstoffe. *Industrie Diamanten Rundschau (IDR)*, **21** (4), 243–7.

8 Krenkel, W. (2003) C/C-SiC composites for hot structures and advanced friction systems. *Ceramic Engineering and Science Proceedings*, **24**, T.4, 583–92.

9 Weinert, K. and Jansen, T. (2005) Faserverstärkte Keramiken – Einsatzverhalten von Schleifstiften bei der Bohrungsfertigung in C/C-SiC. cfi ceramic forum international/Berichte der Deutschen Keramischen Gesellschaft, **82** (2005) 13, Sonderausgabe 2005, Hrsg. Krenkel, W., Göller Verlag, Baden Baden, pp. 34–9.

10 Jansen, T. (2005) Bohrungsbearbeitung hochharter faserverstärkter Keramik mit Schleifstiften, in *Spanende Fertigung, 4. Ausgabe* (Hrsg. K. Weinert), Vulkan Verlag, Essen, pp. 156–62.

11 Weinert, K., Johlen, G. and Finke, M. (2002) Bohrungsbearbeitung an faserverstärkten Keramiken. *Industrie Diamanten Rundschau (IDR)*, **36** (4), 322–6.

12 Weinert, K. and Jansen, T. (2007) Bohrungsfertigung mit Schleifstiften – faserverstärkte Keramik effizient bearbeiten. *Industrie Diamanten Rundschau (IDR)*, **41** (1), 48–50.

13 DIN ISO 286-1: ISO-System für Grenzmaße und Passungen. Grundlagen für Toleranzen, Abmaße und Passungen. Beuth Verlag, Berlin, 1990.

13
Advanced Joining and Integration Technologies for Ceramic Matrix Composite Systems

Mrityunjay Singh and Rajiv Asthana

13.1
Introduction

In recent years, there has been a great deal of interest in research, development, testing, and application of a wide variety of advanced ceramic-matrix composite (CMC) materials and systems. Most advanced CMCs have been developed for applications in thermal structures, exhaust nozzles, turbo pump blisks, combustor liners, radiant burners, heat exchangers, and a number of other applications that involve severe service conditions. For example, C–C composites containing SiC show promise for lightweight automotive and aerospace applications. In the automotive industry, C/C-SiC brake disks, made by the LSI process, are already being used in some models in Europe. The C/SiC composites and its variants are being developed for hypersonic aircraft thermal structures and advanced propulsion components all over the world. Similarly, SiC/SiC composites are being developed and tested for applications in combustor liners, exhaust nozzles, re-entry thermal protection systems, hot gas filters, and high-pressure heat exchangers, as well as components for nuclear reactors.

Among the various technologies available for the manufacture of CMCs, chemical vapor infiltration (CVI) is the most mature and widely used. However, the main drawbacks of this process are the difficulties in achieving uniform and complete densification of large and/or geometrically complex components and lengthy processing time. High initial capital equipment cost coupled with long processing times leads to the high cost of CMCs produced by this process. On the other hand, the polymer infiltration and pyrolysis (PIP) approach requires multiple infiltration and pyrolysis cycles, adding to the significant cost of components. In recent years, a number of reactive melt infiltration-based approaches have been successfully developed in various parts of the world, in an effort to reduce the processing costs. All these fabrication approaches have certain limitations in terms of the size and shape of CMC components, which can be manufactured with appropriate property attributes and at a reasonable cost. This limits the wide-scale global application of these materials in different industrial sectors.

13.2
Need for Joining and Integration Technologies

In order to overcome some of the traditional manufacturing and fabrication challenges, current design strategies for manufacturing large CMC components and complex structures are increasingly looking into utilizing technologies for joining/attaching smaller components with simpler geometries and dissimilar material systems. Thus, ceramic joining and integration are enabling technologies for the successful implementation of CMCs in a wide variety of high temperature applications [1–3].

Advanced ceramics have been typically joined using techniques such as diffusion bonding, adhesive bonding, active metal brazing, brazing (with oxides, glasses, and oxynitrides), and reaction forming. In the case of CMCs, the greatest success in joining carbon and silicon carbide fiber-reinforced-silicon carbide matrix composites (C/SiC and SiC/SiC) has been achieved by using the reaction-bonding approach [4–10]. This joining technology is unique in producing joints with tailorable microstructures and controllable properties. The formation of joints by this approach is attractive because the thermo-mechanical properties of the joint interlayer can be tailored to resemble those of the base materials. In addition, high temperature fixturing is not needed to hold the parts at the bonding temperature. The temperature capability of the ceramic joints in joined CMCs is similar to that of the CMC substrate materials.

For joining and integration of ceramic-metal systems, active metal brazing techniques have been developed [11–15]. The advantages of such techniques are their simplicity, cost-effectiveness, and ability to form strong and hermetically sealed joints capable of withstanding moderately high temperatures. However, the filler material must show good adherence to the substrates, prevent grain growth and long-term thermo-mechanical degradation due to creep and oxidation, have a closely-matched coefficient of thermal expansion (CTE) with the joined materials, and possess liquidus temperatures greater than the operating temperature of the joint, but lower than the substrate's melting temperature. One critical factor in brazing is the wettability of the ceramic by molten braze alloys, which influences braze flow and spreading, and strength of the joint.

We have developed bonding and integration technologies for C/C, C/SiC, and SiC/SiC composites to metallic systems (nickel and titanium based systems and Cu-clad Mo alloys) for potential aerospace and thermal management applications [13–18]. A number of tubular and flat shapes were joined in these systems, test methods were developed, and failure modes were analyzed [19–21].

13.3
Joint Design, Analysis, and Testing Issues

Joint formation involves two fundamental factors:

- chemical nature (wettability) of joined materials and the interlayer
- thermo-mechanical compatibility between joined materials.

Wetting is a necessary but not sufficient condition for joint formation, and thermo-mechanical compatibility (e.g. CTE mismatch) is a factor that may limit the strength and integrity of the fabricated joint. Other important considerations in joining include surface roughness, stress state, joint configuration, and joint stability under service conditions.

13.3.1
Wettability

Wettability is important in CMC joining using a liquid phase, because it controls the spreading and capillary flow processes. Wettability data serve as a valuable preliminary screening tool for the brazeability of materials. Extensive measurements of contact angles of a wide variety of interlayer materials (metals, alloys, glasses) on monolithic ceramics have been made. For example, braze alloys containing the active metal Ti wet oxide, carbide, nitride, and carbon substrates, and in most cases, the equilibrium contact angles approach acute (e.g. near-zero) values from the initially large obtuse values. Titanium is the most commonly used active metal in brazes because it promotes reaction-enhanced wettability and joint strength in ceramic-metal couples. Figure 13.1 shows the effect of Ti content in some alloys on their contact angle with carbon substrates.

In contrast to monolithic ceramics, the multiphase character and chemically and structurally inhomogeneous surfaces of CMCs render prediction of their brazeability from knowledge of the wettability of individual constituents, a difficult task. Surface roughness, fiber architecture, and surface chemical inhomogeneity significantly influence the contact angle and braze flow characteristics. Only limited measurements of contact angle on CMCs currently exist and the data pertain to the specific experimental conditions employed in the studies [22, 23]. The scarcity of dynamic wettability data on CMCs represents a research opportunity in the joining area.

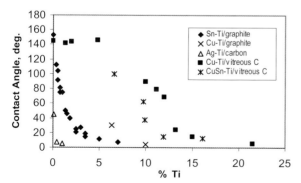

Figure 13.1 Effect of Ti content in different alloys on their wettability with various types of carbon substrates (data are from studies cited in [17]).

13.3.2
Surface Roughness

Surface roughness influences both wettability and stress concentration in joints. Whereas roughness may increase the contact (bonded) area and chemical interactions in a wettable system, besides promoting frictional bonding, it may also enhance stress concentration due to notch effects in brittle substrates. It is difficult to generalise on the effects of roughness because grinding, polishing, and machining not only reduce roughness but also might introduce surface- and sub-surface damage (grain or fiber pull-out and void formation). In addition, residual stresses may be introduced by machining. In monolithic ceramics, damage induced by grinding may be healed by re-firing the ground ceramic part prior to brazing. However, roughening may occur in soft substrates (e.g. graphite, BN) during vacuum heating after polishing. Some porosity is closed by substrate grains detached during substrate polishing. During heating under vacuum these grains are removed along with evacuating gas, resulting in increased surface roughness. Because of diverse and system-specific effect of roughness, it is best to examine its effect in individual systems (Section 13.5).

As-fabricated CMCs are inherently "rough" because of different types of fiber architecture, and this may adversely affect the joining response. There is very little information available on the roughness effects in joining of C/C, C/SiC, and SiC/SiC composites (Section 13.5). For some other CMC systems, the effect of roughness on joint strength has been evaluated. For example, joint strength data on SiC whisker-reinforced Al_2O_3 composite joints made using CuAgTi braze alloys, show [24] that grinding and polishing can lead to substantially different joint strength values; polished composites yielded significantly higher joint strength than ground and as-received SiC/Al_2O_3 composites.

13.3.3
Joint Design and Stress State

Although the use of joining and integration for the manufacture of structural components will simplify the manufacturing process, it will also introduce complexities into the design and analysis of these structures. In the past, joint design and testing activities were principally focused on metal-metal and ceramic-metal systems. Optimum joint designs for these systems accommodate a number of factors, including stress distribution in the joint regions, which are dependent upon joint configuration and chemical and thermal properties mismatch between the joint and substrate materials. In most real-life applications, CMC joints will experience a combination of different types of stresses under operating condition (Figure 13.2) that may be difficult to simulate in lab-scale tests. Therefore, field trials of joined prototype CMC components under actual service conditions are needed. However, lab-scale data on strength and other properties are important for preliminary screening.

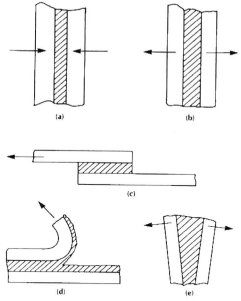

Figure 13.2 Common stress states in CMC joints: (a) compression; (b) tension; (c) shear; (d) peel; (e) cleavage (from Adhesives Technology Handbook, A.H. Landrock, 1985, p. 32, Noyes Publications, Park Ridge, NJ; figure reproduced from the book, Joining of Materials and Structures, R.W. Messler, Jr., Elsevier, 2004, p. 206).

For the polymer-matrix composite systems, various joint designs and design criteria have been established. A wide variety of testing methods have been used to determine the tensile, peel, flexural, shear, and compressive strengths. However, unlike the joining technology for fiber-reinforced polymer-matrix composites, joint design and testing are not well developed and understood for ceramic-ceramic systems, as most of these systems are relatively new. There has been some progress in this area in recent years [25, 26]. Factors such as fiber architecture, interfacial coatings, surface coatings and matrix properties play an important role in joining and integration of CMCs. In addition to the approaches that are under consideration for the design, analysis, and manufacture of single CMC parts, it is essential to develop standardized tests, design methodologies, and life-prediction analysis methods for structures incorporating joints of these materials.

13.3.4
Residual Stress, Joint Strength, and Joint Stability

Joint strength is essentially a convolution of many factors, including wetting, residual stress, joint configuration, phase transformation, and chemical segregation. Residual stresses develop due to a mismatch of thermal expansion coefficients, and may be augmented by external stresses such as those due to machining.

High joining and service temperatures of CMCs can lead to large residual stresses. In addition, thermal stresses that develop between the fiber and the CMC matrix during composite fabrication, processing, and service, may lower the inter-laminar shear stress (ILSS) and result in delamination failure. Frequently, delamination failure within the composite is caused by a weak fiber-matrix bond, which can lead to design efforts for toughening potential in the CMC.

Joint design optimization for real CMC components must consider multi-axial stress distribution as well as stress-rupture behavior because of the time-dependence of the joint strength under isothermal conditions. Prior work [25] has shown that strength loss with time in CMC joints could be rapid and substantial, depending on the testing conditions. The rapid strength loss is often accompanied by undesirable microstructural changes such as formation, growth, and coalescence of cavities, which result in high localized stresses and cracking. Additionally, long-term thermal stability of CMC joints under aggressive environments needs to be evaluated. Limited testing on liquid Si-infiltrated SiC/SiC composite joints shows that long-term thermal stability of CMC joints to 1100 °C is generally good [27]; however, more extensive testing is needed to establish a design-specific database.

Design considerations for CMCs are more complicated than for conventional engineering materials because in most cases these materials will be used at highly elevated temperatures, under stress, in aggressive environments, and for long periods of time. As stated earlier, it would be desirable to evaluate the strength and reliability of joints under states of stress/strain that are comparable to those that will be found in the intended application; more often than not, service conditions involve multi-axial states of stress. Considering the aggressive service conditions that are associated with most of the envisioned applications of CMCs, it is likely that the strength of the joints will be controlled by degradation mechanisms that include slow crack growth, stress rupture, creep, thermomechanical cycling and oxidation/corrosion.

A number of critical issues have been raised [1, 25] in using joint strength data for the design of CMC components. Typically, most of the property data available for design purposes are collected on virgin test coupons. This leads to the following concerns for the designer:

- Are joint properties in the component equivalent to coupon properties?
- Are the initial joint properties retained during the part's lifetime?
- What are the effects of thermal cycles (number and conditions) on the thermomechanical properties of joints, and is the CTE independent of number of thermal cycles?
- Does the material "soften" (lose modulus) and does this change the CTE? If it does lose modulus, can we use this to know when to remove the part?
- Do the oxidation resistance and damping (acoustic signature) change?

If CMC components are to be used on a large scale, at high temperatures, and under extreme operating conditions, these issues must be addressed in the joint design process. Another critical issue highlighted earlier, is the determination of the stress-state at the joint, namely, tensile, shear, or a combination of these two stresses during service conditions. All such considerations require that development of joining technology be integrated into the mechanical design and manufacturing processes.

13.4
Joining and Integration of CMC–Metal Systems

A number of techniques can be used to join CMCs to metals, such as welding, direct bonding, adhesive bonding, and brazing. We focus on brazing which is the most widely used ceramic joining technique. Brazing is a relatively simple and cost-effective joining method applicable to a wide range of ceramics as well as CMCs [13–18]. Ceramic brazing either utilizes tapes, powders, pastes, or rods of metallic or glassy filler alloys containing an active metal such as Ti that wets and spreads in the joint region, and solidifies to form joint interfaces, or pre-metallized ceramic surfaces that promote wettability and bonding.

Brazing is generally done in vacuum furnaces under a high-purity inert atmosphere. Table 13.1 summarizes the CMC/metal joints that have been successfully brazed in simple flat-back configuration. CMCs, such as SiC/SiC, C/SiC, C/C, and ZrB_2-based CMCs, were brazed to themselves and to Ti, Cu-clad-Mo, and Ni-base super-alloys (Inconel 625 and Hastealloy). The braze alloys included Ag, Cu, Pd, Ti, and Ni-base alloys. The brazing process for CMC joining consisted of placing the braze foil or paste between the metal and the composite substrates, heating the assembly to 15 to 20 °C above the braze liquidus in vacuum (10^{-6} torr) under a normal load, isothermally holding for a short time at the brazing temperature, and slowly cooling the joint to room temperature. SEM and EDS analyses and micro-hardness distribution across the joints were done to evaluate the joint quality and the extent of dissolution, inter-diffusion, segregation, and reaction layer formation.

Besides simple flat-back lap joint configuration, more complex tube-on-plate and other non-planar joints were also fabricated. For example, Ti tubes were brazed to graphite-foam (saddle material) and C/C composite plates with different tube-foam contact areas, in order to vary braze contact area and stress applied to joints (Figure 13.3). Mechanical testing of brazed joints showed that failure in tension and shear always occurred in the foam. This was regardless of the type of C/C composite used and whether the Ti tube was brazed to a curved foam plate to maximize the bonded area or to a flat foam surface to maximize stress in the joint [28]. In this example, the maximum shear stresses applied to the braze exceeded 12 MPa, based on the load applied and approximate braze area, and the maximum tensile stresses applied to braze exceeded 7 MPa based on load applied and approximate braze area.

Joint strength testing of several CMC-to-Ti brazed joints was reported [19, 20]. The shear strength values of different joints extracted from the Butt-Strap-Tensile

Table 13.1 General observation on brazing of C/C Composites and CMCs to metals.

Composite	Metallic substrate	Braze	Bonding
C-C[a,f]	Ti	Silcoro-75,[h] Palcusil-15[h]	Weak
C-C and SiC-SiC	Ti	Ticuni, Cu-ABA, Ticusil	Good
C-SiC[i]	Ti and Hastealloy	MBF-20[h]	Good (Ti), Fair (Hastealloy)
C-C[a,f]	Ti and Hastealloy	MBF-20,[h] MBF-30[h]	Good (Ti), Fair (Hastealloy)
SiC-SiC	Hastealloy	MBF-20,[h] MBF-30[h]	Fair (MBF20), Good (MBF30)
C-SiC[i]	Hastealloy	MBF-30[h]	Good
C-SiC[a,i]	Ti, Inconel 625	MBF-20,[h] MBF-30[h]	Good
SiC-SiC	Ti	MBF-20[h]	Good
SiC-SiC	Ti	MBF-30[h]	Good
C-SiC[a,i]	Inconel 625	Incusil-ABA,[h] Ticusil[h]	Good
C-SiC[a,i]	Inconel 625	Cu-ABA,[h] Cusil,[h] Cusil-ABA[h]	Good
C-SiC[a,b,i]	Ti	Cusil-ABA[g,h]	Good
SiC-SiC[a,b,j]	Ti	Cusil-ABA[g,h]	Good,[g] Fair[h]
C-SiC[b,i]	Ti, Inconel 625, Cu-clad Mo[k]	Ticusil[g]	Fair (Ti), Good
SiC-SiC[b,j], C-SiC[a,i], C-C[f] (T-300)	Ti, Inconel 625, Cu-clad Mo[k]	Ticusil[g]	Good
C-C[c,d,e]	Ti, Inconel 625, Cu-clad Mo[k]	Ticusil[g]	Good,[d] Fair[e]
C-C[c,d,e], C-SiC[a,b,i], SiC-SiC[a,b,j]	Cu-clad Mo[k]	Cusil-ABA[g]	Good
C-C[c,d,e]	Ti and Inconel 625	Cusil-ABA[g]	Good
SiC-SiC[a,b,j]	Ti	Cusil-ABA[g]	Good
C-SiC[a,b,i]	Ti	Cusil-ABA[g]	Good,[a] Weak[b]
ZrB$_2$-SiC (ZS), ZrB$_2$-SiC-C (ZSC), ZrB$_2$-SCS9-SiC (ZSS)	Ti, Inconel 625, Cu-clad Mo[k]	Cusil-ABA,[g] Ticusil, Palco, Palni	Good

a Polished.
b Not-polished.
c 3D composite (P120 C-fiber, CVI C matrix), Goodrich Corp., CA.
d Oriented fiber at the joint (3D composite).
e Non-oriented side at the joint.
f T-300 C-fibers in resin-derived C matrix, C-CAT, TX.
g Braze paste.
h Braze foil.
i T300 C-fiber in CVI SiC matrix, G.E. Power System Composites, DE.
j Sylramic SiC fiber in MI-SiC matrix, G.E. Power System Composites, DE.
k H.C. Starck, Inc., MA.

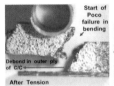

Figure 13.3 Brazed structures with woven K1100 and P120 C/C face sheet/foam/Ti Tube joints. Specimens with different tube-foam contact areas were fabricated in order to vary braze contact area and stress applied to joints [28].

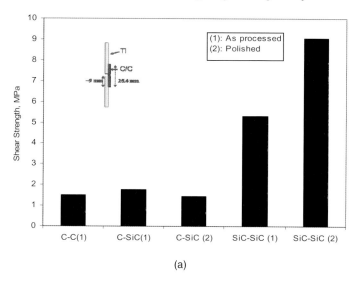

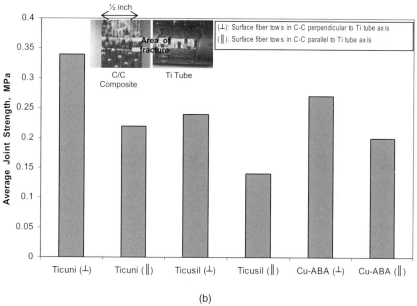

Figure 13.4 Shear strength data on brazed CMC-Ti joints: (a) the effect of polishing the CMC substrate on the shear strength of joints made using Cusil-ABA; (b) the effect of C-fiber orientation on joint strength in C/C plate/Ti tube joints [19, 20].

(BST) test is displayed in Figure 13.4a for a number of CMC-to-metal joints. The average joint strengths are 1.5 to 9.0 MPa. Generally, a polished CMC surface yielded better joints with Ti than as-received CMC. However, the effect of polishing is system-specific because polishing not only reduces surface defects but, as stated earlier, might also introduce surface and subsurface damage. The shear strength of C/C composites brazed to a Ti tube is shown in Figure 13.4b. This figure shows that braze composition and exterior fiber tow orientation influence the C/C-Ti joint strength.

Several variants of conventional brazing techniques have been developed to join CMCs and C/C composites. For example, metal interlayers and liquid infiltration have been used to joint CMCs. For this purpose, transition metals such as Cr and Mo are deposited on C/C composite surfaces via slurry deposition followed by heat treatment to form micrometer-size carbide layers [22]. Metals, such as copper in the form of foil or powder (slurry), are applied to C/C surface followed by heating. This leads to capillary infiltration of a porous carbide layer by the metal, resulting in the formation of a pore-free and crack-free joint. These joined samples can then be directly brazed to other alloys.

Another brazing approach involves use of soft and ductile rheocast Cu-Pb alloys as a thermal bond layer [23] that can absorb CTE mismatch-induced residual stresses. Rheocast alloys have a non-dendritic, globular microstructure, excellent flow properties, and exhibit non-Newtonian behavior. These alloys are obtained via mechanical or electromagnetic (EM) stirring of a partially solidified alloy slurry and prolonged holding of the slurry in the partially solidified state to enable coalescence and globularization of primary solid dendrites. The viscosity of rheocast slurry is shear rate-dependent (pseudo-plastic) and time-dependent at a constant shear rate (thixotropy). Joint strength testing of C/C composites, joined to themselves and to copper using rheocast Cu-Pb monotectic alloys, yielded relatively low (1.5–3 MPa) shear strength values. The main reason for the low shear strength in these systems is the relatively poor wetting of the alloy on C/C composites; contact angle measurements in a hot-stage microscope [23] led to values of 101 to 104 degrees at 710 °C under an argon atmosphere. Such large contact angles will invariably cause poor spreading characteristics. It is conceivable that wettability and joint strength could both be improved with C/C surface modification.

A number of surface modification techniques to promote joining have been developed. For example, joining becomes feasible and joint strength improves when either the metal substrate is oxidized (oxide/ceramic bond) or the ceramic surface is metallized (metal/metal bond). Additionally, a surface film of an active metal may be deposited on the ceramic or CMC substrates using sputtering, vapor deposition, or thermal decomposition. Examples include Ti-coated Si_3N_4, Ti-coated partially stabilized zirconia, Cr-coated carbon, and Si_3N_4 coated with Hf, Ta, or Zr. Pre-coating of the ceramic with a Ti-bearing compound (e.g. TiH_2), which forms a Ti layer on the ceramic upon heat treatment, is also effective in improving the bonding.

A major consideration in all joining processes is the issue of residual stresses that arise from the mismatch of the CTE (Section 13.3.4). Large temperature excursions and a large CTE mismatch are detrimental to the integrity of the joint because of the large residual stresses during cooling. Table 13.2 summarizes the CTE values of some CMCs, C/C composite, and high-temperature metals and

Table 13.2 Representative values of CTE of selected CMCs, metal substrates and metallic brazes.

Material	CTE × 10⁻⁶ K⁻¹
C/C composite	~2.0–4.0 (20–2500 °C)
2D SiC/SiC (0/90 Nicalon Fabric, 40% fiber)	3.0,[a] 1.7[b]
NASA 2D SiC/SiC Panels (N24-C)	4.4[c]
HiPerComp M.I. Si, SiC/SiC (G.E. Global)	3.74,[a] 3.21[b]
ZrB$_2$-SiC and ZrB$_2$-SCS9-SiC (SCS9a vol. fr.: 0.35)	6.59,[a,d] 7.21[b,d]
Cu-clad Mo	~5.7
Titanium	8.6
Inconel 625 (Ni-base superalloy)	13.1
Cusil-ABA®, Ticusil®	18.5
Palco®, Palni®	15

a In-plane value.
b Through-thickness value.
c Average value from RT to 1000 °C, extracted from % expansion data.
d Estimated from Schapery equation; ®registered trademark of Morgan Advanced Ceramics, Hayward, CA.

metallic brazes used to create joints. Very large CTE-mismatch might occur in some systems (e.g. Inconel or Ti joined to C/C composite). Most metallic brazes also have large CTE values (~15 to 19 × 10⁻⁶ K) but fortunately they are usually ductile (20–45% elongation) and might be able to accommodate some of the residual stresses that develop during cooling via plastic flow. However, if bonding at the joint is excellent, and the braze is not ductile (e.g. glass interlayer), then the residual stresses might cause cracking of the ceramic substrate. Judicious selection of braze composition and application of innovative brazing strategies, such as multiple interlayers with graded CTE and transient liquid phase bonding (to lower the joining temperature), are used to reduce the residual stresses in joints.

Effective use of joining techniques also needs to take into consideration the influence of surface coatings, which are applied to many CMCs, on the joining response. Multilayer, multi-functional external coatings provide oxidation and corrosion resistance to the composite. For example, C/C composites are overcoated with a ceramic such as silicon carbide or silicon nitride, and they may also contain internal oxidation inhibitors such as boron, titanium, and silicon. At high operating temperatures, these elements form glassy oxide phases within the C/C composite, which seal any CTE-mismatch induced cracks that might form in the external coating. This creates a barrier to the diffusion of oxygen into the porous composite matrix and thus retards its degradation. One consideration in joining could be whether the external coating should be applied before or after joining. This is especially important for brazed joints because most external coatings are applied at 1000 to 1400 °C and most brazes melt at these temperatures. In addition, the effect of glass forming inhibitors added to the composite on the wettability and adhesion of the braze must be considered.

13.5
Joining and Integration of CMC–CMC Systems

An affordable, robust ceramic joining technology (ARCJoinT) for SiC based ceramics and fiber-reinforced composites (e.g. C/SiC, SiC/SiC, etc.) has been developed at the NASA Glenn Research Center [4–6]. This joining technology permits formation of complex shapes by joining together geometrically simpler shapes. A flow diagram of the technology is given in Figure 13.5. The basic approach involves cleaning and drying of as-machined CMC surfaces prior to joining. A carbonaceous mixture is applied to the joint surfaces followed by curing at 110 to 120 °C for 10 to 20 min. Silicon or a Si alloy (e.g. Si-Mo, Si-Ti) in tape, paste, or slurry form is applied in the joint regions and heated to 1250 to 1450 °C for 5–10 min. Si or Si-alloy reacts with the carbon to form SiC with controlled amounts of Si and other phases (e.g. $TiSi_2$). Joints with good high-temperature strength have been produced for a wide variety of applications by this reaction-bonding approach. In C/SiC composite joints made using the ARCJoinT technology, the shear strength exceeds that of the as-received C/SiC at elevated temperatures up to 1350 °C. This is a positive development for utilizing such joints at elevated temperatures.

Another liquid silicon-based bonding technology has been developed by Krenkel [8, 9] for C/C composites in Germany. A carbonaceous bonding material in the form of a paste (a phenolic precursor containing fine graphite powder) is applied to the C/C composite surface, having a carbon fiber mat or felt as an interlayer. This is followed by liquid silicon infiltration of the bonding materials to yield the required level of conversion.

Other notable attempts to join CMCs include use of calcia-alumina (CA) glass ceramic [29], and zinc-borate (ZBM) glass [30]. A potential advantage of glass as a

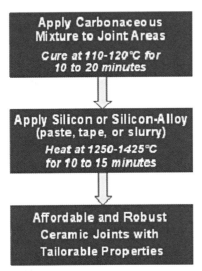

Figure 13.5 Flow diagram of ARCJoinT ceramic joining technology developed at NASA GRC [4].

joining material is that its composition can be designed to achieve the desired level of reactivity, wettability, and closely-matched CTE. In addition, many sintered ceramics develop amorphous phases at their grain boundaries, and good wetting and bonding can be achieved between the glass filler and these amorphous grain boundary phases. The disadvantages of glass-based brazes are their relatively high viscosity and brittle nature. Ferraris et al. [29] reported low-temperature, pressureless joining of HI-Nicalon fiber/SiC CMCs with a eutectic CA glass ceramic (49.77CaO–50.23Al_2O_3). CVI SiC/SiC and polymer infiltration pyrolysis (PIP) SiC/SiC composites were joined using the CA glass as an interlayer. A slurry of CA glass was deposited between the CMC surfaces, and the assembly was heated to 1500 °C for 1 hour under argon. Interestingly, the CA glass ceramic does not wet the PIP CMC (which contained an amorphous SiC phase) but it wets the CVI CMC (which contained a β-SiC phase). No evidence of cracking, porosity, or discontinuity was noted in the CVI SiC/SiC joints, and the fracture path was through the CA glass-ceramic phase and not at the CA/CMC interface; this suggests the joint strength exceeded the fracture strength of the CA glass-ceramic.

In another study [30] on joining C/C, C/SiC, and SiC/SiC composites to themselves, ZBM glass as well as metallic coatings of Al, Ti, or Si were used. Sandwiched, infiltrated, and coated CMC structures were joined. The shear strength values (typically 10–22 MPa) of the CMC joints produced depended primarily on the joining materials. The highest joint strength (22 MPa) was obtained with a silicon interlayer sandwiched between the CMC and low strength (10 MPa) was obtained with an Al interlayer. The glass interlayer between CMC substrates yielded an average shear strength of ~15 MPa in SiC/SiC composite joints. In C/C joints with Si and Al interlayers, SiC and Al_4C_3 phases formed. With Si interlayer, the fracture path in mechanically tested joints was on multiple planes through the Si layer, whereas with the Al interlayer the failure occurred at the interface between the C/C composite and reactively formed layer of Al_4C_3.

Joint microstructures in reaction-formed CVI C/SiC joints with different surface conditions of the starting substrate are shown in Figure 13.6. The CVI C/SiC joints were formed via liquid Si infiltration of a carbonaceous paste applied to the substrate surface, as in the ARCJoinT process. Figure 13.6 shows that microstructurally sound joints form when one C/SiC surface is machined and the other is in an as-received (not machined) state. Inferior quality joints containing voids and microcracks form when either both mating surfaces are machined or neither is machined. Similar results were obtained by Krenkel [9] on C/C joints created using an interlayer of carbon fabric or felt impregnated with a carbonaceous bonding paste and liquid silicon. Highest shear strength was obtained in joints with a combination of a ground and an unground C/C test coupon mated with the carbon fabric (impregnated with the bonding paste), although the greatest improvements in shear strength were achieved when unground specimens were joined using the impregnated fabric as an interlayer. This is because the flexible carbon fabric interlayer can deform to match the surface contour of unground specimens prior to Si infiltration, thus ensuring a void-free and topologically homogeneous interface.

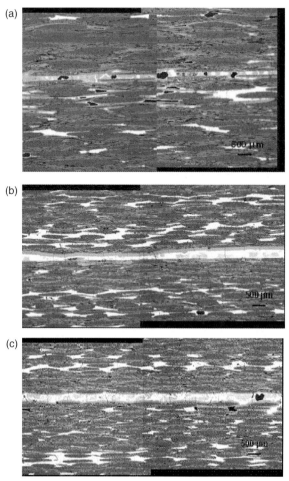

Figure 13.6 Microstructure of reaction-formed CVI C/SiC joints: (a) both surfaces machined; (b) one surface machined and one surface as received; (c) both surfaces as received.

Figure 13.7 shows the stress-strain behavior in CVI C/SiC composite joints made by the ARCJoinT process in a double-notch compression test at different temperatures. The test was performed on CVI C/SiC composite joints according to ASTM C 1292–95a (RT) and ASTM C 1425–99 (HT). These test standards have been recommended by ASTM for determining the interlaminar shear strength (ILSS) of one- and two-dimensional CMCs reinforced with continuous fibers. In this test, shear failure is induced between two notches that are anti-symmetrically located on the opposite sides of the specimen at an equal distance from the mid-plane (Figure 13.7a). While the test is relatively simple to perform and requires small specimens, it suffers from some drawbacks. First, because interlaminar shear failure is induced by the stress concentration at the root of the notches, the

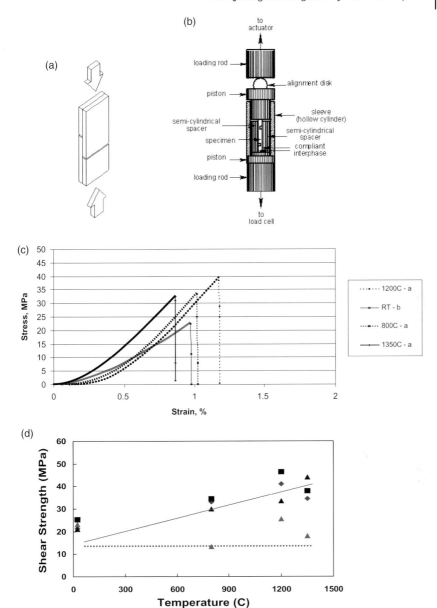

Figure 13.7 (a) Specimen geometry; (b) test fixture used for compression double-notched shear tests; (c) stress-strain behavior in compression double-notched shear test on CVI C/SiC joints at different temperatures; (d) shear strength vs temperature in CVI C/SiC composite joints [25].

shear stress distribution between the notches is not uniform. Second, the specimen does not fail under pure shear as both shear and normal compression stresses are induced. Nevertheless, the test yields reasonable estimates of the joint shear strength.

The dimensions of test coupons for the dataset shown in Figures 13.7c and d were as follows: specimen length (L): 30 mm (±0.10 mm), distance between notches (h): 6 mm (±0.10 mm), specimen width (W): 15 mm (±0.10 mm), notch width (d): 0.50 mm (±0.05 mm), and specimen thickness (t): (adjustable). The test fixture is shown in Figure 13.7b, where the rate of loading was 50 N s^{-1}. Further details on the testing procedure are given in [25]. The shear strength τ of the joints is extracted from the stress-strain data using the following relationship:

$$\tau = \frac{P}{wh}$$

where P is the applied compression load, w is the width of the specimen, and h is the distance between notches.

Figure 13.7 also shows the effect of test temperature on the joint strength. The shear strength of the joints increases with increasing temperature and exceeds the strength of the CVI C/SiC composite substrate. There is minimal effect of surface roughness on the shear strength of the joint, with machined/as fabricated CMC joints showing a relatively higher strength up to 1350 °C, compared to other types of joints.

Measurements of shear strength of CMC joints have been reported in several studies. Table 13.3 summarizes some of the results on C/C, C/SiC, and SiC/SiC composite joints. The strength data were extracted from four-point bend test, single lap test, double-notch compression test, and other methods. There is considerable variation in the shear strength values because the strength depends upon the joining technique, the test method, substrate surface preparation, and the test temperature. Lack of standardized test methodology makes it difficult to utilize such data for actual component design purposes. As a result, significant emphasis is currently placed on the development and performance evaluation of joints in prototype components under actual service conditions in niche application areas. Applications of CMC joining technology in component manufacturing are discussed in the next section.

13.6
Application in Subcomponents

A number of CMC components have been made by integrating sub-elements with the help of advanced joining technology. Some examples are shown in Figure 13.8. The joined CMC components include carbon-carbon composite valves for racing car engines, attachments for sensors [33] for use at elevated temperatures in air

Table 13.3 Representative values of shear strength of CMC joints.

Joined materials and surface preparation	Joint materials	Temp. (°C)	Strength (MPa)	Comments	Reference
C/C composite system	Rheocast Cu-Pb alloy	25	1.5	Strength of Cu/Cu joint with Cu-Pb alloy: 3 MPa at 25 °C	[23]
C/C-SiC composite system[a]	Carbon paste	25	17	Minor improvement in strength, some porosity and free Si	[8]
	Carbon felt	25	18		
	Carbon fabric	25	18.5		
C/C-SiC composite system[b]	Carbon paste	25	11	Significant increase in joint strength	[8]
	Carbon felt	25	19		
	Carbon fabric	25	21		
C/C-SiC composite system[c]	Carbon paste	25	21	No porosity or free Si in the joint region	[8]
	Carbon felt	25	22		
	Carbon fabric	25	26		
C/C-SiC composite system	Liquid Si infiltration of a carbon paste applied to the joint with or without C fabric	25 / 25	21.5 ± 2.28 / 113–198[d]		[9]
CVI SiC/SiC composite system	ARCJoinT (carbonaceous mixture at the joint, cure, infiltrate with Si or Si alloys)	25 / 800 / 1200	65 ± 5 / 66 ± 9 / 59 ± 7	4-point bend test	[27]
MI Hi-Nicalon/BN/ SiC composite system	ARCJoinT (carbonaceous mixture at the joint, cure, infiltrate with Si or Si alloys)	25 / 800 / 1200 / 1350	99 ± 3 / 112 ± 3 / 157 ± 9 / 113 ± 5	4-point bend test; as-fabricated composite strength: 360 MPa at 25 °C	[1]
CVI-SiC/SiC composite system	ARCJoinT (carbonaceous mixture at the joint, cure, infiltrate with Si or Si alloys)	25 / 1200 / 1200	92 / 71 / 17.5[e]	Compression test on double-notched specimens, joints contain SiC + small amounts of Si and TiSi$_2$	[25]
SiC/SiC composite system	CA-glass ceramic (49.77% CaO + 50.23% Al$_2$O$_3$)	25	28		[29]
SiC/SiC composite system	Co-10Ti braze, 1340 °C	25 / 25 / 700	51 in 7.5 min / 55 in 15 min / 75 in 15 min		[31]
SiC/SiC composite system	Si-16Ti, Si-18Cr		71 ± 10 Ti, 80 ± 10 Cr		[32]

a Ground.
b Unground.
c Combination of a ground and an unground specimen.
d Heat treated at 1000 °C.
e After 14.3 h in stress-rupture test at 1200 °C.

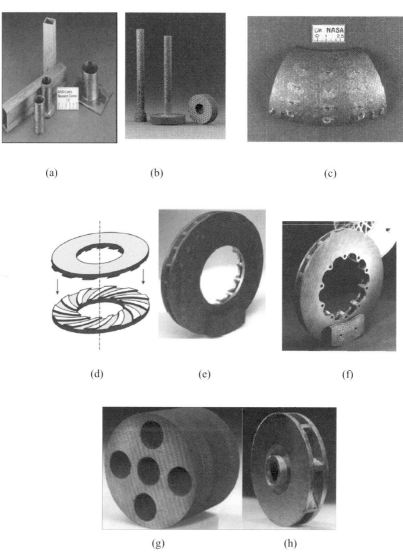

Figure 13.8 Examples of CMC components made by advanced joining technology: (a) joined C/SiC composites; (b) carbon–carbon composite valves for race car engines; (c) attachment for sensors; (d–f) ceramic composite brakes showing cooling channels made of (e) bolts and (f) ribs; (g) heat exchanger substructure; (h) radial pump wheel (Figures d through h by courtesy of Professor W. Krenkel).

and space vehicles, lightweight C/C-SiC brakes with considerably improved thermal stability as compared to standard gray cast iron brake disks, and CMC heat exchangers. CMC brake disks with internal cooling passages between frictional surfaces have been fabricated by joining. Complex arrangements of internal ribs and bolts allow internal ventilation and cooling of the CMC brake. Therefore, these brake disks contain two sub-elements: bolts with longitudinal fiber reinforcement and ring-shaped plates with in-plane reinforcement. These brakes are machined in the C/C stage. This step is followed by bonding with a carbonaceous paste and infiltration with silicon. Similarly, prototype turbine and pump wheels and CMC heat exchangers have been designed and fabricated using advanced ceramic joining technology. Many other types of engineering components have been fabricated by joining of CMC sub-elements and these components have been tested for performance under actual or simulated service conditions.

13.7
Repair of Composite Systems

Many of the joining technologies developed for the manufacture of large and complex components can, with certain modifications, be applied to repair of CMC components damaged in service. A recent innovation has been the development of a technology for advanced in-space repair of reinforced C/C (RCC) composite thermal protection systems of the space shuttle. A new material, called Glenn Refractory Adhesive for Bonding and Exterior Repair (GRABER) has been developed at the NASA Glenn Research Center [34, 35] for multi-use in-space repair of small cracks in the space shuttle RCC leading edge material. The new material has well-characterized and controllable properties such as viscosity, wetting behavior, working life, and so on, and excellent plasma performance as revealed in a number of lab-scale simulations. The material does not require post processing and converts to high-temperature ceramics during re-entry conditions. Initial performance evaluation tests under plasma conditions have been encouraging and further tests are in progress.

In simulated tests, a damaged region of a RCC disk where a crack (0.035' wide and 1.5' long) was created and coatings were spalled using an indenter, was used to evaluate the effect of small damage and coating loss on the performance of RCC center strip during the space shuttle re-entry conditions. Figure 13.9 shows photographs of a test in which GRABER was used to repair the damage. Figure 13.9a shows a GRABER-repaired RCC specimen in an ARCJet test holder, and Figures 13.9b and c show photographs during and after arc jet testing, respectively [34, 35]. The test was conducted for 15 minutes during which the repaired area was exposed to space-shuttle re-entry conditions. It is clearly evident that the repair concept was successful in preventing the hot plasma and other gaseous species getting through the repaired RCC composite. Thus, damage repair using GRABER is a robust repair technique potentially effective under re-entry conditions on small area damage and coatings loss in the RCC material, and it may help increase the

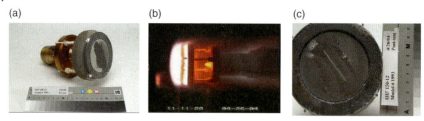

Figure 13.9 (a) GRABER repaired RCC specimen in ARCJet test holder; (b) photograph during testing; (c) post test photograph [34, 35].

safety of vehicles and crew. GRABER is a phenolic based adhesive with silicon as one of the reactive phases. This system has very high viscosity (like peanut butter) and cannot be infiltrated in dry areas of composites. It has been developed and optimized for the repair of damage in RCC composite and further testing will be needed to evaluate its effectiveness in repairing damage in C/SiC and SiC/SiC composites.

13.8
Concluding Remarks and Future Directions

Robust jointing and integration technologies are needed to manufacture large and complex-shaped CMC components and represent an enabling technology in aerospace, power generation, nuclear, and transportation industries. Joint design philosophies, design issues, and joining technologies were discussed along with the relationship of CMC joining to CMC manufacturing. In particular, the role of wettability, surface roughness, and residual stress in joint strength and integrity was highlighted. Mechanical property data on CMC joints were presented together with examples of joined CMC prototype subcomponents. Remarks on recent developments in applying CMC joining technologies to repair of ceramic components damaged in service were also presented.

Clearly, much additional developmental research effort is needed in the joining and integration of CMC components, in order to win the confidence and trust of the designer. At present, a number of critical issues need to be resolved for CMC joints. Some of these include:

1. effectiveness of representing a real component joint by test coupons under laboratory conditions;

2. contact angle and work of adhesion measurements on multiphase CMCs that exhibit considerable chemical and structural inhomogeneity;

3. effect of thermal fatigue and aggressive environments on long-term durability of CMC joints;

4. effect of time on joint properties under various testing conditions, including multi-axial external stresses;

5. effect of oxidation, creep, and thermal shock on joint strength and integrity; and

6. standardization of design and testing methodologies for CMC joints, and development and integration of life prediction models in CMC component manufacture.

Acknowledgments

The authors wish to thank Tarah Shpargel for her help with the microscopy work on joint interfaces. R. Asthana would like to acknowledge the support received from the NASA Glenn Research Center and University of Wisconsin-Stout.

References

1 Singh, M. (2001) Challenges and opportunities in design, fabrication and testing of high-temperature joints in ceramics and ceramic composites, in *Brazing, High-Temperature Brazing and Diffusion welding*, DVS, pp. 55–8.

2 Singh, M. (1999) Critical needs for robust and reliable database for design and manufacturing of ceramic-matrix composites, in *NASA/CR-1999-209316*, September 1999, NASA Glenn Research Center, Cleveland, OH.

3 Lewis, D. III and Singh, M. (2001) Post-processing and assembly of ceramic-matrix composites, in *Materials Handbook*, Vol. 21, Composites, ASM International, Materials Park, OH, pp. 668–73.

4 Singh, M. (1997) A reaction forming method for joining of silicon carbide-based ceramics. *Scripta Materialia*, 37, 1151–4.

5 Singh, M., Farmer, S.C. and Kiser, J.D. (1997) Joining of silicon carbide-based ceramics by reaction forming approach. *Ceramic Engineering and Science Proceeding*, 18, 161–6. .

6 Singh, M. (1999) A new approach to joining of silicon carbide-based materials for high-temperature applications. *Industrial Ceramics*, 19, 91–3.

7 Lewinsohn, C.A., Jones, R.H., Singh, M., Shibayama, T., Hinoki, T., Ando, M. and Kohyama, A. (1999) Methods for joining silicon carbide composites for high-temperature structural applications. *Ceramic Engineering and Science Proceeding*, 20, 119–24.

8 Krenkel, W., Henke, T. and Mason, N. (1997) In-situ joined CMC components, in *Key Engineering Materials, Proceedings of the International Conference on Ceramic and Metal-Matrix Composites, CMMC 96, Part I*, Trans Tech Publications Ltd, Zurich, Switzerland, p. 113.

9 Krenkel, W. and Henke, T. (2000) Modular design of CMC structures by reaction bonding of SiC, in *Proceedings of Materials Solutions Conference '99 on Joining of Advanced and Specialty Materials*, (eds M. Singh et al.), ASM International, Materials Park, OH, pp. 3–9.

10 Martinez-Fernandez, J., Munoz, A., Varela-Feria, F.M. and Singh, M. (2000) Interfacial and thermomechanical characterization of reaction formed joints in SiC-based materials. *Journal of the European Ceramic Society*, 20, 2641–8.

11 Locatelli, M.R., Tomsia, A.P., Nakashima, K., Dalgleish, B.J. and Glaser, A.M. (1995) New strategies for joining ceramics for high-temperature

applications. *Key Engineering Materials*, **111–112**, 157–90.

12 Peteves, S.D., Paulasto, M., Ceccone, G. and Stamos, V. (1998) The reactive route to ceramic joining: fabrication, interfacial chemistry and joint properties. *Acta Materialia*, **46**, 2407–14.

13 Singh, M., Shpargel, T., Morscher, G.N. and Asthana, R. (2005) Active metal brazing and characterization of brazed joints in Ti to carbon-carbon composites. *Materials Science and Engineering*, **412–413**, 123–8.

14 Singh, M., Asthana, R. and Shpargel, T. (2007) Brazing of C-C composites to Cu-clad Mo for thermal management applications. *Materials Science and Engineering A*, **452–453**, pp. 699–704.

15 Singh, M. and Asthana, R. (2007) Joining of advanced ultra-high-temperature ZrB_2-based ceramic composites using metallic glass interlayers. *Materials Science and Engineering A*, **460–461**, 153–62.

16 Asthana, R., Singh, M. and Shpargel, T.P. (2006) Brazing of ceramic-matrix composites to titanium using metallic glass interlayers. *Ceramic Engineering and Science Proceeding*, **27** (2), 159–68.

17 Singh, M., Shpargel, T.P., Morscher, G.N. and Asthana, R. (2005) Active metal brazing of carbon-carbon composites to titanium, in *Proceedings of 5th International Conference on High-Temperature Ceramic-Matrix Composites (HTCMC-5)*, Seattle, WA, September 2004 (eds M. Singh, R.J. Kerans, E. Lara-Kurzio and R. Naslain), The American Ceramic Society, Westerville, OH, pp. 457–62.

18 Singh, M. and Asthana, R. (2007) Brazing of advanced ceramic composites: Issues and challenges, in *Ceramic Transactions*, Vol. 198, (eds K. Ewsuk et al.), John Wiley & Sons, pp. 9–14.

19 Morscher, G.N., Singh, M., Shpargel, T.P. and Asthana, R. (2006) A simple test to determine the effectiveness of different braze compositions for joining Ti tubes to C/C composite plates. *Materials Science and Engineering A*, **418**, 19–24.

20 Morscher, G.N., Singh, M. and Shpargel, T.P. (2006) Comparison of different braze and solder materials for joining Ti to high-conductivity C/C composites, in *Proceedings of 3rd International Conference on Brazing and Soldering*, San Antonio, TX, April 24–26, 2006 (eds J.J. Stephens and K. Scott Weil), ASM International and AWS, pp. 257–61.

21 Singh, M., Shpargel, T.P., Asthana, R. and Morscher, G.N. (2006) Effect of composite substrate properties on the mechanical behavior of brazed joints in metal-composite system, in *Proceedings of 3rd International Conference on Brazing and Soldering*, San Antonio, TX, April 24–26, 2006, (eds J.J. Stephens and K. Scott Weil), ASM International and AWS, pp. 246–51.

22 Appendino, P., Ferraris, M., Casalegno, V., Salvo, M., Merola, M. and Grattarola, M. (2004) Direct joining of CFC to copper. *Journal of Nuclear Materials*, **329–333**, 1563–6.

23 Salvo, M., Lemoine, P., Ferraris, M., Montorsi, M.A. and Matera, R. (1995) Cu-Pb rheocast alloy as joining materials for CFC composites. *Journal of Nuclear Materials*, **226**, 67–71.

24 Moore, A.J., Elliott, W.H. and Kim, H.-E. (1993) Brazing of ceramic and ceramic-to-metal joints, in *Welding, Brazing and Soldering, Metals Handbook*, ASM International, p. 948.

25 Singh, M. and Lara-Curzio, E. (2001) Design, fabrication and testing of ceramic joints for high-temperature SiC/SiC composites. *Transactions of the ASME*, **123**, 288–92.

26 Avalle, M., Ventrella, A., Ferraris, M. and Salvo, M. (2006) Shear strength tests of joined ceramics, in *Proceedings of ESDA06, 8th Biennial ASME Conference on Engineering Systems Design and Analysis*, Torino, Italy, July 2006, ASME International, NY.

27 Singh, M. (1999) Design, fabrication and characterization of high-temperature joints in ceramic composites. *Key Engineering Materials*, **164–165**, 415–20.

28 Singh, M., Morscher, G.N., Shpargel, T.P. and Asthana, R. (2007) Active metal brazing of titanium to high-conductivity carbon-based sandwich structures. *Materials Science and Engineering*, in press.

29 Ferraris, M., Salvo, M., Isola, C., Montorsi, M.A. and Kohyama, A. (1998) Glass-

ceramic joining and coating of SiC/SiC for fusion applications. *Journal of Nuclear Materials*, **258–263**, 1546–50.
30. Salvo, M., Ferraris, M., Lemoine, P., Montorsi, M.A. and Merola, M. (1996) Joining of CMCs for thermonuclear fusion applications. *Journal of Nuclear Materials*, **233–235**, 949–53.
31. Trehan, V., Indacochea, J.E., Lugscheider, E., Buschke, I. and Singh, M. (1998) Joining of SiC for high-temperature applications, in *Proceedings of Materials Conference '98 on Joining of Advanced and Specialty Materials* (eds M. Singh et al.), ASM International, Materials Park, OH, pp. 57–62.
32. Riccardi, B., Nannetti, C.A., Woltersdorf, J., Pippel, E. and Petrisor, T. (2002) High-temperature brazing for SiC and SiC/SiC ceramic-matrix composites, in *Advanced SiC/SiC Ceramic Composites: Developments and Applications in Energy Systems*, Ceramics Trans, Vol. 144 (eds A. Kohyama, M. Singh, H.T. Lin and Y. Katoh), The American Ceramic Society, Westerville, OH, pp. 311–22.
33. Martin, L.C., Kiser, J.D., Lei, J., Singh, M., Cuy, M., Blaha, C.A. and Androjna, D. (2002) Attachment technique for securing sensor lead wires on SiC-based components. *Journal of Advanced Materials*, **34**, 34–40.
34. On-Orbit Shuttle Repair Takes Shape, *Aerospace America*, Aug. 2004, pp. 31–4.
35. NASA One Step Closer to Shuttle Repair in Orbit, *New Scientist*, May 22, 2004, p. 24.

14
CMC Materials for Space and Aeronautical Applications
François Christin

14.1
Introduction

Thermo-structural composites are a unique family of high performance materials. They combine the refractory and structural properties of graphite and ceramics with the high tailorable properties of fibrous reinforcement composites. These materials are of utmost importance for the growth and competitiveness of defense, spatial, aeronautical, and industrial applications. The history of carbon/carbon composites, which are the first thermo-structural composites, can hardly be related without mentioning the leading role that rocket propulsion played in their early research, development, and industrialization, in the 1960s and 1970s.

Over the last 50 years, the temperature capability of metallic super-alloys has never stopped increasing, but the limitation of their melting point, now very close to the operating conditions, is stemming this way.

During the 1980s, Ceramic Matrix Composites) (CMC) were developed to meet the relatively long durations of thermal protection system (TPS) for future reusable launch vehicles. Because of performance needs, economical constraints and, more recently, international environmental regulations evolutions, the design of military and commercial aircraft engines has been driven by two key factors:

- the weight saving
- the increase in gas temperature

These factors are respectively linked to payload gains, efficiency improvement, and NOx and CO emissions reduction. The long life duration of the jet engine component is another requirement, accounting for much of the total lifecycle cost. These needs have necessitated the development of the self-sealing matrix composite concept.

CMCs, with their intrinsic properties (low density, high melting point, high mechanical properties, chemical stability) are well suited to take over from metallic alloys, the ceramic low fracture toughness being deleted due to the fiber reinforcement and tailored fiber/matrix bonding.

Ceramic Matrix Composites. Edited by Walter Krenkel
Copyright © 2008 WILEY-VCH Verlag GmbH & Co. KGaA, Weinheim
ISBN: 978-3-527-31361-7

The main steps of the manufacturing processes will be described and the current space and aeronautical applications will be identified.

14.2
Carbon/Carbon Composites

The history of carbon/carbon composites starts in 1958 in a laboratory of Chance Vought Aircraft, with the pyrolysis, perhaps accidentally, of a phenolic matrix composite [1]. Research in the 1950s, undertaken simultaneously in the United States to develop composites made of a fibrous core embedded in a resin matrix, and in the United Kingdom to develop carbon fibers, formed the basis for development ablative composites in the 1960s. These materials were usually made of carbon, or less frequently silica, fiber reinforcements impregnated with polymerized phenolic resins. After absorbing large amounts of heat, the resins were ultimately charred by the high temperatures of the gas stream expelled from the nozzle of the rocket engine [2].

14.2.1
Manufacturing of Carbon/Carbon Composites

The carbon/carbon manufacturing process can be divided in two main steps:

- fabrication of carbon fiber reinforcement, called preform;
- filling the porosity of the preform by matrix, called densification.

14.2.1.1 n-Dimensional Reinforcement

Three types of carbon fiber precursors can be used: rayon, Polyacrylonitrile (PAN), or pitch fibers, depending on the researched final composites characteristics. For example, ex-pitch carbon fibers will be chosen for high thermal conductivity, ex-PAN carbon fibers for high mechanical properties, and ex-rayon carbon fibers for specific features applications.

By the late 1970s, the main carbon fiber reinforced carbon materials were as follows:

On the one hand, there were the carbon-carbon materials based on two-dimensional reinforcements, whose strong point lay in their low thicknesses capability and simple manufacturing processes. Two ways of processing fabrication were used, the dry route or the liquid route. The dry route consisted of pressing out dry cloth fabrics with special graphite tools for CVI densification. The liquid route was based on the use of prepregs, first shaped and cured, then pyrolysed and heat treated.

The two-dimensional materials suffered from poor interlaminar strength and showed a tendency to delaminate or crack during fabrication. Nevertheless, Snecma Propulsion Solide succeeded in mastering two-dimensional involute

Figure 14.1 Apogee rocket motor (Mage) [3].

carbon-carbon exit cones for rocket motors, since the Mage Apogee boost motor (Figure 14.1).

On the other hand, there were the multidirectional reinforced materials obtained by weaving fibers or by intertwining rods to make n-dimensional reinforcement preforms (three-dimensional, four-dimensional, five-dimensional, and so on) (Figure 14.2). These materials exhibited an improved isotropy, good delamination resistance, and thick part manufacturing capability. When the erosion of the nozzle throat has to be very low, the best reinforcement (but also the most expensive) will be constituted by the four-dimensional material. This preform technology, the development of which started in 1973, is always used. A cubic network is manufactured by inter-twining stiffened rods of carbon fiber leading to a high volumetric fiber ratio construction of 40%, before and after densification.

This four-dimensional material is currently retained for the development of the new generation of French deterrent force solid rocket motors (SRM). It has also been successfully used in the Mage motor series (Figure 14.1) with more than 50 firing tests and 18 flight missions without failures or other unexpected issues. However, only liquid routes (resin or pitch) were applicable for their densification, due to the coarse unit knit. Moreover, they were not suitable for producing complex parts combining both thick and thin areas [3, 5]. These were the reasons why a new texture, combining the advantages of multi-dimensional and two-dimensional preform types, has been developed for application in all propulsion fields, typically integrated throat and exit cones for solid rocket nozzles, hot gas valves, or large size exit cones for liquid rocket engine (LRE).

14.2.1.2 Three-Dimensional Reinforcement Preforms

As previously mentioned, two-dimensional carbon/carbon composites suffer from a weakness in delamination, and also their applications were reduced by thickness limitations. So, three-dimensional preforms have been developed and, because of

(a)

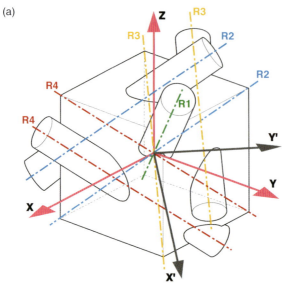

(b)

Figure 14.2 (a) four-dimensional preform construction; (b) four-dimensional Sepcarb after densification [4].

economical considerations, needling is the preferred technology for making carbon preforms by using weaving and needling techniques. At the beginning of the 1980s, SEP (Société Européenne de Propulsion; now Snecma Propulsion Solide), developed a three-dimensional texture called Novoltex. This is a three-dimensional carbon non-woven preform construction made through automatic technology from PAN fibers.

The preform is obtained by needling of a specific cloth. This material is composed of a carbon precursor (PAN precursor) woven cloth combined with a layer of short carbon precursor fibers deposited on it. It is necessary to find out and to master the proper conditioning of the fiber layers on the cloth.

This needling process consists of attaching fabric layers to each other with fibers pushed by hook fitted needles (hooks are designed so that fibers stay where they have been performing when the needles leave the preform). Needling direction is called "Z," while "X" and "Y" represent the two-dimensional cloth layer directions. Z fiber density can be adjusted by mastering needle geometry, needle stroke, or needling density per square inch.

Proper needling tools and parameters were investigated and optimized to obtain the optimal preform characteristics in terms of density, porosity size distribution, and fiber fractions for each type of application (Figure 14.3).

An original aspect of Novoltex is that needling is carried out after each layer deposit so that, at the end of the process, each part of the preform through the thickness has received the same amount of transferred fibers (Figure 14.4). This provides good homogeneity of the Novoltex texture. After this step, a final carbonization treatment is performed to obtain pure carbon fiber. This carbon fiber ratio in the final material is between 20 and 30% by volume.

Specific needling machines were developed to produce different extreme configurations, such as large rectangular blocks (up to 160 mm high), cylinders,

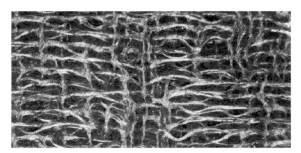

Figure 14.3 Novoltex preform characteristics.

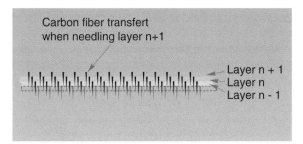

Figure 14.4 Carbon fiber transfer.

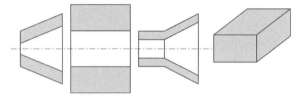

Figure 14.5 Geometric preforms (schematic).

conical forms (2600 mm OD), and thin or thick planar preforms (7000 × 2500 × 60 mm) (Figure 14.5).

After preform manufacturing, it is separated from its needling mandrel and heat treated at elevated temperatures, to transform the carbon precursor into carbon. During the carbonization process, the shrinkage affecting the texture had to be mastered. This tri-dimensional preform, produced by automatic operations, has a regular and fine open pore distribution, well adapted to the CVI densification process (dry route).

Some interest lies in the possibility of adjusting to some extent the orthothropy of the final material by mastering the preform characteristics, from nearly two-dimensional characteristics to a nearly isotropic material. Once densified, these preforms give the material very attractive capabilities, from large and very thin parts (<1.5 mm), where three-dimensional constructions of a coarser-weave is ineffective, to thick parts (>100 mm).

Snecma Propulsion Solide has several needling equipments that are able to build preforms as flat panels up to 2.6 m wide by 6 m long, and cylindrical or

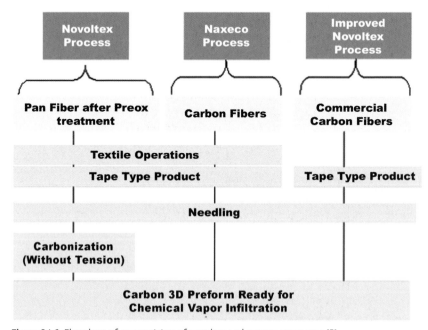

Figure 14.6 Flowchart of comparision of novoltex and naxeco processes [5].

conical parts up to 2.6 m diameter and 3 m length. The preform thickness is adapted to the final size of the parts.

More recently, a new three-dimensional reinforcement called Naxeco has been manufactured by needling directly commercial carbon fibers with a fiber ratio of 35% (Figure 14.6):

- Commercial fiber means that specific raw material purchase at an intermediate manufacturing state ("preoxidized step") and associated drawbacks (long-term supply, price, specific quality control, and process, etc.) are eliminated.

- Carbon fiber means that carbonization treatment with associated cost and drawbacks, such as relatively poor mechanical properties (heat treatment is performed without tensile stress on the fiber for Novoltex) are also eliminated.

The economical Naxeco reinforcement also presents better mechanical performances – at least equal rupture strain in the range of 0.5% and a rupture strength increased by 50 to 100% – that enhances component reliability. The appropriate fiber ratio would lower erosion for propulsion application.

14.2.1.3 Densification

The matrix is obtained either by gas infiltration (CVI), by resin impregnation combined with pyrolysis treatment (PIP), or by pitch impregnation combined with a high isostatic pressure (HIP) carbonization process (Figure 14.7).

CVI (Chemical Vapor Infiltration) The carbon matrix is deposited in the pores of the fibrous carbon structures in the vapor phase by cracking of methane at controlled temperature and low pressure. This process is the most used because it gives a carbon matrix (pure pyrocarbon) with good mechanical properties and is suitable for industrial application such as brake disks for aeronautical applications.

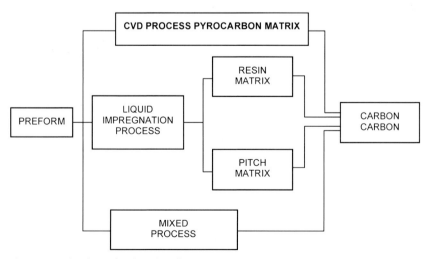

Figure 14.7 Flowchart of carbon densification processes [6].

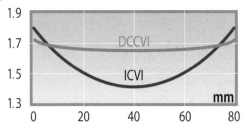

Figure 14.8 Isothermal CVI densification process.

The CVI densification presents many variants such as isothermal, thermal gradient, direct coupling, pressure gradient (forced flow), semi-forced flow, and pulsed flow processes. The common goal of the studies is to decrease the densification duration and the costs.

In the isothermal CVI densification process, the parts are heated by an external graphite susceptor that is itself heated by self induction from an external electric inductor (Figure 14.8).

In the direct coupling process, the axi-symmetrical parts (made of carbon fibers, thus electrically conductive) are self heated by direct Lenz effects from the external inductor. This process favors a central densification with following advantages:

- higher density "at heart"
- reduced densification time
- no coating effect, removing some machining cycles
- initial thinner perform.

The interest of direct coupling vs isothermal CVI process is well demonstrated in Figure 14.8, with comparison of density through the thickness. This process, and also the pressure gradient CVI process, permits the densification cycle duration to be divided by two.

A more rapid densification of carbon/carbon can be obtained by the film boiling process. This process, patented by CEA potential in 1981, can densify parts in less than 10 hours. The part is immersed in liquid hydrocarbon such as toluene or cyclohexane then heated at around 1000 °C by resistive or inductive heating (internal or external radio frequency coil, solenoid, or pan cake or pin coil). Very high thermal gradient and mass transfer fluxes constitute the driving force of this process [7].

These driving forces are induced by vaporization of a liquid precursor, on the external surface of the porous structure to be filled up. This technique presents many advantages: high densification rates (several mm/h), good materials characteristics, reproducibility, and relative simplicity, but it needs high electrical power.

Pitch Matrix Process and Resin Matrix Process The densification of the carbon structures is made by pitch or resin impregnation followed by high-pressure (pitch) or low pressure (resin) carbonization (PIP). The pitch carbon matrix

Figure 14.9 Isothermal CVI densification process [8].

provides carbon-carbon with high thermal conductivity. Conversely, the resin carbon matrix provides carbon-carbon with good mechanical properties and low thermal conductivity.

The three densification processes can be combined as a function of design requirements (cost, material properties, size, shape, etc.). Snecma Propulsion Solide is the only European company with an industrial plant using these densification processes. They have developed specific densification processes for optimized preforms. For example, the pitch impregnation and pressure carbonization is associated with four-dimensional preforms, and CVI is well adapted to needled reinforcements. To make complex parts, the best compromise is to use resin impregnation for the shaping of the preform and then to finalize densification by the CVI process.

Snecma Propulsion Solide owns many installations: CVI furnaces, resin impregnation and pitch impregnation units, and carbonization furnaces at atmospheric and high pressure (up to 1000 bar and 1000 °C) (Figure 14.9). These facilities are able to manufacture complex and large parts with a 2.5 m diameter and 2.5 m length for carbon, as well as silicon carbide matrixes.

14.2.2
Carbon/Carbon Composites Applications

14.2.2.1 Solid Rocket Motors (SRM) Nozzles

The nozzle throat, which is defined as the area where the Mach number is equal to 1, is the most critical part of the SRM. As it has to withstand very high temperatures (>3000 K), high thermal gradient, extreme heat flux, combined with abrasive and chemical erosion, this part needs to be made from a very resistant material, such as tungsten, graphite, or carbon-carbon composites. Ten times lighter than tungsten, high thermal conductivity carbon based materials are excellent candidate materials. But as graphite is brittle, carbon fiber reinforced composites are nowadays standard materials for nozzle throats: they may be "densified" with a phenolic resin system (not carbonized before firing) or with a carbon matrix. For the three largest SRMs in operation, we observe that according to the increasing aluminum content in the propellant grain (Shuttle RSRM, Titan 4 SRMU, Ariane 5 MPS), the three nozzle throats are made of carbon/phenolic (RSRM), graphite/phenolic (SRMU), and finally carbon-carbon composites (MPS).

This selection of materials for large nozzle throats depends also on the capability to produce such large parts: during the Shuttle development phase, it was not possible to produce such large carbon-carbon composites parts at an affordable cost.

Novoltex carbon-carbon composites have been selected, eight years later, as the basic material for the Ariane 5 MPS, because carbon-carbon offered lower ablation rates and higher reproducibility of erosion. This point must be outlined because two boosters are fired at the same time, and need to deliver exactly the same thrust. If not, the thrust imbalance has to be corrected by the thrust vector control system, which is limited to a 6-degree deflection angle. Uncontrolled thrust imbalance drastically reduces the launcher performance, and may jeopardize flight safety. Up to now, development and qualification static firing tests, flight data recorded on 62 MPS, and margins checking after flight recovery all confirm the high level of safety and reproducibility provided by the carbon-carbon Ariane 5 MPS nozzle throat (Figure 14.10).

The exit cones have derived benefit from the thermo-structural capacities of carbon-carbon composites, while at the same time, the designs and components of the nozzles have evolved by the development of these materials.

The designs of exit cones have continuously evolved along with the SEPCARB, due to:

- increase of their mechanical capabilities;
- improvement of their knowledge (more detailed characterization);
- improvement of the prediction of their behavior (mainly with thermo-mechanical nonlinear codes);

Figure 14.10 The carbon-carbon nozzle throat of Ariane 5 Solid Rocket Booster [9].

- designs have been developed in order to meet specific operating requirements, not only for the hot exit cone, but also for the entire nozzle. The design and technology evolution of Snecma Propulsion Solide SRM exit cones is described in [2] and summarized in Table 14.1.

Carbon-carbon composites make it possible to reduce the overall mass of the nozzle, not only because of their low density, but also because of their physical properties adapted to simplified designs. That is why future advanced SRM nozzles feature monolithic nozzle throats – instead of several nozzle throat inserts – and monolithic exit cone replace metallic housing protected with bonded ablative liners (Figure 14.11). This is a good example of where an apparently "expensive" material does indeed help to cut overall cost, due to:

- simplification of the design;
- reduction of parts number;
- lowering of assembly and quality insurance manpower, and so on;
- increasing performances and reliability.

14.2.2.2 Liquid Rocket Engines (LRE) [9]

Traditionally based on high refractory metallic alloys, the technology evolved in the mid-1990s to carbon-carbon composites introduction in LRE design. The

Table 14.1 Reduction of weight for upper stages SRM exit cones.

Years	Thermal protection	Housing/thickness (mm)	Mass per unit area (kg m^{-2})
1970–1980	Carbon/phenolic (16 mm)	Metallic (4 mm)	40 to 35
1980–1997		C/epoxy	30
1985–2000	Carbon/Carbon (9 mm to 2.3 mm)		15 to 10

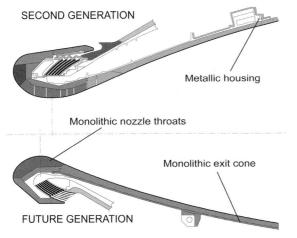

Figure 14.11 Nozzle design evolution of advanced SRMs.

Figure 14.12 RL10 B2 EEC during qualification test.

pioneer is the RL10-B2 cryogenic engine, which powers the Boeing Delta 3 (and future Delta 4) upper stage. Based on the Extendible Exit Cones (EEC) technology successfully demonstrated on several operational SRMs (ICBM Upper Stages, Shuttle IUS), and also on the RL10 A4 (with a small extendible ring made of thin Columbium), Pratt and Whitney decided to drastically enhance the performance of the RL10 engine by using a larger EEC for the RL10B2, and increasing the expansion ratio from 83 for RL10A4 up to 285 for RL10B2.

The resulting specific impulse is 464s, as demonstrated by two flights of the Boeing Delta 3 launcher. It may be underlined that this huge specific increase (>30s) is based on the EEC weighing only 92kg. This extendible nozzle is the largest carbon-carbon nozzle ever produced in the western hemisphere: and is now in current production at Snecma Propulsion Solide (Figures 14.12 and 14.13).

14.2.2.3 Friction Applications [10]

Carbon-carbon materials have proven superior to conventional materials for high performance brakes. For this application their main advantages are lightness, good friction characteristics even at high temperature, very good thermal capacity, high dimensional stability, high resistance to heat cycling, and low wear. The Sepcarb Novoltex C-C is widely used for brake disks, such as for the Airbus or Formula 1 (Figure 14.14).

14.3
Ceramic Composites

Because of the carbon sensitivity to oxidation, ceramic matrices have been developed, since the middle of the 1970s, to replace carbon matrices, in order to obtain materials of long lifetime capability, withstanding high thermal fluxes and mechanical loads, under oxidative environments.

14.3 Ceramic Composites

Figure 14.13 Extendible nozzle in production [9].

Figure 14.14 Carbon-carbon brake disks composites [10].

14.3.1
SiC-SiC and Carbon-SiC Composites Manufacture

Until 1992, SiC-SiC material families developed by Snecma Propulsion Solide were made of a two-dimensional reinforcement, using CG (Ceramic Grade) Nicalon from Nippon Carbon, with a pyrocarbon interphase [11], and a SiC matrix deposited by CVI.

14.3.1.1 Elaboration

For applications in oxidative environments, Snecma Propulsion Solide selected C-SiC or SiC-SiC composite material, obtained by:

- CVI process; and
- densification with combination of liquid impregnation and pyrolysis (PIP).

CVI-SiC is preferred to LPI-SiC (Liquid Phase Infiltration) because it provides a pure SiC matrix. LPI-SiC is not pure enough, so thermal stability is affected, as for SiC fibers. Thermal stability is higher with CVI-SiC (proved over 1800 °C) than with LPI-SiC.

The silicon carbide matrix is obtained by cracking of gas (methyltrichlorosilane) circulating through the pores of the perform, which is maintained at high temperature in a specific infiltration furnace.

A combination has been found using liquid densification and CVI processes, making CMC composites cost efficient materials. This combined process takes advantage of the liquid route (shape and simplified tooling) and the CVI route (mechanical characteristics and thermal stability of the pure SiC matrix). It is suitable to manufacture components such as flaps for aircraft engines, Thermal Protection Systems (TPS), and hot structures for re-entry vehicles.

14.3.2
SiC-SiC and Carbon-SiC Composites Applications

14.3.2.1 Aeronautical and Space Applications

These materials offered good strength at room temperature, of about 300 MPa, and a non-brittle behavior, with an enhanced failure strain of about 0.5 %.

In doing so, the opening of a wider range of applications were expected. The feasibility of different aero-engine parts (Figure 14.15), hot gas valve parts, thermal

Figure 14.15 SiC-SiC and C-SiC aeronautical applications [10].

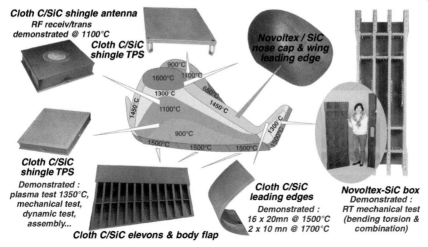
Figure 14.16 Hermes TPS [10].

structures, and TPSs based on carbon-ceramic or ceramic-ceramic materials, have been already demonstrated (Figure 14.16). Qualification of SEPCARBINOX material (carbon-SiC material with an enhanced finishing surface treatment) was pronounced for the Rafale M88-2 engine outer flaps in 1996, and its lifetime objective has been fully demonstrated and the flaps are in the series production phase (Figure 14.17).

14.3.2.2 Liquid Rocket Engines Applications [12]

Thermo-structural composites provide several advantages for LRE, due to their physical properties. Particularly, as exit cones, thermal protections, or thermal insulators are no longer necessary. Film cooling or dump cooling can be suppressed, thus increasing performances and simplicity.

That is why, for non-cooled LRE exit cones, which may be exposed to very hot and oxidative combustion gases (cryogenic or storable-propellant engines), oxidation resistant composites such as coated carbon-carbon composites, or CMC are preferred.

For example, Snecma Propulsion Solide has developed and successfully tested (in extra-nominal conditions) an experimental SEPCARBINOX exit cone for the HM7 cryogenic engine, which powers the third stage of Ariane 4 launcher (Figure 14.18). This exit cone is made of Novoltex Carbon reinforcement, associated with a carbon matrix and a silicon carbide (SiC) matrix. The weight is 24 kg compared to 84 kg for the metallic exit cone, including cooling equipment: this represents a 70% weight reduction, and 2 s more of specific impulse owing to the suppression of liquid hydrogen cooling flow mass. It has been tested successfully 10 times for 1650 s, and twice the second firing test has been 20% longer than the nominal burn time. The maximum measured temperature was over 2000 K in the forward part of the exit cone. After post-test expertise, no degradation was observed, thus

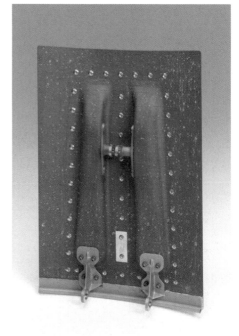

Figure 14.17 SEPCARBINOX flaps for rafale M88-2 engine.

Figure 14.18 HM7 Sepcarbinox C-SiC exit cone as fabricated and fitted on HM7 (Ariane 4 third stage).

demonstrating the reusability of such a SEPCARBINOX exit cone. From samples cut out on several locations of the exit cone, it was observed that the mechanical resistance was unaffected, and no oxidation detected even at microscopic scale. So these two full-scale firing tests have left unchanged the safety mechanical and thermal margins, and have given high confidence in the long-life capability of this kind of CMC.

Seven firings cumulating 900 s, including a 610 s duration test, were successfully performed. The chamber walls were heated to 1723 K (2640 °F) without any problems (no crack, no erosion, no leakage, etc.).

14.3.3
A Breakthrough with a New Concept: The Self-Healing Matrix

In the early 1990s, development of advanced ceramic materials was initiated with the ambitious objective to develop composites, carbon, or SiC fiber reinforced, useable above their mechanical yield point for a temperature range up to 1373 K, and capable of an operating life exceeding 1000 h in an oxidative atmosphere integrating 100 h in severe conditions (stress level up to 120 MPa).

However, testing in more severe conditions, such as tensile/tensile Low Cycle Fatigue LCF tests performed at a stress level of 120 MPa led to early failures due to the insufficient sealing of the matrix micro-cracks, leading to premature oxidation of the interphase. This was illustrated by a lifetime duration of about 10 to 20 h at 873 K and less than 1 h at 1123 K.

The material concept was mainly based on the use of the novel self-healing matrix and the use of a multilayer woven reinforcement to reduce delamination sensitivity during the manufacturing process and during material operating conditions involving high shear stresses.

14.3.3.1 Manufacturing of Ceramic Composites

Reinforcement   Planar multilayer reinforcements called Guipex preforms have been developed (Figure 14.19). The considered technology is applicable either to carbon or to Hi-Nicalon fiber reinforcements. Guipex are made of layers linked together. The number of layers is adjustable, according to the required composite thickness, and can be typically between 2 and 7 mm.

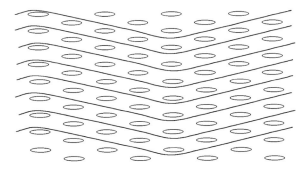

Figure 14.19 A Guipex performs.

14.3.3.2 The Self-Healing Matrix

The self-healing matrix is obtained by the cracking of gas containing boron-silicon-carbon molecules circulating through the pores of the perform, which is maintained at a high temperature in a specific infiltration furnace. In the first development phase [13, 14], the concept, based on a novel self-healing matrix, was applied to a SiC multilayer reinforcement (CERASEP A410). The improvement in the behavior under oxidative atmospheres, induced by the self-healing matrix, has made carbon reinforcement use also possible (SEPCARB-INOX A500), without anti-oxidation finishing treatment. In addition to its thermo-mechanical potential, the carbon reinforcement economical impact makes this material very attractive [15].

14.3.3.3 Characterization

CERASEP A410 and SEPCARB-INOX A500 Description The reinforcement of both A410 and A500 materials uses a multilayer fabric with a link between the different layers. In order to avoid the delamination sensitivity, this link is completed during the weaving process, to obtain the GUIPEX preform (Figure 14.19) available with carbon and ceramics fibers and a 1.5 to 7 mm thickness range. The matrix is deposited by the CVI process. A hardening first step in graphite toolings assumes the shape, including a tailored interphase treatment to guarantee the thermo-mechanical behavior and the duration life in oxidative environment. The self-healing matrix, combining specific phases of CVI process deposited carbide layers, eliminates the finishing treatment. By its volumic aspect, the self-healing matrix approach enhances the damage tolerance of the composite, in comparison with the finishing treatment, the oxidative species being trapped in the whole thickness matrix by a glassy phase formation sealing the micro-cracks (Figure 14.20).

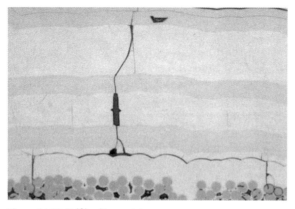

Figure 14.20 Self-healing matrix sequences.

A410 and A500 Mechanical and Thermal Database [16] The A410 and A500 characteristics are measured on test coupons machined from standard $200 \times 200\,\text{mm}^2$ plates; the thermo-mechanical behavior of these self-healing matrix composites is similar to SiC/SiC and C/SiC materials, obtained with a monolithic SiC matrix, the improvement being about the duration life under oxidative environment (Figure 14.21, Tables 14.2 and 14.3).

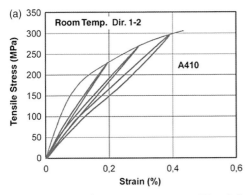
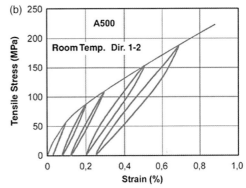

Figure 14.21 Tensile stress-strain curves at RT and dir. 1 and 2: (a) Cerasep A410; (b) Sepcarbinox A500.

Table 14.2 A410 and A500 physical properties.

Materials	Fiber vol. fraction	Density (g cm^{-3})	Porosity (%)
A410	35	2.20–2.30	12–14
A500	40	1.90–2.10	12–14

Table 14.3 Main characteristics ruling thermomechanical behavior.

CERASEP A410		SEPCARBINOX A500	
σ1 (MPa)	315	σ1 (MPa)	230
ε1 (%)	0.50	ε1 (%)	0.80
E1 (GPa)	220	E1 (GPa)	65
α1 (E-6/°K)	4–5	α1 (E-6/°K)	2.5
λ1 (W/m.k)	10	λ1 (W/m.k)	10
λ3 (W/m.k)	3	λ3 (W/m.k)	3
E.α (at 1000 °C)	~800	E.α (at 1000 °C)	~170

The combination of thermal properties and elastic modulus of A500 material reflects a lower sensitivity to thermal gradients compared with A410 material, attractive for components sustaining heterogeneous thermal fields.

A410 and A500 Lifetime Duration in Oxidative Environment [17]   The tensile fatigue and creep behavior have been characterized under air (Figures 14.22–14.24).

The improvement induced by the self-healing matrix is illustrated by the gap between the performances of the Cerasep A400 (Nicalon fibers and self-healing

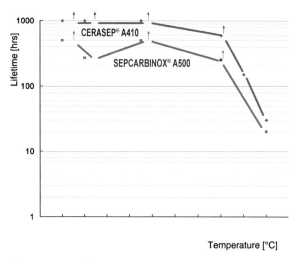

Figure 14.22 T/T fatigue at 120 MPa in air (0.25 Hz).

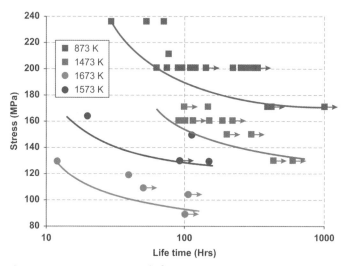

Figure 14.23 Cerasep A410 tensile fatigue (0.25 Hz) and creep in air.

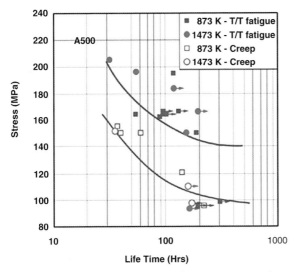

Figure 14.24 Sepcarbinox A500 tensile fatigue (0.25 Hz) and creep in air.

matrix) and those of the Cerasep A373 (Nicalon fibers and monolithic SiC matrix); the last enhancement is obtained by Hi-Nicalon fibers introduction in Cerasep A410. The self-healing matrix effect is also noteworthy with the A500 Sepcarbinox (carbon fibers and self-healing matrix).

A high potential for the Cerasep A 410 is demonstrated at 850 °C (several thousands of hours with stresses up to 150 MPa without rupture); for the Sepcarbinox A500, the lifetime duration is above 100 h for 160 MPa stress level in fatigue and 110 MPa in creep.

14.3.4
Representative Applications of These New Materials

The initial objective of the CMC development was to enlarge the gas turbine engine operating temperature field, replacing the refractory alloys limited by their melting point. Another approach is carried out, the Cerasep A410 and Sepcarbinox A500 breakthrough standing on their weight saving capability and their long life duration performances under high thermal gradients. The following representative applications illustrate this orientation.

14.3.4.1 Military Aeronautical Applications

These new CMCs have been identified as an excellent choice for replacing the metallic hardware in exhaust nozzle, such as the F100-PW-229 engine, which powered F15 fighters [18] [19].

Divergent flap and seals are typically manufactured from nickel-based superalloys, such as René 41 or Waspalloy. The severe thermo-mechanical environment produces extensive cracking in the metallic components, while the high temperatures produce excessive creep deformation. Six seals for the F100-PW-229, with two different cross-sections, were made: two of A410 with variable cross-section, two of A410 with a constant cross-section and two of A500 with constant cross-section (Figure 14.25).

After various testing phases carried out at several locations, the different seals installed on the engines in hot streak locations (i.e. in the worst conditions) have accumulated between 4600 and 6000 Total Accumulated Cycles (TACs), involving 1300 to 1750 engine operating hours, and including approximately 100 hours of afterburner operation and more than 15 000 afterburner lights. Testing was well beyond the full-life objective of 4300 TACs, and significantly exceeds the life of the current metal hardware. The seals experienced no delamination, no wear, and appeared to be in excellent condition after ground testing. One seal each of A500 (4600 TACs) and A410 (4850 TACs) were cut into tensile specimens for comparison with the as-processed materials data base. This residual strength testing showed no loss for A410 and only 6% loss for A500 (Tables 14.4 and 14.5). These are excellent results and highlight their exceptional durability in an exhaust nozzle environment.

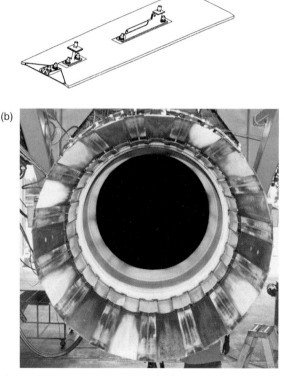

Figure 14.25 F100-PW-229 CMC seal schematics.

Table 14.4 RT tensile properties [20].

Plate number	UTS (MPa)	Elongation (%)	Modulus (GPa)
1324	200	0.73	84
1234	243	0.81	84

Table 14.5 A500 CT seal (4609 TACS) post engine test mechanical properties [20].

Sample No.	UTS (efficient) (MPa)	Strain to failure (%)	Young's modulus (GPa)
1	181	0.69	66
2	183	0.59	78
3	178	0.77	68
4	189	0.89	61
5	For morphological analysis		
6	192	0.74	65
7	197	0.95	62
8	198	0.68	69
9	197	0.89	73
10	205	0.68	82
11	198	0.71	85

The next step is flight demonstration. In parallel, work is continuing to reduce the cost of CMC materials by a factor of two.

14.3.4.2 Commercial Aeronautical Applications

A recent application of the SiC/Si-B-C composites concerns civil planes with mixed flow (BR 710–715, RB 211, PW 500, SaM146 engines) or separated flow (CFM56-7 or CFM56-2-3 engine) nozzle. These parts meet the aircraft manufacturers' expectations for high performances and competitive nacelles.

They are made of Inconel 625 or 718 and have a 700 to 1000 mm diameter and 400 to 500 mm length. A first demonstration has been made with a CFM56–5C engine showing good behavior of the SiC/Si-B-C composite after 600 cycles and 200 engine operating hours, including 70 take-off hours. The mixer (Figure 14.26) permits a weight gain of 35%; this gain can reach 60% in the case of TP400 engine for A400 M plane.

The complex shape of this kind of part can necessitate the use of liquid way hardening with a pre-ceramic resin associated with the CVI process for Si-B-C matrix. Parallel to the technology demonstration with scale one part, studies are in progress to satisfy economical and technical requirements.

(a)

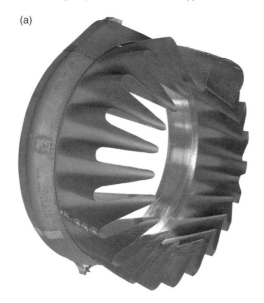

(b)

Figure 14.26 (a) CMC mixer; (b) integrated on CFM56-C engine [8].

References

1 Cavalier, J.C. and Christin, F. (1992) Les Techniques de l'Ingénieur A 7 804.
2 Cullerier, J.L. (1992) GEC Alsthom Technical Review N° 8, p. 23.
3 Choury, J.J. (1976) C-C materials for nozzles of solid propellant rocket motors. AIAA paper 76-609.
4 Maistre, M.A. (1976) Development of a 4D reinforced carbon-carbon composite. AIAA paper No. 76-607.
5 Broquere, B. (1997) C-C nozzle exit cones: SEP's experience and new developments, AIAA paper 97-2673.
6 Michel, B. (2006) Advanced technology SRM nozzles – history and future, AIAA 2006-4596.
7 Gachet, C., Louarn, V., Blein, J. and David, P. (2005) US, France, Japan, Carbon-Carbon Composites Technical Exchange (C3TEX), Washington DC, June 15–17.

8. Cavalier, J.C., Berdoyes, I. and Bouillon, E. (2006) Composites in aerospace industry, CIMTEC.
9. Broquère, B. H. and Fischer, R. (2001) Composites for propulsion: improvement of performances in rocket propulsion by the use of carbon-carbon materials, National workshop on advanced composite material and processing, Thiruvananthapuram—February 10.
10. Christin, F. (2004) A global approach to fiber n-D architectures and self sealing matrices: from research to production; HTCMC5.
11. Jouin, J. M., Cotteret, J. and Christin, F. (1991) Designing ceramic interfaces: understanding and tailoring interface for coating, composite and joining applications, Second European colloquium, November 11–13, Institute for Advanced Materials, Commission of the European Communities, Petten, The Netherlands.
12. Pouliquen, M.F., Soler, E. and Melchior, A. (1987) Thermostructural composite materials for liquid propellant rocket engines, AIAA-paper 87-2119
13. Bouillon, E., Abbé, F., Goujard, S., Pestourie, E. and Habarou, G. (2000) Mechanical and thermal properties of a self-healing matrix composite and determination of the life time duration, 24th Annual Conference on Composites. *Advanced Ceramics, Materials and Structures*, 21, 459–67.
14. Bouillon, E., Lamouroux, F., Baroumes, L., Cavalier, J.C., Spriet, P. and Habarou, G. (2002) An improved long life duration CMC for jet aircraft engine applications, ASME paper GT-2002-30625.
15. Bouillon, E., Spriet, P., Habarou, G., Louchet, C., Arnold, T., Ojard, G.C., Feindel, D.T., Logan, C.P., Rogers, K. and Stetson, D.P. (2004) Proceedings of ASME Turbo Expo Land, Sea, and Air GT2004 Vienna.
16. Rougès, J.M., Bertrand, S. and Bouillon, E. (2006) Long life duration to CMC materials, ECCM12, Biarritz.
17. Bouillon, E., Spriet, P., Habarou, G., Arnold, T., Ojard, G., Feindel, D., Logan, C., Rogers, K., Doppes, G., Miller, R., Grabowski, Z. and Stetson, D. (2003) Engine test experience and characterization of self sealing ceramic matrix composites for nozzle applications in gas turbine engines, June 16–19, Atlanta, Georgia, USA, GT2003-38967.
18. Bouillon, E., Spriet, P., Habarou, G., Louchet, C., Arnold, T., Ojard, G.C., Feindel, D.T., Logan, C.P., Rogers, K. and Stetson, D.P. (2004) Proceedings of ASME Turbo Expo Land, Sea, and Air, GT2004-53976, June 14–17, Vienna.
19. Zawada, L., Richardson, G. and Spriet, P. (2005) HTCMC-5-28.
20. Zawada, L., Bouillon, E., Ojard, G., Habarou, G., Louchet, C., Feindel, D., Spriet, P., Logan, C., Arnold, T., Rogers, K. and Stetson, D. (2006) *Manufacturing and flight test experience of ceramic matrix composite seals and flaps for the F100, gas turbine engine*, Proceedings of ASME Turbo Expo 2006, Power for Land, Sea, and Air, GT2006-90448, May 8–11, Barcelona, Spain.

15
CMC for Nuclear Applications

Akira Kohyama

15.1
Introduction

Fiber reinforced composite materials have been historically and classically tailored structural materials appearing in many old documents. Some still remain on the Earth, such as the Great Wall of China. The concepts of fiber reinforced materials as advanced tailored materials were developed in the last half of the twentieth century and many extensive R&D efforts have been performed during this period [1, 2]. The major part of these efforts has been in aerospace application and we see many spin-offs in our daily life.

Many outstanding accomplishments with fiber reinforced composite materials have been expanding their areas in technological applications, which start from mild conditions to very severe environments. One example is the quest for very or ultra high temperature applications, where materials are to be changed from plastics and metals into inter-metallics and ceramics. Over the last few decades, R&D efforts on ceramic composites have been very extensive, especially in the fields of aerospace and energy. Among them, carbon fiber reinforced carbon composite (C/C) and silicon-carbide fiber reinforced silicon-carbide composite (SiC/SiC) R&Ds have been much emphasized in nuclear energy research [3, 4].

In order to provide sufficient energy to our society, without strong impacts on the environment, economically competitive and stable core energies are crucial needs. Nuclear fission and fusion are believed to be very attractive options [5, 6] and many efforts have been extensively carried out worldwide. In the field of fission, reactor R&D activities have a wide spectrum from near term reactors to generation IV reactors [7]. Although there has been much progress in conceptual design study, supporting activities from engineering and materials have been premature and insufficient.

The materials R&D methodology has been unique in the case of ceramic fiber reinforced ceramic matrix composite (CMC) materials, such as C/C, C/SiC, and SiC/SiC, where ceramic fibers, ceramic matrix, and interphase connecting fiber and matrix are the three key components. Materials design is carried out by optimizing these components to meet the material's requirements.

Ceramic Matrix Composites. Edited by Walter Krenkel
Copyright © 2008 WILEY-VCH Verlag GmbH & Co. KGaA, Weinheim
ISBN: 978-3-527-31361-7

In this chapter, starting with the brief introduction of ceramic fiber reinforced CMCs, material design, process engineering, and performance evaluation methodologies toward the application to GFR are presented.

15.2
Gas Reactor Technology and Ceramic Materials

The Generation I.V. Reactor International Forum (GIF) is the international activity for developing the generation IV reactors [7]. There are six reactor types, including the very high temperature reactor (VHTR) and gas cooled fast reactor (GFR), which are gas reactors using gas as coolant and nuclear heat as transporting media.

Figure 15.1 shows the ongoing R&D strategies and issues resulting from the development of the prismatic modular reactor (PMR) and the pebble bed reactor (PBR), to VHTR and GFR as the final goal. In the case of PMR and PBR, reactor materials and He system technology are available from industrial metallic materials and current technologies. But for the VHTR and GFR, very or ultra high temperature materials are essential for them to remain attractive as advanced core energy systems. The key materials for these reactors are reactor core structural materials, where neutron absorption, thermal conductivity, and fusion (melting or decomposition) temperatures are key requirements. Intermediate heat exchanger (IHX) is another important area for materials R&D, where radiation damage tolerance is not an important issue. However, in these applications there are many similarities to R&D, and many R&D activities have been coordinated, both for core structure and IHX applications.

Table 15.1 indicates the characteristics for forming potential ceramic materials for GFR and VHTR. The table shows that neutron absorption is required to allow

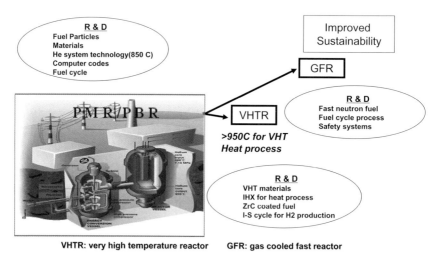

Figure 15.1 Gas reactor technology and R&D pathway [8].

15.2 Gas Reactor Technology and Ceramic Materials

Table 15.1 Choice of materials for GFR and VHTR.

	Material	Neutron absorption	Thermal conductivity	Fusion temp. (°C)		Material	Neutron absorption	Thermal conductivity	Fusion temp. (°C)
Carbides	Sic (α + β)	○	○	○ (2972)	Silhicides	MoSi$_2$	×	○	○ (2050)
	ZrC	○	○	○ (3400)		TaSi$_2$	×	○	○ (2200)
	Tic	○	○	○ (3100)		WSi$_2$	×	○	○ (2165)
	VC	○	○	○ (3810)		TiSi$_2$	○	○	× (1540)
	TaC	×	○	○ (3800)		ZrSi$_2$	○	○	× (1520)
	WC	×	○	○ (2900)		HfSi$_2$	○	○	× (1750)
	HfC	×	○	○ (3800)		VSi$_2$	○	○	× (1660)
Oxides	Al$_2$O$_3$	○	×	○ (2050)	Nitrides	ZrN	○	○	○ (2952)
	MgO	○	×	○ (2832)		T N	○	○	○ (2950)
	ZrO$_2$	○	×	○ (2370)		AN	○	○	○ (2227)
	Y$_2$O$_3$	○	×	○ (2427)		TaN	×	○	○ (3087)
	SiO$_2$	○	○	× (1470)		Si$_3$N$_4$	○	○	× (1827)

Note: ○, acceptable; ×, unacceptable.

for a sufficient fission reaction and especially for GFR breeding reactions, which should be larger than unity. Thermal conductivity is an important parameter for reactor core design and has a strong impact on the attractiveness of VHTR and GFR, not simply because of this property. From Table 15.1, SiC, ZrC, TiC, VC, ZrN, TiN, and AlN can be recognized as potential candidates, but nitrides are not preferable from a high induced radioactivity viewpoint.

Furthermore, for the application for reactor core reliability with a reasonable safety margin is desirable and monolithic ceramics have an inherent character of brittleness. The proposed method to overcome this very brittle nature, which causes catastrophic fracture or sudden loss of system performance, is to manufacture fiber reinforced composite materials.

15.3
Ceramic Fiber Reinforced Ceramic Matrix Composites (CFRC, CMC)

Ceramic fiber reinforced composites are usually known as CFRC or CMC. Carbon fiber reinforced carbon matrix composites are a kind of CMC, but are usually called CC or CC composites. In this chapter, the main emphasis is on SiC/SiC composite materials. CMC has three major components, which are fiber, interphase, and matrix. The brief introduction of these components and major processing are shown in Figure 15.2.

In the case of reinforcing fibers, there are whisker, short fiber, and long (continuous) fiber. In this chapter, only long fiber reinforcement is introduced, for this type of CMC is supposed to be the potential candidates for nuclear fission and

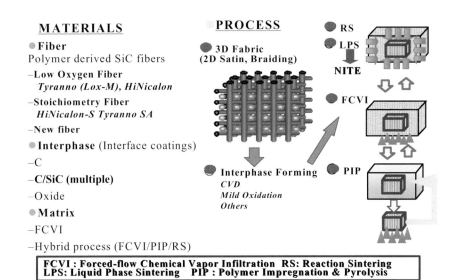

Figure 15.2 CMC: components and processing.

fusion. To make flexible fiber architecture, flexibility of the fiber, which enables it to be woven or to be blazed to make fabrics or textures, is essentially required. Polymer derived SiC fibers and C-fibers meet this requirement, thus these fibers are mainly used for advanced CMCs. There has been much progress in advanced fibers and now new fibers such as Tyrano-SA and Hi-Nicalon-S have been produced (Table 15.2). Now a new fiber from large continuous production line, Cef-NITE, is also available, which retains its integrity at more than 1400 °C.

Interphase is another important component to give pseudo-ductility to CMCs. The Chemical Vapor Deposition (CVD) process is mainly used for interphase coating, including the multilayer coating method, which is a repeating CVD process to form SiC and C multilayers. Other potential methods to make C-enriched surface layers of SIC fiber are polymer pyrolysis and thermal treatment. Oxide or nitride interphase formation is another option. However, in the case of high temperatures and under severe nuclear environments, oxide and nitride are not adequate for carbide ceramics from micro-structural stability and induced radioactivity viewpoints.

Matrix formation is the final stage in the process of making CMC. To fill the open space within the fiber structure with a fiber coating (with interphase at the surface of the fibers) by different methods, is the so-called densification process of matrix. As shown in Figure 15.2, the FCVI process has been recognized as the most reliable high-quality method for densification. The CVI method is the most established with high crystallinity, high purity, and near stoichiometry, but limitations in shape, geometry, and porosity are issues for many applications. Other methods, such as RS, LPS, PIP, and their hybrid processes, are attractive options in the manufacture of various kinds of CMCs. PIP has issues in baseline properties, crystallinity, and stoichiometry, but process improvement and near-stoichiometry polymer development are in progress. The melt infiltration (MI) method or RS method has issues in phase control and uniformity control, but improvement is in progress by increasing its attractiveness. LPS has similar issues to the MI/RS method.

The typical example of the process development is based on the liquid phase sintering (LPS) process modification, where the new process called Nano-powder Infiltration and Transient Eutectic (NITE) process has been developed [8].

Table 15.2 Characteristics of representing SiC fibers.

SiC fiber	C/Si atomic ratio	Oxygen content (wt.-%)	Tensile strength (GPa)	Tensile modulus (GPa)	Elongation (%)	Density (g cm^{-3})	Diameter (μm)
Tyranno SA Gr.3	1.07	<0.5	2.6	400	0.6	3	7
Hi-Nicalon Type-S	1.05	0.2	2.6	420	0.6	3.1	11
Hi-Nicalon	1.39	0.5	2.8	270	1.0	2.74	14

15.4
Innovative SiC/SiC by NITE Process

Because of its superior chemical stability at very high temperature, inherent heat resistance, and irradiation stability, and low activation properties under radiation environment, SiC/SiC has been considered as an attractive option for advanced nuclear energy systems. The extensive recent works are making their way toward high crystallinity constituents. Due to improvements in reinforcing SiC fibers and availability of fine nano-SiC powders, the well-known LPS process has been drastically improved to become a new process called the NITE Process.

As indicated in Figure 15.3, slurry of SiC nano-powders and additives is infiltrated into SiC fabrics and dried for making prepreg sheets. After the lay-up of the sheets, hot pressing is applied to make NITE SiC/SiC. For the near-net-shaping process to make tubes, pipes, turbine blades, and blanket for fusion reactors, this basic process is modified to include prepreg wire production, filament winding, or three-dimensional fabric weaving and pseudo-HIP processes. To keep the advantage of the NITE process, the following are essential and to satisfy these requirements in industrial fabrication, line production is still under way. These are:

- use near stoichiometry SiC fiber with high crystallinity;
- make protective interface by fiber coating of carbon and SiC;
- use SiC nano-powders with appropriate surface characteristics [9].

One of the greatest advantages of the NITE process is its flexibility in shape and almost no limitation on size.

Figure 15.4 gives some examples of composite materials made by the NITE process. About one liter in volume of the two-dimensional SiC/SiC composite

NITE (Nano-Infiltration Transient-Eutectic Phase) Process
Dense and robust structures (cf. PIP, CVI, …)
Excellent Gas Tightness
Fairly high thermal conductivity, Chemical stability
Flexibility in Size and Shape
Applicability of existing net-shaping techniques
Low production cost

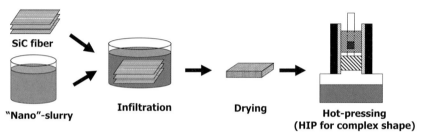

Figure 15.3 The concept of NITE process.

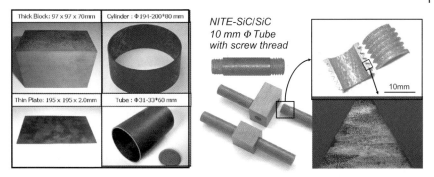

Figure 15.4 Shape flexibility of NITE SiC/SiC composites.

cubic blocks was successfully produced (Figure 15.4, upper left), where no cracks or cavities were detected by the naked eye. The upper right of Figure 15.4 shows the real size model of a 100 kw gas turbine combustor liner. The lower left of the figure shows a 2 mm thin plate of two-dimensional SiC/SiC. In these materials, basic properties have been measured. They all presented excellent results in high density, high crystallinity, high thermal conductivity, and basic mechanical properties. Process improvement and optimization with the emphasis on maintaining sound protection interface are current technical challenges.

15.5
Characteristic Features of SiC/SiC Composites by NITE Process

The outstanding total performance of NITE SiC/SiC composite material is based on the highly crystalline and highly dense microstructure.

Figure 15.5 shows a low magnification SEM image of a cross-section perpendicular to the fiber axis (top left image) and a high magnification TEM image of a fiber-interface matrix. These images present the full dense and small crystalline microstructure of SiC with carbon interface. The selected area diffraction images (right) clearly define that fiber and matrix are highly crystalline beta-SiC and interface is pyrolytic carbon. These micro-structural features provide excellent thermal stress values in merit and high hermetic property presented by helium permeability, as shown in Figures 15.6 and 15.7, respectively.

Figure 15.6 represents the potential of thermal stress tolerance, M, the thermal stress value of merit, defined by the equation in the figure, where σ_U = Ultimate tensile strength, K_{th} = Thermal conductivity, v = Poisson's ratio, E = Young's modulus, and α_{th} = Thermal expansion coefficient.

A higher M value indicates the excellence in thermal stress tolerance and pure Al comes out the worst in this figure in almost all temperature ranges. The conventional and commercially available CVI-SiC/SiC is better than Cerasep N3-1, which is similar to Titanium alloy 318 at lower temperatures. 8Cr-2W steel (F82H) is one of the candidates to reduce activation ferritic steels for fusion reactor. F82H

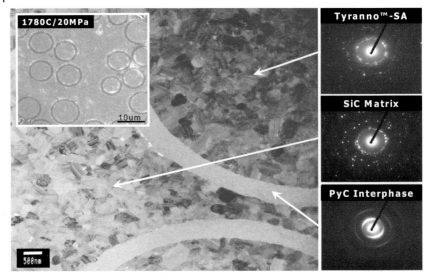

Figure 15.5 Microstructure of NITE SiC/SiC.

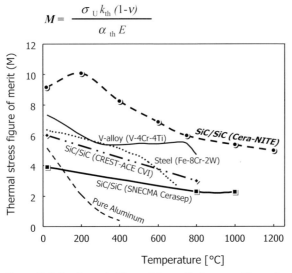

$$M = \frac{\sigma_U k_{th} (1-v)}{\alpha_{th} E}$$

Figure 15.6 Excellence in Thermal Stress Resistance – Thermal stress figure of merit: M.

steel presents excellent properties, especially below 600 °C [7]. V-alloy is similar to steel below 500 °C, but above that temperature the excellence becomes clear with other materials. The SiC/SiC composite produced by CVI is similar with steel, but SiC/SiC made by the NITE process, Cera-NITE, has a very high M value from room temperature to 1300 °C. This high M value has an enormous impact on the design of high temperature components, such as the gas turbine combustor liner,

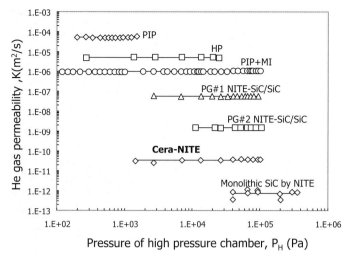

Figure 15.7 Improvement in hermeticity of NITE-SiC/SiC.

turbine blade, fuel pin, reactor core components, and heat exchangers. For high temperature gas system application, gas leak tightness or hermetic property is important, but unfortunately ceramics are well known as inferior materials from a hermeticity viewpoint, especially for ceramic composite materials with high porosity and micro-cracks. The NITE SiC/SiC is becoming the first leak-tight ceramic fiber reinforced matrix composite.

To be used for the shield fuel pin of GFR, gas tightness, hermeticity, and shielding capability of FP are essential. This property is also essential for gas cooling system or heat exchange components using gas, such as fusion reactor blankets and the IHX for VHTR. Ceramic composite materials have been widely recognized as poor gas tightness materials and cannot be used for gas systems unless there are applied gas shield coatings or cladding. Owing to the excellent near theoretical density character of NITE-SiC/SiC, He gas permeability was measured. Figure 15.7 indicates the progress of gas tightness improvement during the pilot grade production of NITE-SiC/SiC. The more than seven orders of magnitude superiority of He gas permeability of monolithic NITE-SiC, compared with other SiC or C by other methods, has not been attributed. The number 3 pilot grade products of NITE-SiC/SiC are more than five orders of magnitude better than other ceramic composites [10].

There still remains one large concern about the gas tightness of ceramic composite materials, even if the gas tightness satisfies the design requirement from the beginning. This is the well-known long-term concern about the weak characteristics of the ceramic materials to be easily destroyed under service conditions. In order to verify the stability of the gas tightness feature, the effect of heat cycling was investigated [10] and further effects of tensile deformation at higher than the proportional limit strength (PLS) have been studied [11]. Figure 15.8 shows the

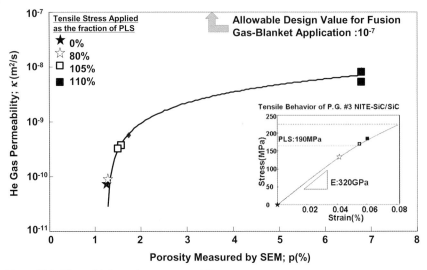

Figure 15.8 Effect of deformation on permeability.

result of the tensile deformation effect study, where porosity enhanced or introduced by tensile deformation was measured and the He gas permeability was plotted against the porosity.

The porosity dependence behavior is similar to the data for the porous ceramics as the function of porosity. This may suggest that the micro-cracks induced by plastic deformation being slightly higher than the PLS may not have a great contribution. In these materials, the inter-bundle porosity was about the order of 0.01% and these values stayed below 0.02%, even after the deformation by 1.2 × PLS where intra-bundle porosity of 0.075% before deformation reached almost 6%. Those data are encouraging, showing that the NITE-SiC/SiC is stable enough to keep He permeability, even under deformation such as micro-crack initiation and propagation of proportional limit stressing.

15.6
Effects of Radiation Damage

There have been many research activities into SiC since the early stage of Fast Breeder Reactor R&D projects, where SiC was recognized as easily degraded by even the smallest amount of neutron radiation. These results have been discouraging to ceramic material researchers in the past.

Due to progress in high purity ceramic production and in highly crystalline and stoichiometry fiber production, radiation damage studies have been enhanced in fusion reactor materials programs in the past few decades.

15.6 Effects of Radiation Damage

The major trends recognized are:

- Irradiation-induced strengthening is observed in most experiments.
- Irradiation-toughening is apparent, despite the elastic modulus decrease.
- Increase in data scatter is very significant after irradiation.
- Enhanced fracture toughness and blunting are both contributing to the apparent strengthening.

15.6.1
Ion-Irradiation Technology for SiC Materials

An ion irradiation experiment is a powerful tool, where irradiation conditions and high temperature flexibility and controllability are excellent. High damage rate capability is another advantage for ion irradiation. The first systematic research of the point-defect swelling of SiC was performed by the ion-irradiation method at DuET Facility, Kyoto University in Japan [12, 13]. Figure 15.9 introduces the DuET facility and evaluation methods of swelling and three-dimensional crack development. Because the charged particles rapidly lose their kinetic energy by the electric and nuclear stopping power in a material, the irradiated ions cannot penetrate deeply into a material [14]. The damaged range is limited near the surface, thus the development of investigation techniques is also important. The depth profiles of displacement damages and ion concentration are normally calculated by TRIM codes (Figure 15.10) [15], where features of radiation damage characteristics are also indicated. A nano-indentation technique combined to the ion irradiation experiments is a new method to investigate the crack propagation and fracture toughness of SiC [16].

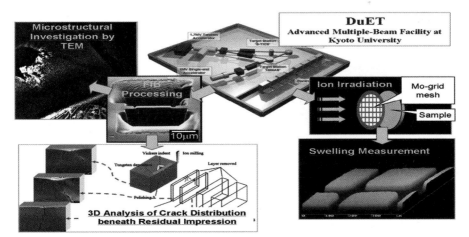

Figure 15.9 DuET facility and investigation techniques for ion irradiation researches at Kyoto University.

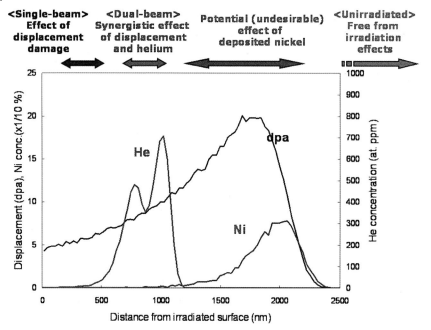

Figure 15.10 A depth profile calculation of displacement damage and ion concentration in SiC using TRIM.

15.6.2
Micro-Structural Evolution and Swelling

Accumulation of point defects introduced by displacement damage is the origin of swelling in SiC. Though the neutron irradiation data had been accumulated since 1969, some data relating to the swelling of SiC was misleading. It was difficult to understand the correct trend of swelling from only neutron irradiation data, due to their relying on irradiation condition data. The ion-irradiation research clarified the trend of point-defect swelling of (3C) CVD-SiC up to 1400 °C. Figure 15.11 shows the temperature dependence of saturation (3 dpa irradiated) swelling by ion irradiation superimposed on the neutron published irradiation data [17].

The swelling of SiC is divided to three regimes, which are the amorphization range below 200 °C, the point-defect swelling regime between 200 and 1000 °C, and the void swelling regime over 1000 °C. At lower temperatures, as the point-defect accumulation is much heavier than the annihilation of defects, the strain by defect accumulation causes amorphization [18]. The amorphization of SiC depends on the irradiation temperature, dose, and irradiated particles. Heavier charged particles, such as Si ions and Xe ions, are easier to amorphize SiC than electrons (Figure 15.12) [19].

The volume swelling by amorphization increases to over 10% at room temperature. In the point-defect swelling regime, the point defects accumulate and cause

15.6 Effects of Radiation Damage

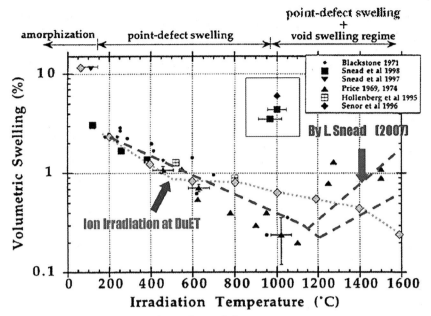

Figure 15.11 Temperature dependence of point-defect swelling investigated by ion irradiation experiments [17].

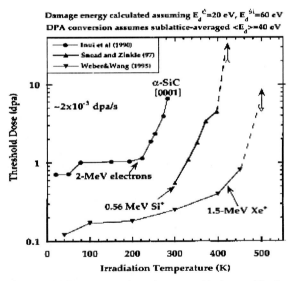

Figure 15.12 Temperature dependence on critical amorphization for SiC [19].

the swelling. A balance of volume changes by an interstitial and a vacancy causes the swelling in this temperature range. The dose dependence of point-defect swelling was investigated by ion irradiation experiments at DuET using 5.1 MeV Si ions. The point-defect swelling increases with increasing dose and saturates at 1 to 3 dpa

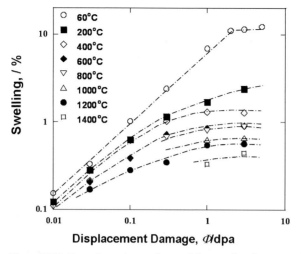

Figure 15.13 Dose dependence of point-defect swelling by ion irradiation on SiC [20].

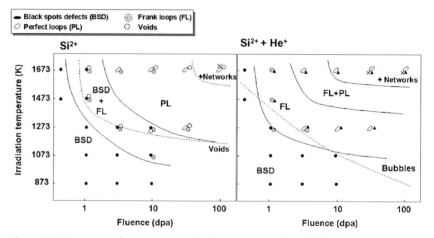

Figure 15.14 Summary of microstructure developments in irradiated SiC.

(Figure 15.13) [20]. The saturation swelling of irradiated CVD-SiC decreases due to the annihilation of defects and clustering of interstitials.

Based on the systematic ion irradiation study at Kyoto University, under the collaboration with IEA, the Fusion Materials Working group for SiC/SiC and the Japan-US collaboration program for fusion materials, the micro-structural evolution with irradiation temperature and dose is summarized (Figure 15.14) [21].

15.6.3
Thermal Conductivity

Thermal conductivity is also an important key property for nuclear applications. Although the swelling and the mechanical properties under irradiation environments have been studied extensively, research into the radiation effects on thermal properties has been limited. The discussion of thermal conductivity for nuclear applications might be complex because a SiC/SiC composite has many factors that affect its thermal conductivity. In properties of the composite bases on the solid-state physics of SiC, recent studies have revealed that the physics of thermal conductivity of pure-SiC changes by irradiation [22, 23]. The thermal conductivity of a covalent crystal, SiC, depends on a phonon which relates to the lattice vibration in crystals. The phonon is scattered by other phonons, defects, dislocation, grain boundaries, and so on. In the case of ceramics designed to have high thermal conductivity, the existence of even a few irradiation defects drastically affect phonon transport at low temperature, and saturate the thermal conductivity. Figure 15.15 shows the thermal conductivity at room temperature of CVD-SiC neutron irradiation.

The neutron irradiation was performed at HFIR in ORNL with the dose of 4 to 8 dpa. The thermal conductivity of non-irradiated CVD-SiC was 327 ± 25 W/mK. Though the post irradiation thermal conductivity increases with irradiation temperature up to 1468 °C, the highest value is ~111 W/mK at 1468 °C, which is 34% of the un-irradiated SiC [22]. The thermal conductivity reduced to about 10 W/mK at low temperatures, the defect drastically affected by the reduction of thermal conductivity. The thermal defect resistance, defined as $1/K_{rd} = 1/K_{irr} - 1/K_{unirr}$, where $1/K_{rd}$ is the thermal defect resistance, and K_{irr} and K_{unirr} are the values of thermal conductivity of irradiated and un-irradiated materials, respectively. The thermal defect resistance clearly relates to the swelling in the point-defective regime (Figure 15.16).

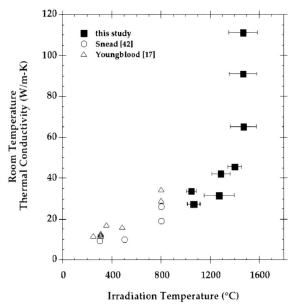

Figure 15.15 Thermal conductivity measured at room temperature of neutron irradiated CVD-SiC [24].

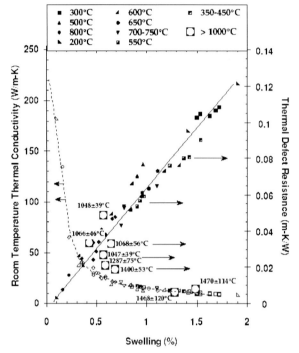

Figure 15.16 The relation among thermal defect resistance, thermal conductivity at RT and swelling of CVD- SiC [24].

The correlation of swelling and the thermal defect resistance does not go above 1100 °C, which is the void swelling regime of SiC. The voids are less effective on phonon scattering than the point-defect strain. The thermal conductivity of SiC/SiC composites becomes worse than the monolithic CVD-SiC because the SiC/SiC composites include many reinforcing fibers, pores, interface of matrix and fiber, interphases, or coating of fibers, which provide a pseudo-ductility to the composite. These imperfections of the material scatter the phonons and reduce the thermal conductivity. The properties of SiC/SiC composites show anisotropy caused by the fabrication of reinforcements. Figure 15.17 shows the temperature dependence of thermal conductivity of various CVI SiC/SiC composites. Because the relationship between the thermal conductivity and the mechanical strength is the trade-off, the highest thermal conductivity material normally does not have the best mechanical property. The highest limit of the thermal conductivity of a SiC/SiC composite is that of high-purity SiC having a theoretical density. Though the strongest factor that scatters phonons is the irradiation defects, it is able to improve the thermal conductivity of SiC/SiC composites by increasing the density of the matrix and the contrivance of fabrication of reinforcements [25].

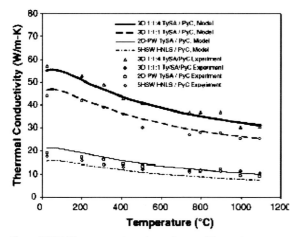

Figure 15.17 The temperature dependence of thermal conductivity of various CVI SiC/SiC composites [26].

15.6.4
Mechanical Property Changes

SiC is a ceramic which has covalent bonds. Although the property of monolithic CVD-SiC looks brittle, the mechanical properties are changed by the accumulation of irradiation defects. The irradiation temperature dependence of the properties is summarized by Snead and Nozawa *et al.* [27]:

- The irradiation reduces the elastic modulus at lower temperatures and the elastic modulus change by irradiation becomes negligible at about 1000 °C.

- The point-defect accumulation and lattice relaxation are believed to cause the elastic modulus change.

- The nano-indentation hardness slightly increases after the irradiation, and the change of hardness is almost uniform against the irradiation temperature.

- The fracture toughness significantly increases in the intermediate temperature range of 300 to 1000 °C.

- Though the number of data of Chevron notched beam testing for the irradiated CVD-SiC is not sufficient, the fracture toughness of 800 to 1100 °C is apparently higher than the un-irradiated SiC.

A significant degradation in the mechanical strength of SiC/SiC composites was a serious issue in the early years of development. The cause of this problem was the shrinkage of SiC fibers [28, 29]. Early SiC fibers, such as Tyranno-TE and Nicalon CG fibers, have an amorphous phase. This shrinkage causes the decrease of interfacial shear strength and friction resistance of fiber against matrix, thus the ultimate strength significantly reduces after neutron irradiation [28]. Because the low crystallinity of SiC fibers with high oxygen content brought into low irradiation makes them resistant to the composites, new SiC fibers have been developed with stoichiometric composition and high crystallinity. Tyranno-SA reduced oxygen and C/Si ratio came close to 1. A SiC/SiC composite consists of many fibers, matrix, interphases, and the other elements such as residual silicon or carbon, sintering additives, and impurities. The dimensional change of a SiC/SiC composite is strongly affected by the fabrication of fibers, and shows strong anisotropy.

The CVI matrix behaves similar to the CVD-SiC under the neutron irradiation environments because of high purity and highly crystallized SiC microstructure. Normalized flexural strengths of CVI SiC/SiC composites irradiated by neutron are summarized in Figure 15.18.

Though the flexural strength of Nicalon and Hi-Nicalon composites tends to decrease with increasing neutron dose, the Hi-Nicalon Type-S composites show very high irradiation resistance up to 800 °C and 10 dpa [30]. The tensile property changes by neutron irradiation at high temperatures of advanced SiC/SiC composites (UD, Hi-Nicalon Type-S reinforcements, and CVI matrix) are shown in Figure 15.19 [31].

Higher temperature and heavier dose irradiation experiments are in progress for advanced SiC/SiC composites. Relatively heavier irradiation was also performed at the fast test reactor, JOYO, in Japan, where Tyranno-SA and Hi-Nicalon Type-S composites (two-dimensional, CVI matrix) were irradiated up to 12 dpa at 760 °C [32]. The stress-strain curves of three irradiation conditions did not indicate significant changes, the advanced SiC/SiC composites showing high irradiation

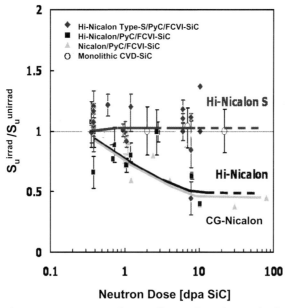

Figure 15.18 Neutron irradiation effects on flexural strength of SiC/SiC composites.

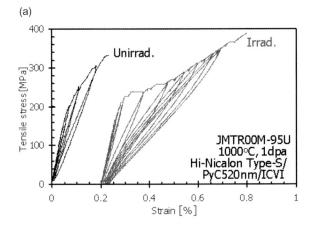

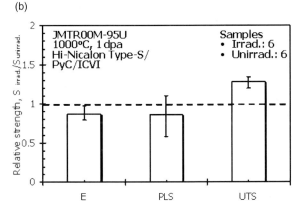

Figure 15.19 Tensile property changes of Hi-Nicalon Type-S composites by neutron irradiation at 1000 °C, 1 dpa [31].

resistance under these conditions. However, as major degradation of graphite strength is expected to take place at about this dose level, irradiation to even higher doses will be essential to determine the very high dose effects on the composite strength [21, 32].

15.7
Mechanical Property Evaluation Methods

CMCs have a continuation of the basic unit structure with a finite size, the size and structure of which is varied by the choice of architecture. Therefore, the composites exhibit anisotropy, providing various types of failure modes: tensile, compressive, and shear. For engineering design, the mechanical tests of the individual fracture mode are needed. Many test methods for CMCs are presently standardized (Table 15.3). However, considering these issues, the existing test methods are

Table 15.3 List of standard test methods for CMCs.

Failure mode		ASTM	JIS	ISO
Tensile	Room temp.	C1275	R1656	15 733
	High temp.	C1359	–	–
	Off-axis	D3518(PMC)	–	–
	Tran-thickness	C1468	–	–
	Creep	C1337	–	–
	Fatigue	C1360	–	–
Compressive	Room-temp.	C1358	R1673	20 504
Flexural		C1341	R1663	–
Shear	Interlaminar	C1292	R1643	20 505
	Iosipescu	D5379m	–	20 506
Fracture energy	In-plane mode-I	–	R1662	–
	Interlaminar mode-I	D5528(FRP)	–	–
		D6671m(FRP)		

still incomplete. Of particular importance is consideration of the physical meaning of the measured value.

In the development of a material for nuclear applications, irradiation resistance needs to be proven, where a small specimen test technique (SSTT) has to be developed due to the strict limitation of irradiation volume from availability, uniformity, and irradiation cost, and the strict requirements to reduce radioactive materials for testing, and post-test radioactive waste management. One of the difficulties for SSTT development is a strong demand for statistically reliable data results in an increasing number of valid tests. There have been many efforts to develop SSTT for metallic materials, but no SSTT for CMCs has been established so far. Therefore, most of the standardized test methods, which generally require a large specimen, need to be updated. For that purpose, miniaturization of a specimen is an urgent need, where specimen size effect has to be clarified.

15.7.1
Impulse Excitation Method for Young's Modulus Determination

An impulse excitation method is used to evaluate a dynamic Young's modulus of the material (ASTM-C1259, JIS-R1644, and ISO-17561). The dynamic Young's modulus can be estimated by detecting the specific mechanical resonant frequency in the flexural mode of vibration. The experimental error of the impulse excitation method is exceptionally small with high reproducibility. This technique is often applied in evaluation of composites. Due to the potential non-uniformity of the composites, special care is necessary in discussing the physical meaning of the data.

15.7.2
Bulk Strength Testing Methods for Ceramics

A flexural test method is applied to quantify the strength of the ceramics (ASTM-C1161, JIS-R1601, and ISO-14704). Carefully considering the surface condition of the specimen, a miniature bend specimen with the volume of $20 \times 1 \times 1 \, mm^3$ can be successfully applied to the irradiation experiment study [33]. The strength of ceramics depends significantly on the surface finish of the specimen. Specifically, the recent result by Byun *et al.* [34] indicates that the statistical strength of the ceramics depends significantly on the surface area rather than the volume. The information about the specimen size and surface finish therefore need to be reported in advance of the test.

An internal pressurization test and a diametral compression test were applied to evaluate fracture strength of miniature cylindrical and hemispherical SiC specimens [35]. Due to the uniform pressurization anticipated, the internal pressurization method is believed to provide much reliable and reproducible data. In contrast, the results by the diametral compression test were somehow questionable unless considering the influence of stress concentration on the contact area. The finite element study to correlate the influence of stress concentration on the load section enables the more reasonable result [36].

The statistical strength of SiC was measured by flexure and internal pressurization [37]. Both tests consistently demonstrated the key finding, that is, the Weibull strength slightly increases by irradiation, while the Weibull modulus decreases.

Another attempt to measure the fracture strength of ceramics was conducted using a C-sphere flexural specimen [38].

Fracture toughness of ceramics is one of the important basic mechanical properties and is measured by various test methods: micro- and nano-indentation, surface crack in flexure, double cantilever beam, double torsion, single-edge notched beam, Chevron notched beam, and fractography (ASTM-C1421, JIS-R1607, -R1617, ISO-15732, -18756, -24370). In addition, a spiral notch torsion test has been recently proposed as a pure mode-I testing method [24]. In any test methods, the stable crack extension needs to be achieved to validate the test.

An indentation test technique is often used to evaluate hardness, elastic modulus, and fracture toughness. This test was successfully applied to evaluate the local crack propagation behavior after ion irradiation [39]. However, careful examination is required since the information obtained from this technique is limited to a small area near the surface and the data depends significantly on the surface condition. Together with the indentation test, other techniques should be applied to generate reliable data.

Strength of ceramic fibers for reinforcement of CMC is another important property to be appropriately evaluated, but standardization of the test methods is still insufficient. A single filament tensile test has been developed to evaluate tensile properties of a mono filament (ASTM-D3379 and JIS-R1657). Highly reliable data of a fiber diameter and a tensile strain are needed to determine the strength. However, non-uniformity of fiber diameter and roundness observed in many ceramic fibers and the difficulty in measuring the fiber size at fracture by post test measurement make it difficult to obtain accurate results. So far, by applying this technique to the neutron irradiation studies, the excellent stability of tensile strength of advanced SiC fibers against neutron irradiation up to ~8 dpa has been verified.

The Bend Stress Relaxation (BSR) test is a unique technique to evaluate creep behavior, originally developed for thermal creep studies on ceramic fibers [39] and recently advanced for evaluation of irradiation creep deformation [40]. By measuring the radius of the elastically bent specimen after the test, the creep strain is estimated. Applying this developmental technique to irradiation creep study, the steady state irradiation creep compliance of CVD-SiC was identified up to 0.7 dpa. The transient creep behavior was also compared between polycrystalline and single crystal SiC. There is an attempt to apply this technique to the composites but further discussion on validity is required.

15.7.3
Test Methods for Composites

A flexural test is conventionally used to measure composite strength (ASTM-C1341 and JIS-R1663). Contrary to simplicity in testing, the analysis in flexure is complicated since the stress distribution of the material cannot be symmetrical because of the co-existing failure modes of tensile, compression, shear stresses, and the stress gradient. Moreover, the crushing and buckling at the contact point invalidates the test.

A tensile test method is standardized in ASTM-C1275, JIS-R1656, and ISO-15733. Since a large test coupon is defined in test method standards, development of the SSTT is important. To design a miniature tensile specimen, the specimen size effect on tensile properties was investigated and key features are summarized:

- very minor size effects on tensile properties, if the fiber volume fraction in the loading direction in a unit structure is unchanged;
- the size dependency of off-axis tensile properties, probably due to the size-relevant change of fracture modes;
- very minor effect of specimen geometry [26, 41, 42].

Two types of miniature specimens: a face-loaded specimen for room-temperature test and an edge-loaded miniature specimen for high-temperature applications (Figure 15.20) are proposed [43] and have been widely used in many neutron irradiation experiments. An advantage in using the face-loaded specimen is good applicability to the composites with weak inter-laminar shear strength. By contrast, the edge-loaded specimen requires strong shear strength to prevent failure at the gripping section. No systematic difference in reliability of data between these two methods was reported.

A mini-composite tensile technique is another effective testing method. Specifically, by applying the mini-composite tensile test method, interfacial parameters

15.7 Mechanical Property Evaluation Methods | 375

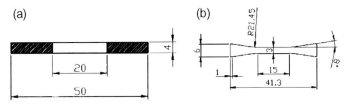

Figure 15.20 Miniature tensile specimens: (a) face-loaded straight-bar specimen; (b) edge-loaded contoured specimen [17].

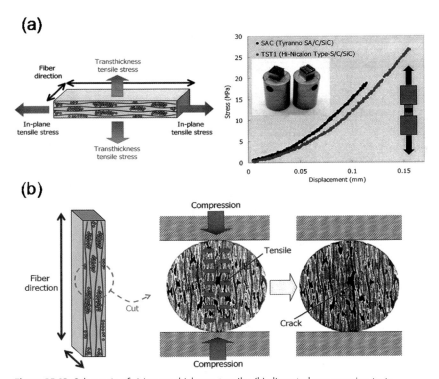

Figure 15.21 Schematic of: (a) trans-thickness tensile; (b) diametral compression tests.

can be estimated by measuring a matrix crack space and a density [44]. Because of the difficulty in determining a crack extension path, this technique is not recommended for interfacial shear evaluation of the multilayered composites.

A trans-thickness tensile testing method is standardized in ASTM-C1468 (Figure 15.21). No quasi-ductility without fiber pullouts resulted in the brittle-like fracture at the maximum applied stress [45]. Since the specimen is adhesively bonded on the fixtures, it is not easy to induce a crack within the gauge section of the specimen and so this method does not look appropriate for post-irradiation test methods, concerning the difficulty in handling the irradiated specimen and the high temperature applicability.

A diametral compression test (Brazilian test) is another option to investigate the trans-thickness tensile strength of the composites. The conceptual design of this technique was first proposed for fracture toughness evaluation of concrete and recently rearranged for the composites. There still remains the concern of the negative effect of stress concentration at the contact area with the fixture. Nevertheless, the diametral compression test has an advantage in miniaturization and high temperature applications. According to the specimen size effect study [46], a specimen diameter of 3.2 to 9.5 mm and a thickness of 1.7 to 6.0 mm are recommended. For neutron irradiation, a 3 mm-thick, 6 mm-diameter disk specimen was proposed. The diametral compression test technique is now under standardization in the ASTM subcommittee C28.07.

Ceramics, in general, are strongly resistant to compression compared with tension. However, the compressive strength of the composites are generally inferior due to the presence of a weak F/M interface. It is difficult to distinguish the microscopic fiber fracture and the macroscopic composite buckling. The compressive test (ASTM-C1358, JIS-R1673, and ISO-20504) is undoubtedly one of the most difficult testing methods and therefore needs to be developed more technically and analytically.

Shear strength is categorized by the correlation between directions of the working force and the aligned fibers (Table 15.4). The in-plane shear stress (IPSS) works equally in each plane of the stacked sheets perpendicular to the stacking direction. In contrast, the inter-laminar shear stress (ILSS) works in the stacking direction of fabric sheets. A rail shear test, an asymmetric four-point flexural test, an Iosipescu shear test, and an off-axis tensile test are usually applied for the evaluation of IPSS, while a short beam test and a double notch shear (DNS) test are applied for ILSS. The individual test methods are standardized in ASTM-C1292, -D5379M, JIS-R1643, ISO-20505, and -20506.

Although the off-axis tensile test method is well established, the co-existing mixed fracture modes depending on the off-axis angle make the analysis difficult. Specifically, due to the intrinsic size effect, the off-axial strength further decreases with decreasing specimen width [47], although a high-density SiC/SiC composite seems to exhibit a very minor size effect [48].

The Iosipescu shear test and the asymmetrical 4-point bend test are believed to be of best valuable, since a nearly pure shear stress field can be achieved between notches by carefully considering the loading alignment. A crushing failure with a fixture, which often invalidates the test, must be avoided. Besides, a crack extension behavior depending on a textile structure needs to be carefully discussed in analysis.

In DNS testing, stress concentration differs, depending on the distance between the two notches [49]. In ASTM-C1292, the notch separation of 6 mm is recommended. The simple compression test configuration enables a high-temperature test.

Of many interfacial shear testing methods, a single fiber push-out method is the most promising. One end of the fiber embedded in the matrix is stressed by

Table 15.4 List of shear testing methods for CMCs.

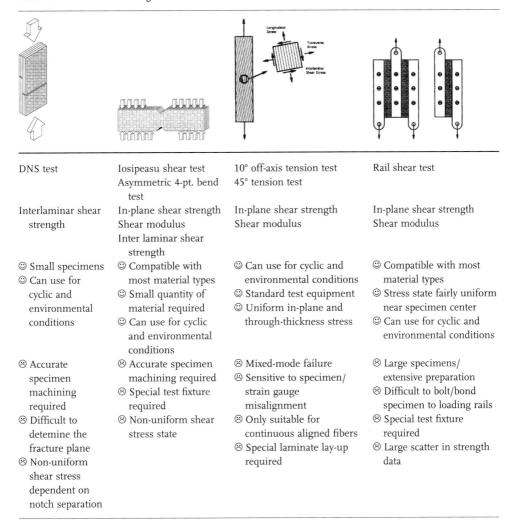

DNS test	Iosipeasu shear test Asymmetric 4-pt. bend test	10° off-axis tension test 45° tension test	Rail shear test
Interlaminar shear strength	In-plane shear strength Shear modulus Inter laminar shear strength	In-plane shear strength Shear modulus	In-plane shear strength Shear modulus
☺ Small specimens ☺ Can use for cyclic and environmental conditions	☺ Compatible with most material types ☺ Small quantity of material required ☺ Can use for cyclic and environmental conditions	☺ Can use for cyclic and environmental conditions ☺ Standard test equipment ☺ Uniform in-plane and through-thickness stress	☺ Compatible with most material types ☺ Stress state fairly uniform near specimen center ☺ Can use for cyclic and environmental conditions
☹ Accurate specimen machining required ☹ Difficult to detemine the fracture plane ☹ Non-uniform shear stress dependent on notch separation	☹ Accurate specimen machining required ☹ Special test fixture required ☹ Non-uniform shear stress state	☹ Mixed-mode failure ☹ Sensitive to specimen/strain gauge misalignment ☹ Only suitable for continuous aligned fibers ☹ Special laminate lay-up required	☹ Large specimens/extensive preparation ☹ Difficult to bolt/bond specimen to loading rails ☹ Special test fixture required ☹ Large scatter in strength data

the small indenter, and load and indenter displacement are monitored during the test [50].

Since sharp indenter tips, such as a Vickers or a Berkovich, cause severe deformation of the fiber, it is difficult to apply on a low shear strength fiber, such as carbon fiber with an onion structure and a multilayer coating interface. In such a case, a flat-bottom indenter can be applied. There are two important events addressed in the push-out process:

- the crack initiation at the interface;
- the complete debonding and following sliding at the maximum load (Figure 15.22).

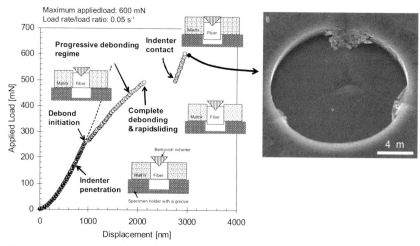

Figure 15.22 Schematic of the push-out test.

By applying the non-linear shear-lag model, intrinsic shear parameters at the F/M interface can be obtained:

- an interfacial debond shear strength (IDSS);
- an interfacial friction stress (IFS) [51].

Recent results revealed that both IDSS and IFS had a dip of around 1 to 2 dpa and recovered to the un-irradiated level with increasing neutron dose, that is, "turn-around" behavior [52].

In order to evaluate friction stress at the F/M interface, the fiber push-back test is also employed. A friction coefficient is monitored during the push-back process.

Evaluation of the fracture energy (often called "fracture resistance"), that is, the energy to initiate the crack propagation from the intrinsic or machined flaws, becomes more important. Although the importance of failure evaluation has been recognized, the existing test methods cannot necessarily be standardized for that purpose. Fatigue and creep are important for practical system design. However, the existing methods do not focus on the "failure" behavior. Moreover, the test development for joint and coating should be more encouraging. The current standards (ASTM-C1469, -D905, -F734, JIS-R1624, and -R1630) cannot simply apply to the very strong joints, the strength of which is equivalent to that of a composite substrate [53].

15.7.4
Development of Materials Database

The latest material handbook of SiC [27] has had many updates:

- physical, thermal, and mechanical property database for un-irradiated and irradiated high-purity CVD-SiC;

- summary of irradiation enhanced microstructure, swelling, thermal conductivity, and mechanical properties over the wide temperature and neutron fluence range;
- introduction of new test techniques.

This handbook is still incomplete but believed the best available database for high-quality SiC, which solves many problems in the early handbook [37]. Recent progress in irradiation data of CVI-SiC/SiC composites reveals the excellent irradiation tolerance to intermediate neutron dose (~10 dpa) [54]. Besides, irradiation-induced change of shear strength at the F/M interface of the composites is being clarified [52]. The manufacturing technology is developed, hence standard SiC/SiC composites are available. The compilation of a materials database becomes a key for industrialization and accordingly many irradiation campaigns are in progress.

15.8
New GFR Concepts Utilizing SiC/SiC Composite Materials

Although there are many conceptual designs of GFR, the fuel types can be categorized into:

- Coated Particle Fuel
- Shield Fuel Pin
- Composite Ceramics Fuel

Control rods and reflectors are other core structural components. Other hot gas circuits for energy conversion system, including the gas turbine and internal heat exchanger, are of great importance. In all these areas, high temperature ceramics have a strong impact on improving the attractiveness of reactors. Applications of the SiC/SiC composites to coated particle fuel type and shield fuel pin type have been explored, intending to improve energy conversion efficiency, to reduce reactor core size and to improve reactor safety margins.

Figure 15.23 is the He cooled fast reactor core and fuel concept using coated particle fuel, where horizontal flow cooling concept with directly cooling system is applied. For this purpose, tubes with diameters of 8.4 and 20 cm have to be developed with 5 and 40% porosity, respectively. In this case, the SiC/SiC composites can provide excellent safety margins from high temperature stability and radiation resistance to neutrons.

Figure 15.24 is the He cooled fast reactor core and fuel concept using shield fuel pin. Also in this case, SiC/SiC composite materials can provide higher energy conversion efficiency and reduction of core size from their high temperature properties and neutron damage tolerance. R&D of fabrication technology of fuel pins and core components are ongoing.

The final goal of the fabrication is to make fuel pins of 3 m in length, 10 mm in inner diameter, and 1 mm wall thickness. Currently, a 300 mm length tube with

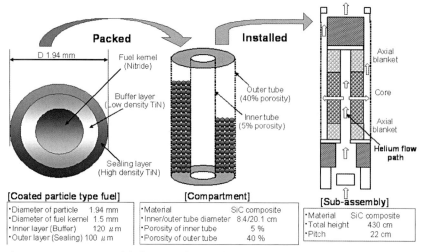

Figure 15.23 GFR core and fuel concept for coated particle fuel [55].

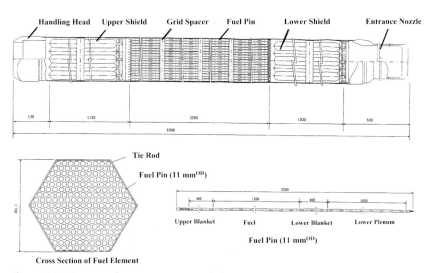

Figure 15.24 GFR core design concept using fuel pin.

the goal dimensions of diameter and wall thickness has been successfully fabricated by HIP processing.

For the application of SiC/SiC in GFR, other than core structural components and fuel, many potential applications can be found. This is also similar in the case of VHTR. Figure 15.25 shows the potential application for VHTR designed by JAEA (Japan Atomic Energy Agency). Although the current HTTR (High Temperature Test Reactor in JAEA-Oarai) uses SiC/SiC instead of SiC in these areas, the VHTR is much enhanced.

Component	HTTR	R&D
Fuel block Reflector	Isotropic reactor grade graphite (IG-110)	Irradiation data Life extension (Non-destructive method)
Control rod cladding	Alloy 800H	Irradiation data Design code
Upper shield Core barrel	Graphite Metals	Irradiation data Design code

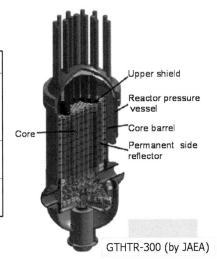

GTHTR-300 (by JAEA)

Figure 15.25 Potential application areas of SiC/SiC in VHTR.

15.9
Concluding Remarks

There are many ongoing efforts for R&D in advanced SiC/SiC toward the near-term application to energy systems, especially for nuclear energy systems. GFR is one of the important targets, but even current light water systems, other generation IV reactor systems than GFR and even fusion reactor systems, are potential and attractive candidates.

For these applications, development of joining and coating technology is important. For the design of those systems, basic material properties such as a material database is an essential need and a database of neutron damage effects on material performance is inevitable.

References

1 Advanced Composite Materials (2002) (MRS).
2 Snead, L.L., Hinoki, T., Taguchi, Y., Jones, R.H. and Kohyama, A. (2003) Silicon carbide composites for fusion reactor application. *Advances in Science and Technology*, **33**, 129–40.
3 Kohyama, A., Katoh, Y. and Jimbo, K. (2004) *Materials Transactions*, **45**, 51–8.
4 Kohyama, A., Singh, M., Lin, H.T. and Katoh, Y. (eds) (2002) Advanced SiC/SiC Ceramic Composites, in *Ceramic Transactions*, Vol. 144, American Ceramic Society.
5 Energy White Paper (2003) (TSO, UK).
6 Bloom, E.E., Zinkle, S.J. and Wiffen, F.W. (2004) Materials to deliver the promise of fusion power-progress and challenges. *Journal of Nuclear Materials*, **329–333**, 13–19.
7 A Technology Roadmap for Generation IV Nuclear Energy Systems and Supporting Documents (2002) (INEEL, US/DOE).
8 Katoh, Y. and Dong, S.M. (2002) Thermo-mechanical properties and microstructure

of silicon carbide composites fabricated by nano-infiltration and transient eutectoid process. *Fusion Engineering and Design*, **61–62**, 723–31.

9 Shimoda, K., Eto, M., Lee, J.K., Park, J.S., Hinoki, T. and Kohyama, A. (2004) Influence of surface micro-chemistry of SiC nano-powder on the sinterability of NITE-SiC. Proceedings of HTCMC-5.

10 Hino, T., Hirohata, Y., Yamauchi, Y., Hashiba, M., Kohyama, A., et al. (2004) Plasma material interaction studies on low activation materials used for plasma facing or component. *Journal of Nuclear Materials*, **329–333**, 673–81.

11 Toyoshima, K., Schaffron, L., Hinoki, T., Kohyama, A., Yamauchi, Y., Hirohata, Y. and Hino, T. (2006) *Effects of GFR/fusion environments on helium gas permeability of NITE-SiC/SiC.* Presented at Spring Meeting of JIM, Tokyo, Japan.

12 Kishimoto, H., Ozawa, K., Kondo, S. and Kohyama, A. (2005) Effects of dual ion irradiation on the swelling of SiC/SiC composites. *Materials Transactions*, **46**, 1923–7.

13 Kohyama, A., Katoh, Y., Ando, M. and Jimbo, K. (2000) New multiple beams-material interaction research facility for radiation damage studies in fusion materials. *Fusion Engineering and Design*, **51–52**, 789–95.

14 Kulcinski, G.L., Brimhaqllo, J.L. and Kissinger, H.E. (1972) Production of voids in pure metals by high energy heavy ion bombardment, in *Radiation Induced Voids in Metals*, AEC 26. US Atomic Energy Commission, pp. 449–78.

15 Biersack, J.P. (1980) A Monte Carlo computer program for the transport of energetic ions in amorphous targets. *Nuclear Instruments and Methods*, **174**, 257–69.

16 Park, K.H., Kishimoto, H. and Kohyama, A. (2004) 3D analysis of cracking behaviour under indentation in ion-irradiated β-SiC. *Journal of Electron Microscopy*, **53**, 511–13.

17 Snead, L.L., Osborne, M.C., Lowden, R.A., Strizak, J., Shinavski, R.J., More, K.L., Eartherly, W.S., Bailey, J. and Williams, A.M. (1998) Low dose irradiation performance of SiC interphase SiC/SiC composites. *Journal of Nuclear Materials*, **253**, 20–30.

18 Gao, F., Weber, W.J. and Devanathan, R. (2001) Atomic-scale simulation of displacement cascades and amorphization in β-SiC. *Nuclear Instruments and Methods in Physics Research. Section B*, **180**, 176–86.

19 Snead, L.L., Zinkle, S.J., Hay, J.C. and Osborne, M.C. (1998) Amorphization of SiC under ion and neutron irradiation. *Nuclear Instruments and Methods in Physics Research. Section B*, **141**, 123–32.

20 Katoh, Y., Kishimoto, H., Jimbo, K. and Kohyama, A. (2002) The influences of irradiation temperature and helium production on the dimensional stability of silicon carbide. *Journal of Nuclear Materials*, **307–311**, 1221–6.

21 Katoh, Y., Snead, L.L., Henager, C.H., Jr Hasegawa, A., Kohyama A., Riccardi, B. and Hegeman, H. (2007) Current status and critical issues for development of SiC composites for fusion applications. *Journal of Nuclear Materials*, **367–370**, 659–71.

22 Snead, L.L., Katoh, Y. and Connery, S. (2007) Swelling of SiC at intermediate and high irradiation temperatures. *Journal of Nuclear Materials*, **367–370**, 677–84.

23 Youngblood, G.E., Senor, D.J. and Jones, R.H. (2004) Effects of irradiation and post-irradiation annealing on the thermal conductivity/diffusivity of monolithic SiC and f-SiC/SiC composites. *Journal of Nuclear Materials*, **329–333**, 507–12.

24 Wang, J.A. (2003) Oak ridge national laboratory (ORNL) spiral notch torsion test (SNTT) system. *Practical Failure Analysis*, **3**, 11–15.

25 Katoh, Y., Nozawa, T., Snead, L.L., Hinoki, T. and Kohyama, A. (2006) Property tailorability for advanced CVI silicon carbide composites for fusion. *Fusion Engineering and Design*, **81**, 937–44.

26 Nozawa, T., Katoh, Y., Kohyama, A. and Lara-Curzio, E. (2003) Specimen size effect on the tensile and shear properties of the high-crystalline and high-dense SiC/SiC composites. *Ceramic Engineering and Science Proceedings, Section B*, **24**, 415–20.

27 Snead, L.L., Nozawa, T., Katoh, Y., Byun, T.S., Kondo, S. and Petti, D.A. Handbook of SiC properties for fuel performance

modeling. *Journal of Nuclear Materials*, in printing.

28. Snead, L.L., Steiner, D. and Zinkle, S.J. (1992) Measurement of the effect of radiation damage to ceramic composite interfacial strength. *Journal of Nuclear Materials*, **191–194**, 566–70.

29. Hollenberg, G.W., Henager, C.H., Youngblood, Jr., G.E., Trimble, D.J., Simonson, S.A., Newsome, G.A. and Lewis, E. (1995) The effect of irradiation on the stability and properties of monolithic silicon carbide and SiCf/SiC composites up to 25 dpa. *Journal of Nuclear Materials*, **219**, 70–86.

30. Jones, R.H., Giancarli, L., Hasegawa, A., Katoh, Y., Kohyama, A., Riccardi, B., Snead, L.L. and Weber, W.J. (2002) Promise and challenges of SiCf/SiC composites for fusion energy applications. *Journal of Nuclear Materials*, **307–311**, 1057–72.

31. Ozawa, K., Hinoki, T., Nozawa, T., Katoh, Y., Maki, Y., Kondo, S., Ikeda, S. and Kohyama, A. (2006) Evaluation of fiber/matrix interfacial strength of neutron irradiated SiC/SiC composites using hysteresis loop analysis of tensile text. *Materials Transactions*, **47**, 207–10.

32. Ozawa, K., Nozawa, T., Katoh, Y., Hinoki, T. and Kohyama, A. (2007) Mechanical properties of advanced SiC/SiC composites. *Journal of Nuclear Materials*, **367–370**, 713–18.

33. Katoh, Y. and Snead, L.L. (2005) Mechanical properties of cubic silicon carbide after neutron irradiation at elevated temperatures. *Journal of ASTM International*, **2**, 12377.

34. Byun, T.S., Lara-Curzio, E., Lowden, R.A., Snead, L.L. and Katoh, Y. (2007) Miniaturized fracture stress tests for thin-walled tubular SiC specimens. *Journal of Nuclear Materials*, **367–370**, 653–8.

35. Byun, T.S., Hong, S.G., Snead, L.L. and Katoh, Y. (2005) *Influence of specimen type and loading configuration on the fracture strength of SiC layer in coated particle fuel*. Presented at the 30th Annual International Conference on Advanced Ceramics & Composites, 2005, Cocoa Beach, FL, USA.

36. Hong, S.G., Byun, T.S., Lowden, R.A., Snead, L.L. and Katoh, Y. (2007) Evaluation of the fracture strength for silicon carbide layers in the tri-isotropic-coated fuel particle. *Journal of the American Ceramic Society*, **90**, 184–91.

37. Snead, L.L., Nozawa, T., Katoh, Y., Byun, T.S., Kondo, S. and Petti, D.A. Handbook of SiC properties for fuel performance modeling. *Journal of Nuclear Materials*, in printing.

38. Wereszczak, A.A., Jadaan, O.M., Lin, H.T., Champoux, G.J. and Ryan, D.P. (2007) Hoop tensile strength testing of small diameter ceramic particles. *Journal of Nuclear Materials*, **361**, 121–5.

39. Morscher, G.N. and DiCarlo, J.A. (1992) A simple test for thermo-mechanical evaluation of ceramic fibers. *Journal of the American Ceramic Society*, **75**, 136–40.

40. Katoh, Y., Snead, L.L., Hinoki, T., Kondo, S. and Kohyama, A. (2007) Irradiation creep of high purity CVD silicon carbide as estimated by the bend stress relaxation method. *Journal of Nuclear Materials*, **367–370**, 758–63.

41. Nozawa, T., Katoh, Y. and Kohyama, A. (2005) Evaluation of tensile properties of SiC/SiC composites with miniaturized specimens. *Materials Transactions*, **46**, 543–51.

42. Shinavski, R.J., Engel, T.D., Lara-Curzio, E., Porter, W.D., Radovic, M. and Wang, H. (2007) *Mechanical and thermal evaluation of near-stoichiometric SiC fiber-reinforced SiC with braided architectures for nuclear applications*. Presented at the 31st Annual International Conference on Advanced Ceramics & Composites, 2007, Daytona Beach, FL, USA.

43. Nozawa, T., Katoh, Y., Kohyama, A. and Lara-Curzio, E. (2002) *Specimen size effects in tensile properties of SiC/SiC and recommendation for irradiation studies*. Proceedings of the 5th IEA. Workshop on SiC/SiC Ceramic Matrix Composites for Fusion Structural Applications, 74–86.

44. Lamon, J., Rebillat, F. and Evans, A.G. (1995) Microcomposite test procedure for evaluating the interface properties of ceramic matrix composites. *Journal of the American Ceramic Society*, **78**, 401–5.

45 Hinoki, T., Lara-Curzio, E. and Snead, L.L. (2004) *Effect of interphase on transthickness tensile strength of high-purity silicon carbide composites*. Presented at the 28th International Conference of Advanced Ceramics and Composites, 2004, Cocoa Beach, FL, USA.

46 Hinoki, T., Lara-Curzio, E. and Snead, L.L. (2003) Evaluation of transthickness tensile strength of SiC/SiC composites. *Ceramic Engineering and Science Proceedings*, **24**, 401–6.

47 Evans, A.G., Domergue, J.-M. and Vagaggini, E. (1994) Methodology for relating the tensile constitutive behavior of ceramic-matrix composites to constituent properties. *Journal of the American Ceramic Society*, **77**, 1425–35.

48 Nozawa, T., Lara-Curzio, E., Katoh, Y. and Shinavski, R.J. (2007) *Tensile properties of braided SiC/SiC composites for nuclear control rod applications*. Presented at the 31st Annual International Conference on Advanced Ceramics & Composites, 2007, Daytona Beach, FL, USA.

49 Lara-Curzio, E. and Ferber, M.K. (1997) Shear strength of continuous fiber ceramic composites. *ASTM STP*, **1309**, 31–48.

50 Lara-Curzio, E. and Ferber, M.K. (1994) Methodology for the determination of the interfacial properties of brittle matrix composites. *Journal of Materials Science*, **29**, 6152–8.

51 Nozawa, T., Snead, L.L., Katoh, Y., Miller, J.H. and Lara-Curzio, E. (2006) Determining the shear properties of the PyC/SiC interface for a model TRISO fuel. *Journal of Nuclear Materials*, **350**, 182–94.

52 Nozawa, T., Katoh, Y. and Snead, L.L. (2007) The effects of neutron irradiation on shear properties of monolayered PyC and multilayered PyC/SiC interfaces of SiC/SiC composites. *Journal of Nuclear Materials*, **367–370**, 685–91.

53 Hinoki, T., Eiza, N., Son, S., Shimoda, K., Lee, J. and Kohyama, A. (2005) Development of joining and coating technique for SiC and SiC/SiC composites utilizing NITE processing. *Ceramic Engineering and Science Proceedings*, **26**, 399–405.

54 Katoh, Y., Nozawa, T., Snead, L.L. and Hinoki, T. (2007) Effect of neutron irradiation on tensile properties of unidirectional silicon carbide composites. *Journal of Nuclear Materials*, **367–370**, 774–9.

55 Konomura, M. *et al.* (2003) *A promising gas-cooled fast reactor concept and its R&D plan*. Proceedings of Global 2003 (American Nuclear Society, 2003).

16
CMCs for Friction Applications
Walter Krenkel and Ralph Renz

16.1
Introduction

The development of carbon fiber reinforced SiC-composites (C/SiC, C/C-SiC) for their use in friction systems started in the early 1990ies and represents a successful spin-off from space technologies to terrestrial applications [1]. These lightweight and thermally stable materials based on the silicon melt-infiltration of modified carbon/carbon composites and manufactured by the LSI-process, overcame the drawbacks of C/C (influence of temperature and humidity on the coefficient of friction) and grey cast iron (high weight, insufficient thermal stability), the two standard materials for brake disks in cars and aircraft. C/SiC composites show superior tribological properties in terms of high and stable coefficients of friction (CoF) and extreme wear resistance. First prototypes of full-sized brake disks for high speed trains and passenger cars were developed between 1994 and 1997 from the German Aerospace Center in Stuttgart [2–5]. Currently, about 50,000 C/SiC brake disks are produced commercially per year by different manufacturers and CMCs are going to substitute grey cast iron brake disks in premium and high performance cars [6, 7]. Also in emergency brake systems of elevators and cranes, where sintermetallic and resin-bonded (organic) brake pads have reached their thermal limits, C/SiC materials are increasingly used because of their high impact and abrasive resistance [8–10]. C/SiC brake pads and disks have found their niche applications so far, a future breakthrough of this technology to mass markets, however, requires a further reduction of the still high manufacturing costs. In various parts of the world a number of institutional as well as industrial activities exist to scale-up the technology and to reduce the costs [11–25].

16.2
C/SiC Pads for Advanced Friction Systems

Originally, LSI-C/SiC composites have been developed for aerospace applications as hot structures in thermal protection or propulsion systems. Therefore, at the beginning of the development, only fabric-reinforced 2D-composites were available. Disk-on-disk tests with this first-generation C/SiC composites showed high surface temperatures of more than 1000 °C, high but unstable coefficients of friction and a considerable low wear resistance due to oxidation effects. This insufficient tribological behaviour required a consequent modification of these initial CMC composites to improve their applicability in high performance brake systems (Figure 16.1).

The main reason for the unstable coefficient of friction was the low transverse thermal conductivity of the standard 2D-C/SiC composites (9 W/mK).

Higher transverse conductivities were achieved by different modifications:

- Use of carbon fibers with a high thermal conductivity, e.g. graphite fibers
- Increasing amount of fibers oriented perpendicular to the friction surface
- Increase of the ceramic fraction.

Whilst the technical and economical efforts grow considerably for the first two methods, the increase of the silicon and silicon carbide content within the composite is a more cost-efficient way. Higher ceramic contents can be achieved easily by reducing the fiber content resulting in a higher silicon uptake and a more pronounced SiC formation and consequently a higher density of the C/SiC composite material. Therefore, the lower the fiber volume content and the higher the density, the higher the transverse thermal conductivity of the CMC material. Nevertheless, higher densities coincide with a decrease in strength and fracture toughness. The resulting CMC composites must therefore be designed according to the individual requirements of the brake system as a compromise between sufficient thermal and mechanical properties.

The following figures demonstrate these relationships for some representative C/SiC materials. Figure 16.2 shows the transverse thermal conductivity for short fiber reinforced C/SiC composites in dependence on the fiber volume content in the green body (carbon fiber reinforced plastic, CFRP). The values vary from 23–29 W/mK for fiber contents of 55–30 Vol.-%.

Figure 16.2 also shows the relationship between the material density and the transverse thermal conductivity of C/SiC composites, reinforced with HT-fibers.

Figure 16.1 Bidirectionally reinforced brake disks and pads made of C/SiC (first generation) [9].

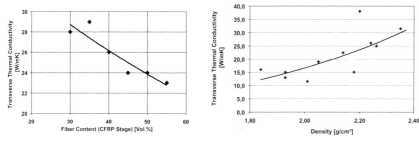

Figure 16.2 Thermal conductivity at 50 °C versus fiber volume content and versus density of short fiber reinforced C/SiC composites (HT-fibers) [9].

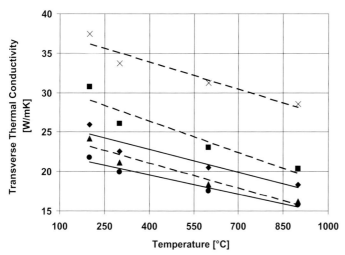

Figure 16.3 Transverse thermal conductivity of different C/SiC composites versus temperature [9].

Generally, the lower values correspond with continuous fiber reinforcements (i.e. fabrics), whereas the highest densities have been measured particularly with short fiber architectures.

As for most other materials, the thermal conductivity of C/SiC composites decreases with higher temperatures. Figure 16.3 demonstrates this relationship for different material modifications which have been manufactured by varying the fiber architecture and the processing conditions. The average decrease of transverse conductivity between 300 °C and 900 °C, which represents the surface temperature range occurring during high performance brakings, amounts to approximately 30%. As a consequence, suitable C/SiC materials for brake pads and disks must be selected by their thermal conductivity at high temperatures.

16.2.1
Brake Pads for Emergency Brake Systems

C/SiC composites have been successfully applied in rotors and stators of emergency brakes which are used in mechanical engineering and conveying systems. Conventional brake systems with metallic disks and organic friction linings have reached their thermal limits in particular under extreme power input.

In screening tests, CMC composites with different transverse thermal conductivities were tested by pressing one rotating brake disk against two stationary disks of the same material. Figure 16.4 shows the results of different C/SiC modifications and their influence on the stability of the CoF.

The wear rates are only slightly decreased by higher transverse thermal conductivities as shown in Figure 16.5, but this tribological parameter is of inferior importance for emergency brake systems. Figure 16.6 shows an emergency brake system during a high energy stop braking and a set of three brake disks made of LSI-C/SiC.

16.2.2
C/SiC Brake Pads for High-Performance Elevators

The most powerful drive units of elevators are designed for buildings of up to 500 m in height and can accelerate a mass of up to 45,000 kg up to a speed of 10 m/s (Figure 16.7). If the speed exceeds this value, the elevator is automatically brought to a controlled emergency stop. Thereby, the brake pads can reach

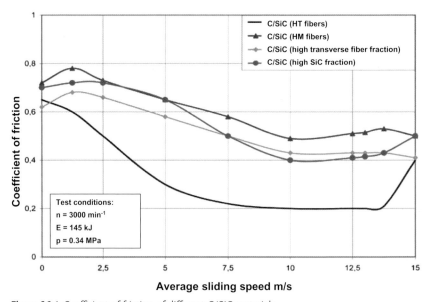

Figure 16.4 Coefficient of friction of different C/SiC materials.

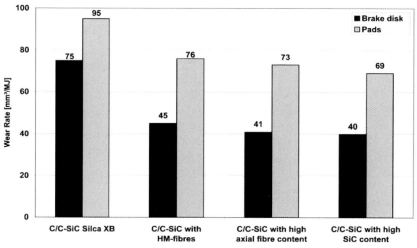

Figure 16.5 Wear rates of C/C-SiC with different transverse thermal conductivities (test conditions see Figure 16.4).

Figure 16.6 Emergency brake system during high energy stop braking (left) and a set of one rotating brake disk and two stationary disks, each made of 2D reinforced C/SiC with an outer diameter of 110 mm and a thickness of about 8 mm (Mayr Antriebstechnik, Germany).

temperatures of up to 1200 °C. At such high temperatures, conventional braking materials like organic or sintermetallic pads are far beyond their limits of thermal stability and wear resistance.

For this application, LSI-C/SiC composites were adapted by varying the type of carbon fiber, the fiber architecture, the type of precursor and the manufacture process of the CFRP green body. These modifications lead to significantly different tribological and mechanical properties (Figure 16.8). Graded C/SiC composites,

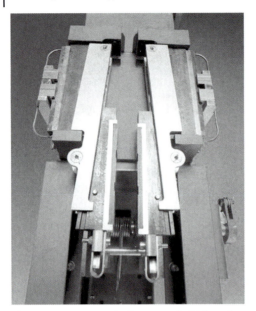

Figure 16.7 Emergency brake system for high performance elevators with C/SiC brake pads [8].

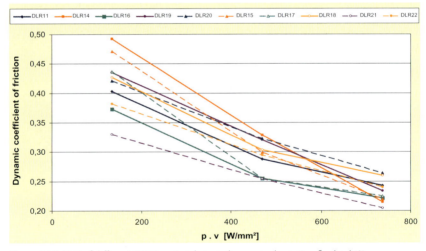

Figure 16.8 CoF of different C/SiC materials tested on a steel rotating flywheel [8].

achieved by a thermal pre-treatment of the carbon fibers prior to the CFRP manufacture, with highly abrasive-resistant SiC-rich surfaces in combination with a carbonaceous core material of high mechanical strength, show almost no wear, high CoF and sufficient mechanical properties. After intensive screening tests, these graded composites were chosen for the serial production of brake pads (dimensions $142 \times 34 \times 6\,mm^3$) which started in 2003.

16.3
Ceramic Brake Disks

High performance cars need a powerful brake system. This simple equation turns out to be more complex than it appears. The brake system and especially the brake disks of modern cars not only have to fulfill the task to decelerate the car failure safely under every condition e.g. dry or wet conditions, but also guarantee acceptable wear rates, high braking comfort and low weight. Under consideration of technical and economical aspects, the use of grey cast iron brake disks presently offers a compromise. Today, more or less all brake disks for automotive use are made of this material. However, its corrosion resistance is limited and its density of $7.2\,g/cm^3$ is comparatively high. The weight of the brake disks influences the unsprung and the rotating mass of the car and has a tremendous influence on the driving performance, agility, handling, etc. Therefore, the weight of the brakes should be as low as possible.

The use of metal matrix composites (MMC) based on an aluminum matrix reinforced with ceramic particles with densities of $2.8–3.0\,g/cm^3$ leads to essentially lighter disks, but the temperature stability is too low for high performance cars, where brake disk temperatures of about 800 °C may occur. Therefore their use as brake disks and brake drums is limited to small and very lightweight cars.

In very weight-sensitive racing cars carbon/carbon (C/C) brake disks are state-of-the-art. C/C brakes were introduced into Formula 1 motor racing in the early 1980s by Brabham [26] and are still the first choice for race applications in the Formula 1 and various other race series (Figure 16.9).

Due to the low density of the carbon material of about $1.8\,g/cm^3$ and its excellent temperature stability up to 1000 °C these brakes offer great racing performance. Despite these outstanding properties, C/C brakes cannot be used for series-production vehicles. The strong dependence of the coefficient of friction on temperature and moisture as well as extensive wear rates caused by the oxidation of carbon above 400 °C prohibit their use under normal driving conditions [27].

Carbon fiber reinforced ceramic composite materials, originally developed for the thermal protection of spacecrafts [28, 29], combine the high temperature

Figure 16.9 Carbon/Carbon brake system for race application e.g. used in American Le Mans Series race cars (Porsche RS Spyder).

stability and the low weight of C/C with an excellent wear behavior and stable friction values. Ceramic brakes were presented for the first time at the 1999 IAA (International Motor Show, Frankfurt, Germany) to the public and were introduced in 2000 in the Mercedes CL 55 AMG F1 Limited Edition [30] and in 2001 in the Porsche 911 GT2 [31]. Since their first application they have become more and more common for sport cars (Porsche, Ferrari, Lamborghini, Bugatti, etc.). In 2005 the first use of ceramic brake disks in an upper class car, the Audi A8-W12, could be observed [32]. Furthermore the use in a high performance middle class car (e.g. Audi RS4) in 2006 underlines the ongoing market penetration of ceramic brake disks not only in the field of sport cars.

16.3.1
Material Properties

As Figure 16.10 demonstrates, a ceramic brake disk consists of a ceramic friction rotor mounted on a metallic bell. The friction rotor is made of a carbon fiber reinforced ceramic composite material (C/SiC). Unlike in aerospace applications, where typically multidirectional woven fabrics are used for fiber reinforcement, for ceramic brakes chopped or milled short fibers are processed. The matrix of C/SiC consists of silicon carbide (SiC) and small amounts of free silicon (Si). While carbon fibers possess outstanding mechanical properties at high temperatures in air, their stability in an oxidative environment is very low. The oxidation of carbon starts at a temperature of about 400 °C in air and leads to an ongoing destruction of the brake disk at higher temperatures. This limits the use of carbon/

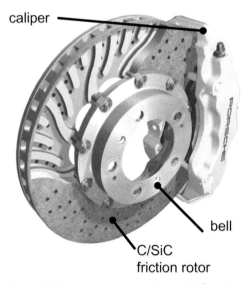

Figure 16.10 Ceramic composite brake disk from Porsche (PCCB, Porsche Ceramic Composite Brake) introduced in 2001 in a Porsche 911 GT2.

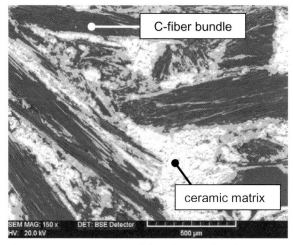

Figure 16.11 Microstructure of short fiber reinforced C/SiC.

Table 16.1 Mechanical and thermophysical properties of carbon fiber reinforced silicon carbide (C/SiC) in comparison to a typical grey cast iron brake material GJS-200 [27].

Material		GJS-200	C/SiC, SGL
Density	g/cm^3	7.2	2.3
Reinforcement	–	–	short fiber
Tensile Strength	MPa	150–250	20–40
Fracture Strain	%	0.3–0.8	0.3
Young's Modulus	GPa	90–110	30
Coefficient of Thermal Expansion	10^{-6} 1/K	9; 12 (RT; 300 °C)	1; 2 (RT; 300 °C)
Thermal Conductivity (⊥)	W/mK	54	40
Heat Capacity	J/kgK	500	800
	J/dm^{3}K	3600	1800

carbon brakes besides racing. As Figure 16.11 shows, the carbon fiber bundles of the C/SiC brake material, which are still sensitive to oxidation, are embedded in a protective ceramic matrix. The result is a more oxidation resistant material than normal carbon/carbon [33].

Generally, the properties of C/SiC depend strongly on its composition, the fiber content and the fiber orientation. The carbon fibers decrease the brittleness of SiC considerably so that the damage tolerance lies in nearly the same order of magnitude as for grey cast iron. To give an impression of some relevant mechanical and thermophysical properties, Table 16.1 compares the material data for a typical grey cast iron brake material GJS-200 with C/SiC. The low density of 2.3 g/cm^3 from C/SiC offers a great lightweight potential. However, the specific heat capacity of GJS-200 is lower than the specific heat capacity of C/SiC, the absolute heat capacity of the friction ring with the same dimensions is higher for GJS-200. Normally, this leads to a larger brake disk diameter for ceramic brakes of about 1″ and reduces the weight savings from 65% (resulting from the lower density of C/SiC) to approximately 50% despite of using a metallic bell. The strength of the ceramic material depends on the fiber orientation and fiber length and is constant until the maximum temperature is reached during braking. GJS-200 shows at temperatures above 400 °C an tremendous decrease in strength, resulting in crack formation under thermal shock conditions. The high coefficient of thermal expansion of GJS-200 favors the crack formation and is the reason for brake distortion and judder at high braking temperatures. Due to the lower thermal conductivity of C/SiC perpendicular to the friction surface of the brake disk, the brake disks temperatures are generally higher than in conventional metallic brakes. More importance has to be attached to the heat protection of surrounding parts as well as the brake fluid overheating must be impeded.

16.3.2
Manufacturing

The manufacturing process for C/SiC brake disks is principally split into three main steps: The shaping of the brake disk rotor, the pyrolysis and the siliconizing (Figure 16.12). In the first step special treated short PAN (Poly Acrylic Nitrile) based carbon fibers, additives and phenolic resin are mixed to a homogeneous mass and a carbon fiber reinforced plastic (CFRP) green body is made using conventional warm press technique. The shaping takes place under exactly defined temperature-pressure-conditions in column presses with several cavities. Depending on the functional requirements, brake disks are made of C-fibers containing 3000–420000 filaments based on different recipes. The fiber length influences

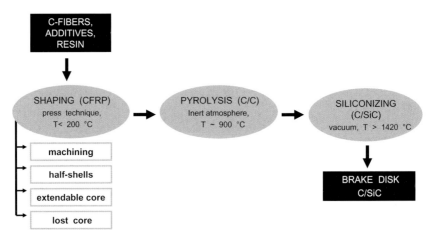

Figure 16.12 Manufacturing process for ventilated brake disks, generally split into three main steps: shaping, pyrolysis and siliconizing.

mostly the fiber strength, the oxidation behavior and the manufacturing. Long fibers lead to higher strength values but lower oxidation resistance and are more difficult to process. To get balanced properties and to guarantee an effective manufacturing, fiber fractions, meaning different fiber length-width-distributions, can also be processed.

Normally high performance brake disks are not massive but vented. To obtain a green body in a cost efficient way, it is common to use lost cores or a compression mould with extendable cores. The advantage of the lost core technique is that complex formed cooling channels can be realized without additional machining, while the use of extendable cores is limited to simple undercut-free geometries. In the early stage of the ceramic brake development half-shells were pressed and fixed together to form the cooling channels. The most expensive way to produce vented brake disks is by mechanical machining comparable to the manufacturing process of carbon/carbon brakes. In the second step the green body is temperature treated at up to 900 °C in inert atmosphere to protect the carbon from oxidation. The shrinkage of the phenolic resin during carbonization, the conversion to amorphous carbon and the stiffness of the embedded carbon fibers lead to a typical microcracked carbon/carbon microstructure with high porosity. The pyrolysis is characterized by a weight loss of the green body and changes in dimension. Where lost core technique is used, during carbonization the core disintegrates and can be easily removed without residue. For a time and energy efficient carbonization a through-type automatically temperature and velocity controlled furnace is used. To exclude an oxygen penetration special sensors permanently control the atmosphere inside the furnace.

In the last high temperature process step, the porous body is put together with liquid silicon under vacuum at temperatures above the melting point of silicon (1420 °C). After its infiltration into the porous microstructure, the liquid silicon reacts with the carbonaceous matrix to form silicon carbide, surrounding the fiber bundles as an inner oxidation protection. In the ideal case, the silicon only reacts with the carbon matrix, whereas the fibers remain unchanged. In practice, the carbon fibers are coated with additional carbon before mixing the raw materials to protect the carbon fibers against conversion to SiC. After the siliconizing process remains, beside the silicon carbon matrix and the carbon of the fiber bundles, free silicon as a further matrix component.

Because of the high hardness of the ceramic matrix diamond grinding tools are required for further machining steps. For an efficient tooling the preliminary materials stages are more suitable. For example the CFRP and the C/C can easily be machined with standard carbide tools and at high feed speeds to obtain high removal rates. Under economical aspects, a near net shape manufacturing of the brake disks from the beginning is essential and extensive machining of the ceramic product should be avoided.

Before assembling the friction rotor with the bell, an optional oxidation protection can be applied in order to avoid an early burnout of the carbon fibers lying unprotected on the disk surface where temperatures above the oxidation temperature of carbon occur quite often during high performance braking.

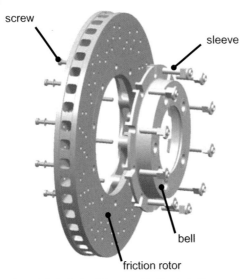

Figure 16.13 Brake disk assembly (front disk).

In the last step the friction rotor and the metallic bell are fixed together by special shaped sleeves to guarantee an unhindered thermal expansion of the ceramic rotor and especially the metallic bell (Figure 16.13). According to the requirements concerning thickness, axial run-out and flatness tolerance, unbalance etc. the brake disk has to be finished after mounting.

Beside the technical and economical aspects of manufacturing, the quality assurance of ceramic brake disks plays an important role in avoiding the delivery of faulty parts. Therefore, a lot of process parameters are used for evaluating the brake disks quality as well as batch tests (e.g. frequency analysis and optical verification) are carried out. Additional destructive strength tests guarantee reliability for the produced brake disks.

16.3.3
Braking Mechanism

Ceramic brake disks offer substantial improvements in several fundamental areas [34] and set a new benchmark in decisive user benefit aspects as shown in Figure 16.14 in comparison to conventional grey cast iron disks [35].

The outstanding property of ceramic disks, their low material density, offers a tremendous weight advantage (about 50%) over grey cast iron components. In case of a Porsche 911 Carrera S equipped with ceramic brakes, the weight saving compared to the grey cast iron model is over 15 kg. This reduces the weight of both the unsprung and rotating mass of the car with various improvements of driving performance, agility, handling, shock absorber response, road-holding comfort and economy of fuel.

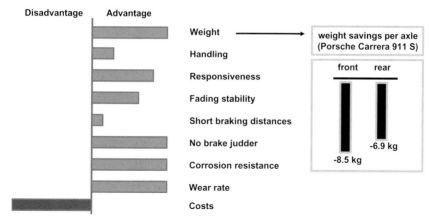

Figure 16.14 Advantages and disadvantages of ceramic brakes compared to grey cast iron disks.

The braking ability of the ceramic brakes is every bit as impressive as its weight. In splits of a second, the maximum brake force is built up, the driver has an excellent pedal feeling and the car a spontaneously responding behavior. The ceramic brake disks achieve not only high, but also constant friction coefficients, maintaining them independent of temperature. This excellent fading resistance ensures greater balance and powerful precise braking when slowing down from high speeds, driving downhill or under racing conditions as proved in a range of race series, including the Porsche Michelin Supercup [36].

The high and constant friction coefficient of ceramic brake disks provides shorter braking distances in even the toughest road and race conditions. The fact that the friction coefficient between the tire and the street is limited, the braking distances of cars equipped with ceramic brake disks is only little shorter compared to cars with grey cast iron disks. However, the spontaneous response gives the driver a feeling of safety and in critical moments the crucial centimeters can be won.

Even at high operating temperatures, the high thermal resistance and the low thermal expansion of the ceramic material (Table 16.1) ensure extraordinary dimensional stability resulting in excellent hot brake judder characteristics. Brake judder is perceived by the driver as minor to severe vibrations during braking. Hot judder is usually produced as a result of longer moderate braking. Commonly, it occurs when decelerating from higher speeds. The vibrations are the result of uneven thermal distributions so called Hot Spots. Hot Spots are concentrated thermal regions that alternate between both sides of a disk and cause, due to the materials distortion, sinusoidal waviness felt by the driver as vibrations.

Another critical phenomenon during braking is cold judder. Cold judder is caused by an incorrect finishing of the brake disk or by disk thickness variations (DTV) due to wear. These variations in the disk surface are usually the result of extensive vehicle road usage. Due to the ceramic matrix, the hardness of the

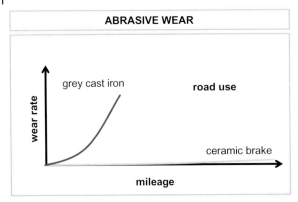

Figure 16.15 Idealized wear behavior of ceramic brake disks under normal braking conditions.

friction surface compared to grey cast iron is very high and extremely low wear rates under normal road conditions can be measured (Figure 16.15). This leads to a high durability of the ceramic brakes without losses in comfort. At last, ceramic brake disks are designed to last the entire lifetime of a car. Thanks to the non-iron based surface, corrosion is totally avoided, increasing the life span of the brake disk.

The main disadvantage of ceramic disks are their costs. The ceramic brake production is characterized by comparative low volumes and labor intensive and time consuming manufacturing steps. Also the technology requires totally new machining and automation concepts. To overcome these restrictions the level of automation must be increased considerably, the valuable carbon fiber must be used as effectively as possible and the production time must be shortened [37]. Efforts have been made but further improvements have to be done. Therefore a close cooperation between the manufacturer and the automotive industry is helpful, because the design of the brake disk and the system requirements have a tremendous influence on the brake disk costs.

16.3.4
Design Aspects

The brake rotor design plays a very important role in the effective use of the specific advantages of the ceramic composite material. Therefore the designer must keep the peculiarity of the material in mind and must adapt the brake disk to the requirements of the whole braking system. Obviously, this is a very complex task which requires a lot of experience and fundamental knowledge.

The maximum kinetic energy, the brake-power distribution and the mechanical and thermal loads caused by braking substantially determine the needed dimensions for the front and rear brake disk. To analyze the stress and temperature distribution FEM software is used. Therefore, the mechanical and thermal loads have to be known and entered in the FE calculation of the brake disk model. As

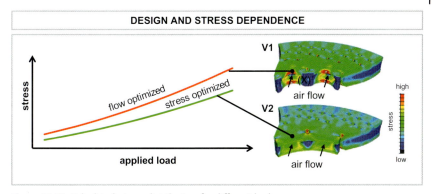

Figure 16.16 Calculated stress distribution for different brake disk designs while braking: V1: flow optimized, V2: stress optimized.

Table 16.2 Typical dimensions of ceramic brakes compared to the metallic version from selected Porsche cars.

		Carrera GT Ceramic Brakes	911 Carrera S Ceramic Brakes	911 Carrera S Metallic Brakes
Top Speed		330 km/h	293 km/h	293 km/h
Maximum Kinetic Energy		6.9 MJ	6.0 MJ	6.0 MJ
Front Brake Disks	Dimension	380 × 34 mm	350 × 34 mm	330 × 34 mm
	Weight	4.9 kg	5.8 kg	10.5 kg
Rear Brake Disks	Dimension	380 × 34 mm	350 × 28 mm	330 × 28 mm
	Weight	4.9 kg	5.7 kg	9.0 kg
Bell Material		Aluminum	Stainless steel	Stainless steel

Figure 16.16 shows, the highly stressed areas of the brake rotor have to be designed very carefully to avoid mechanical over-loads and to guarantee a maximum of air flow.

For better braking under wet conditions and for an efficient cooling brake disks are often perforated. The bores drilled in axial direction influence the friction coefficient, the wear of the friction pads, the stress and temperature distribution as well as the brake-noise behavior. Therefore the distribution of the bores has to be designed under consideration of these aspects.

To give an idea about typical brake disks dimensions, Table 16.2 shows an overview of several models from Porsche with ceramic and metallic brake disks. The weight of the ceramic disks is relatively low and is influenced by the material used for the bell. In case of a high strength aluminum bell, additionally more than 1 kg can be saved instead of using a bell made of steel.

As mentioned, oxygen attacks the carbon in the C/SiC both on the surface and inside the composite. The oxygen diffuses into the interior via an interconnected pore and crack network and can significantly reduce the strength of the brake disk,

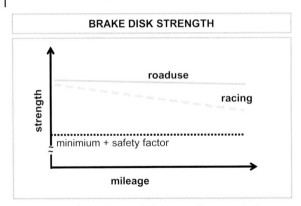

Figure 16.17 Idealized strength behavior of ceramic brake disks under road and racings conditions.

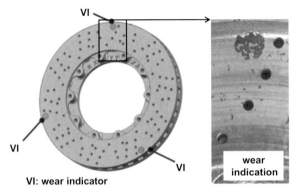

Figure 16.18 Special inserts to visualize the degree of oxidation.

particularly under racing conditions (Figure 16.17). To minimize the loss of strength, additional friction layers and an effective cooling design are used.

Additional friction layers with an oxidation optimized composition are often used to hinder the burn out of exposed carbon fibers on the friction surface. Also the brake disk's body is protected from high temperatures and from high oxygen attack. The manufacturing of the friction layers and their application to the brake rotor is completely integrated in the production process. Regarding the high lifetime of the ceramic disks, the oxidation status of the ground body has to be easily visible. Therefore special wear inserts are integrated in the friction surface (Figure 16.18). These inserts visualize the end of the disk life time.

The inner design of the brake rotor is a crucial point in the brake system's cooling optimization. Therefore since the introduction in 2001 a systematic effort has been made to develop a high flow rate brake rotor by using analytical and computational fluid mechanics software. As Figure 16.19 shows, the number of cooling channels has recently been doubled, while a new vent geometry offers a better airflow through the disk.

16.3 Ceramic Brake Disks | 401

Figure 16.19 Ventilated ceramic brake disk with an involute cooling channel design (right) in comparison to a more efficient channel design.

Figure 16.20 Cooling air flow for the front and rear brakes to avoid high braking temperatures to minimize wear, protect surrounding parts from overheating and guarantee a maximum of fading stability (Porsche 911 Turbo).

With more cooling channels, there are more internal walls creating greater structural stability. Externally, these modifications are clearly visible in the form of a modified drill-hole pattern. The resulting design of the brake rotor proves its high aerodynamics performance on the fluid test rig: the cooling flow rate through the rotor is increased by more than 20% compared to the original design [38]. With the increased cooling flow rate of the brake rotor, the cooling performance of the brake system has been considerably improved. To provide the required cooling air flow rate, the air flows through an air intake in the front of the car in a special air duct for the brake ventilation and over a spoiler directly on the brake disk (Figure 16.20).

16.3.5
Testing

As the braking procedure is very complex to simulate, various dynamometer and vehicle tests are carried out before delivery to guarantee the functionality of the brake system. Due to the fact, that the tribological behavior of brake disks is strongly influenced by the used brake pad material as well as the calipers it is necessary to test the whole brake system under realistic conditions on an inertia dynamometer. To get an impression about the friction behavior normally standardized brake tests are carried out. An example of a standardized test is the AK-Master: It is a dynamometer test developed by automotive engineers in Europe. This test is used to determine the general performance of friction materials. Both, the coefficient of friction and wear rates are recorded in dependence of brake pressure, deceleration, velocity and temperature. The experimental results obtained during these tests answer whether a brake pad material has stable and constant friction values or whether it shows a strongly influenced friction behavior or high wear rates. A specialty of Porsche is the fading test. This fading test is a high-load brake test and tests the high-end performance of a brake system. The boundary conditions are: 25 sequenced brake stops from 90% v_{max} to 100 km/h with a deceleration of $0.8\,m/s^2$ equal one fading. This sequence leads to brake disk temperatures above 700 °C (Figure 16.21) which have to be withstood by the brake disk as well as by the brake pads. Obviously, ceramic disks are very insensitive to these temperatures whereas the brake pads composition has to be chosen very carefully. Armed with metal containing linings, ceramic disks maintain their frictional coefficient regardless of temperature for the ultimate performance in fade-free stopping power.

Besides dynamometer tests with their advantages considering quickness, complexity and repeatability of the results an intensive vehicle testing is necessary, especially because some phenomena appeared for the first time in vehicle testing. The validation of the dynamometer test results take place in special endurance and race track vehicle tests with different load profiles and environmental conditions, such as arctic temperatures or tropic heat. Whereas in endurance tests a customer road using is primarily in focus, special track tests give information about the ultimate performance of the brake system, stopping distance from several speeds and the cooling effectiveness, as well as the temperature behavior of the surrounding parts including the brake fluid.

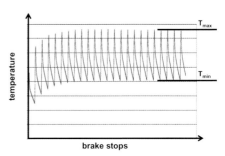

Figure 16.21 Porsche dynamometer test to analyze the high performance brake potential (left) and a typical fading temperature profile.

16.4 Ceramic Clutches

Carbon/Carbon clutches are well known from racing and some high performance cars [26]. Carbon is very temperature resistant and lightweight and is often used under severe conditions and leads to small, compact and lightweight clutches. Racing clutches are normally designed with a focus on:

- minimum weight,
- low moment of inertia,
- minimum dimensions,
- maximum strength,
- optimal performance,
- maximum heat resistance [39].

For the use in series-production vehicles the wear resistance of carbon is too low and the friction characteristic is too aggressive to offer sufficient start-comfort. Organic or sintered friction materials used in a single disk clutch construction with a metallic friction disk, offer good friction behavior but due to the limited heat resistance and the high density, these clutches are comparatively heavyweight and big.

To overcome theses restrictions, Porsche has developed, especially for the Carrera GT, a ceramic composite clutch [40, 41]. The double-disk dry ceramic composite clutch (Figure 16.22) meets the typical racing requirements of a smaller diameter and low weight combined with long service intervals and good start-up behavior. As no appropriate clutch system for the Carrera GT was available on the market, Porsche and selected Partners (Sachs and SGL) have developed a

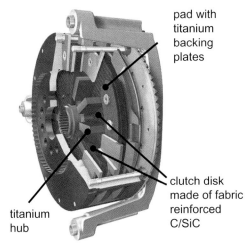

Figure 16.22 Ceramic composite clutch from Porsche (PCCC) combining low weight, small diameter and good wear and friction behavior.

completely new clutch using an innovative ceramic composite material for the friction disk and a special metal containing pad material as counterpart. Both, the disks and the pads show friction values over 0.4 with a degressive friction path i.e. insensitive to grabbing. The maximum power transmission lies over 1000 Nm.

The ceramic material consists of a carbon fiber reinforced silicon carbide material (C/SiC) with fabric reinforcement. In comparison to brake applications, clutch disks are more thin-walled and have to withstand mechanical loads resulting from up to 20,000 rpm. Instead of a short fiber reinforcement the ceramic material of the clutch disk is built up with different layers of carbon fiber fabrics to guarantee the necessary strength. The manufacturing is similar to the process of the brake disks: Forming a CFRP green body, pyrolysis and siliconizing. Due to the fabric reinforcement a near net shape manufacturing couldn't be realized and flat plates were made. After the siliconizing step the clutch disks are cut out of the resultant ceramic plates with water jet at around 3000 bar (Figure 16.23).

Thanks to the material's heat resistance and the high friction values a clutch plate diameter of only 169 mm could be realized. The compact dimensions of the clutch, helps to give the engine and transmission a very low center of gravity (Figure 16.24). The result is a very sportive character of the car improving agility and handling. The weight reduction e.g. over the 7.6 kg conventional clutch of the Porsche 911 Turbo is more than fifty percent for the 3.5 kg PCCC and lead to an excellent rev-up behavior of the engine and a very spontaneous acceleration.

Comprehensive tests on the test bench and extensive vehicle testing, also under extreme loads showed that the PCCC is able to at least match the service intervals of conventional clutches [42].

Figure 16.23 Water jet cutting of the clutch disks at about 3000 bar.

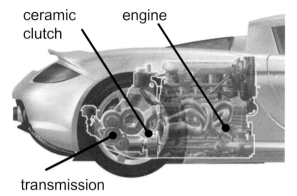

Figure 16.24 The small diameter of the clutch enables a low center of gravity for the engine.

Beside the outstanding properties, the price for the PCCC (Porsche Ceramic Composite Clutch) is roughly ten times higher than from conventional clutches. Therefore the use is actually limited for supersport cars and racing, particularly for long distance and off-road races. Further improvements in the material's production technology to lower the costs possess excellent future prospects for high performance applications in mass production vehicles.

References

1 Krenkel, W. (1994) CMC Materials for High Performance Brakes, in Proceedings of the 27th International Symposium on Automotive Technology and Automation (ISATA), Conference on Supercars, Aachen, Germany, 31 Oct.–4 Nov.
2 Friction Unit, European Patent EP 0797555, 1994.
3 Brake Disk, European Patent EP 0777061, 1994.
4 Krenkel, W. (1996) Keramische Verbundwerkstoffe für Hochleistungsbremsen, in *Verbundwerkstoffe und Werkstoffverbunde* (Eds.: G. Ziegler), DGM-Verlag, Oberursel, Germany, pp. 329–332.
5 Pfeiffer, H., Krenkel, W., Heidenreich, B. and Schlenk, L. (1997) Bremsscheiben aus keramischen Verbundwerkstoffen für Schienenfahrzeuge, in *Werkstoffe für die Verkehrstechnik* (Ed.: U. Koch), DGM Informationsgesellschaft, Frankfurt, Germany, p. 275.
6 SGL Carbon setzt auf Forschung und Entwicklung, cfi/Ber. DKG 83, No. 11–12, 2006.
7 Carbon/Keramik Bremssysteme: Leistungsfähig–Sicher–Komfortabel, cfi/Ber. DKG 83, No. 11–12, 2006.
8 Abu El-Hija, H., Krenkel, W. and Hugel, S. (2005) Development of C/C-SiC Brake Pads for High-Performance Elevators, *International Journal of Applied Ceramic Technology*, Volume 2, Issue 2, pp. 105–113.
9 Krenkel, W., Heidenreich, B. and Renz, R. (2002) C/C-SiC Composites for Advanced Friction Systems, *Advanced Engineering Materials*, Volume 4, No. 7, pp. 427.
10 Renz, R. and Krenkel, W. (2000) C/C-SiC Composites for High Performance Emergency Brake Systems, 9th European Conference on Composite Materials (ECCM-9), Brighton, UK, June 4–7.
11 Krenkel, W. (2000) Development of a Cost Efficient Process for the Manufacture of CMC Components, Doctoral Thesis,

University of Stuttgart, DLR Forschungsbericht 2000–2004.

12 Krenkel, W. (2002) Design of Ceramic Brake Pads and Disks, American Ceramic Society, *Ceramic Engineering and Science Proceedings*, Volume **23**, Issue 3, pp. 319.

13 Krenkel, W. (2003) Ceramic Matrix Composite Brakes, in *Advanced Inorganic Structural Fibre Composites IV* (Eds.: P. Vincenzini, C. Badini), TECHNA, Faenza, Italy, pp. 299–310.

14 Krenkel, W., Abu El-Hija, H. and Kriescher, M. (2004) High Performance C/C-SiC Brake Pads, *American Ceramic Society, Ceramic Engineering and Science Proceedings* (Eds.: E. Lara-Curzio, M.J. Readey), Volume **25**, Issue 4, pp. 191–196.

15 Krenkel, W. and Berndt, F. (2005) C/C-SiC Composites for Space Applications and Advanced Friction Systems, *Materials Science & Engineering*, **A412**, Elsevier B.V., pp. 177–181.

16 Krenkel, W. (2004) Carbon Fiber Reinforced CMC for High-Performance Structures, *International Journal of Applied Ceramic Technology*, Volume **1**, Issue 2, pp. 188–200.

17 Vaidyaraman, S., Purdy, P., Walter, T. and Horst, S. (2001) C/SiC Material Evaluation for Aircraft Brake Applications, in: Krenkel, W., Naslain, R., Schneider, H. (Eds.), *High Temperature Ceramic Matrix Composites*, Wiley-VCH, Weinheim, Germany, pp. 802–808.

18 Krupka, M. and Kienzle, A. (2000) Fiber Reinforced Ceramic Composites for Brake Discs, SAE Technical Paper Series 2000-01-2761.

19 Weiss, R. (2001) Carbon Fiber Reinforced CMCs: Manufacture, Properties, Oxidation Protection, in: Krenkel, W., Naslain, R., Schneider, H. (Eds.), *High Temperature Ceramic Matrix Composites*, Wiley-VCH, Weinheim, Germany, pp. 440–456.

20 Gadow, R. and Speicher, M. (2001) Manufacturing of Ceramic Matrix Composites for Automotive Applications, *Ceramic Transactions*, Volume **128**, pp. 25–41.

21 Rak, Z. (2001) CF/SiC/C Composites for Tribological Application, in: Krenkel, W., Naslain, R., Schneider, H. (Eds.), *High Temperature Ceramic Matrix Composites*, Wiley-VCH, Weinheim, Germany, pp. 820–825.

22 Heidenreich, B., Renz, R. and Krenkel, W. (2001) Short Fibre Reinforced CMC Materials for High Performance Brakes, in: Krenkel, W., Naslain, R., Schneider, H. (Eds.), *High Temperature Ceramic Matrix Composites*, Wiley-VCH, Weinheim, Germany, pp. 809–815.

23 Krenkel, W. (2003) C/C-SiC Composites for Hot Structures and Advanced Friction Systems, *Ceramic Engineering and Science Proceedings*, Volume **24**, No. 4, pp. 583–592.

24 Zhang, J., Xu, Y., Zhang, L. and Cheng, L. (2007) Effect of Braking Speed on Friction and Wear Behaviour of C/C-SiC Composites, *International Journal of Applied Ceramic Technology*, Volume **4**, Issue 5, pp. 463–469.

25 Data Sheet Starfire Systems "Ceramic Composite Brake Rotors", www.starfiresystems.com. 06/2005.

26 Savage, G. (1990) *Carbon/Carbon Composites*, Chapmann+Hall, London.

27 Breuer, B. and Bill, K.H. (2006) *Bremsenhandbuch, Grundlagen, Komponenten, Systeme, Fahrdynamik*, ATZ-MTZ-Fachbuch, Vieweg Verlag, ISBN 978-3-8348-0064-0.

28 Kochendörfer, R. (1987) *Heiße tragende Strukturen aus Faserverbund-Leichtbauwerkstoffen*, DGLR-Jahrbuch.

29 Krenkel, W. and Hald, H. (1988) Liquid Infiltrated C/SiC – An Alternative Material for Hot Space Structures, Proceedings of The ESA/ESTEC Conference on Spacecraft Structures and Mechanical Testing, Noordwijk.

30 DaimlerChrysler: Coupé im Rennanzug, DaimlerChrysler Times, 09.06.2000.

31 Porsche: Porsche Ceramic Composite Brake (PCCB), Die Serienfertigung hat begonnen, Porsche Pressemitteilung Nr. 72/00, 23.08.2000.

32 Audi: Keramikbremsen für den Audi A8 12-Zylinder, Audi Pressemitteilung, 07.06.2005.

33 Kienzle, A. and Meinhardt, J. (2006) Kurzfaserverstärkte C/SiC-Keramiken für den Fahrzeugbau, Technik in Bayern Nr. 5.
34 Neudeck, D., Martin, R. and Renzow, N. (2002) Porsche Bremsenentwicklung – "Vom Rennsport auf die Straße", Fortschritt-Berichte VDI Reihe 12, Nr. 514.
35 Renz, R. and Seifert, G. (2007) Keramische Hochleistungsbremsscheiben im Sportwagenbereich, Berichtsband DVM TAG Reifen, Räder, Naben, Bremsen.
36 Porsche, Überzeugungs-Kraft, Porsche Christophorus Magazine 316, 2005.
37 Güther, H. M. and Grasser, S. (2007) Carbon Fibers: Entering Mass-Produced Cars? Auto Technology No. 1.
38 Porsche, Hart im Nehmen, Porsche Christophorus Magazine 311, 2004.
39 ZF Sachs, Formula Clutch Systems, Delivery Program, 2006.
40 Porsche, Little, Strong, Black, Porsche Christophorus Magazine 306, 2004.
41 Steiner, M., Erb, T., Hölscher, M. et al. (2004) Innovative Material Usage in the Porsche Carrera GT, Proceedings of FISITA, Barcelona.
42 Porsche Engineering Group, The ceramic clutch – a world first from Porsche, Porsche Engineering Magazine No. 02, 2004.

Index

a

active filler controlled pyrolysis (AFCOP)
    process 178
advanced friction system 385f.
aeronautics 132
    – application 327ff., 340ff.
    – military aeronautical application 346
    – commercial aeronautical
      application 348
air permeability 220f.
Almax fabrics 196ff.
alumina 191ff., 207ff., 219ff.
aluminate
    – rare-earth 197
aluminum hydroxide 223
application
    – aeronautical 327ff., 340ff.
    – C/SiC 340ff., 385ff.
    – CERASEP 345
    – CMC 153ff., 172
    – friction 338, 385ff.
    – hot structure 385
    – liquid rocket engine 329ff., 341
    – propulsion system 385
    – SEPCARBINOX 345
    – space 131, 327ff., 340ff.

b

bend stress relaxation (BSR) 374
bending stiffness 41
bending test 248
bond 57
bond concept 293
    – galvanic mounted point 293
    – sintered mounted point 293
bonding 50
    – metallic 291
boost motor 329
BoraSiC® 90
boron carbide 90

braid 27
    – angle 27
    – biaxial 28
    – pattern 27
    – triaxial 28
    – UD 32
braided structure
    – three-dimensional 34
braiding 173
    – four-step 34
    – square 34
    – technology 32
    – three-dimensional rotary 34
brake disk 385ff.
    – assembly 396
    – carbon/carbon (C/C) 391
    – ceramic 391ff.
    – grey cast iron 391
    – mechanism 396ff.
    – metal matrix composite (MMC) 391
    – ventilated 394
brake judder 397
brake pad 385ff.
    – organic 389
    – sintermetallic 389
brake rotor design 398
brake test 402
braking material 389
braze alloy 309
bridging 208
brittle failure 246
brittle fiber 43
brittle rigid material 300
BSR, *see* bend stress relaxation
bulk strength testing 373
burner 220ff.
    – nozzle 206
    – pore 227

Index

c

cantilever test 42
carbon 80f., 122
– fabrics 72
– matrix 73
– pore filling mechanism 79
– pyrolytic (pyC) 175
carbon fiber 12, 172
  – reinforced plastics (CFRP) 117ff., 176, 280ff.
  – reinforcement 71
  – transfer 331
  – type 71
  – structure 14
carbon fiber reinforced carbon composite (C/C) 69ff., 81ff., 97, 134, 245f., 309ff., 328ff., 353
  – brake disk 93ff., 391
  – bulk protection system 83ff.
  – bundle 116ff.
  – delamination detection 280
  – felt susceptor 108
  – fixture 102ff.
  – industrial application 82, 92
  – insulation 107
  – load carrying 102
  – manufacturing 69ff.
  – multilayer coated 88
  – non-destructive testing 267
  – oxidation 84
  – oxidation protection 83
  – preform 118
  – tray 103
  – unidirectional composite 71
carbon fiber reinforced silicon carbide (C/SiC) 86, 114ff., 130ff., 179, 247ff., 280ff., 304ff., 385ff.
  – application 340ff., 385ff.
  – brake disk 385
  – brake pad 385
  – ceramics 276
  – coating 175
  – composite 136, 276, 304ff., 339ff.
  – feature 179
  – fracture behavior 245
  – LPI 246
  – LSI-based 131
  – material 177
  – non-destructive testing 267
  – nozzle extension 182
  – precursor 176
carbon fiber reinforced silicon carbide composite (C/C-SiC) 86ff., 114ff., 130ff., 287ff., 385
  – CVI/LSI 245
  – LSI-based 131
cellulose
  – regenerated 17
ceramic brake disk 391ff.
ceramic clutch 403f.
ceramic composite 338ff.
  – clutch 403
  – manufacturing 343
  – material 391
ceramic fiber reinforced composite (CFRC) 356
ceramic matrix composite (CMC) 1ff., 114ff., 141ff., 280ff., 287, 303ff., 356
  – all-oxide 205ff.
  – application 153ff., 172
  – CMC system 314
  – combustion chamber 155
  – corrosion 151ff.
  – data 150
  – fiber 1
  – friction application 385ff.
  – fusion reactor 156
  – gas turbine 155
  – high temperature property 151
  – hot structure in space 153
  – joining and integration 303ff.
  – joint 308
  – low-cycle-fatigue (LCF) 151
  – manufacturing 175
  – material 113ff., 170ff.
  – mechanical property 148
  – metal system 309
  – micrograph 147
  – mini 208
  – non-destructive testing/evaluation/inspection (NDT/NDE/NDI) technique 261ff.
  – nut 153
  – oxidation 152
  – property 121, 146ff.
  – screw 153
  – stress-strain behavior 149f.
  – thermal shock 151f.
ceramics 376
  – bulk strength testing 373
ceramization 176
Cera-NITE 360
CERASEP
  – A373 345
  – A400 345
  – A410 344ff.
  – application 345

– lifetime duration in oxidative environment   345
CESIC material   114
chemical bond type   5
chemical vapor deposition (CVD)   88, 166, 191, 357
  – coated C/C   106
  – coating   87
  – process   7
  – SiC coating   88ff.
chemical vapor infiltration (CVI)   75f., 113ff., 136, 333f., 357
  – CMC   143, 173
  – coated C/C   106
  – densification   79, 332
  – fiber   116
  – gradient   145
  – impregnation   79
  – isothermal   76
  – precursor   166
  – pressure gradient   77
  – process   141ff.
  – rapid   77
  – SiC   370
  – SiC/SiC   370ff.
  – thermal gradient   76
chip transportation   295
chromia
  – matrix porosity   195
  – toxicity   195
clutch system   403
CMC, see ceramic matrix composite
coating   60f.
  – engineered interface   188
  – environmental barrier (EBC)   92, 130, 223
  – fugitive   235
  – porous   61
  – protective   223
  – thickness   191
coefficient of friction (CoF)   385ff.
coefficient of thermal expansion (CTE)   130ff., 304ff.
combustion   223
  – chamber   155
combustor liner   189, 206, 227
composite
  – damage tolerant   49
  – fabric-reinforced 2D   386
  – graded   390
computed tomography (CT)   280
  – defect detection   280
  – functional principle   278
  – micro-structural   282

– process accompanying   284
– X-ray   278
conductivity
  – thermal   130
  – transverse   386f.
coolant   291
cooling air flow   401
cooling channel   395ff.
core
  – extendable   395
  – lost   395
  – roughness depth   295
crack   49ff.
  – arrest   55
  – bridging   141, 205ff.
  – deflection   59, 189, 205ff., 219
  – deviation   49ff., 62
  – formation   142
  – interface   53ff.
  – interfacial   62
  – matrix   49ff.
  – micro   267
  – resistance   141ff.
  – transverse   50ff.
creep   217, 248
  – resistance   1, 57
CT, see computed tomography
CTE, see coefficient of thermal expansion
cutting   39
  – grain   291
CVD, see chemical vapor deposition
CVI, see chemical vapor infiltration   75
cyclic stress   253
cyclic test   250

d

damage   234ff.
  – detection   272
  – tolerance   49ff., 61
debond energy   65
debonding   205ff., 233, 299
  – progressive   233
defect detection   280
  – computed tomography   280
deformation
  – inelastic   241
  – twin   195
delamination   215ff.
DEN (double end notch)   239ff.
densification   79, 332
design aspect   398
deviation   66
diamond   291, 395
  – polycristalline (PCD)   291

direct metal oxidation (DIMOX)   114, 232
displacement damage   366
Ditrichlet cell   214
DLR (Deutsches Zentrum fuer Luft- und Raumfahrt)   209
double notch shear (DNS)   376
drapability   42
draping   39
drilling   287ff.
– blind hole   298
– carbon fiber reinforced silicon carbide composite (C/C-SiC)   287
– hollow drilling tool   291
– machining   287ff.
dynamometer test   402

e
echo-pulse analysis   265f.
electron-beam physical vapor deposition (EB-PVD)   224
electrophoretic deposition (EPD)   191
electroplated surface   298
elevator   388
elongation   41
emergency brake system   385
environmental barrier coating (EBC)   92, 130, 223
EPD   194
European Apogee Motor (EAM)   183
evaluation, *see* non-destructive testing

f
fabrics   210
– knitted   28
– non-crimp   29
– reinforcement   404
– three dimensional woven   30
fading resistance   397
fading test   402
failure
– brittle   246
fatigue   216, 251ff.
FE model   239f.
fiber   1ff.
– architecture   21, 387ff.
– brittle   43
– bundle   216
– carbon   1ff., 71
– ceramic   1
– ceramic matrix composite   1
– coater   191
– continuous   194
– CVI   115f.
– degradation   195
– fabrics   216
– inorganic   1
– monazite coated   196
– non-oxide   172
– oxide   172
– polymer impregnation   115
– property   4f.
– protection   115
– pull-out   141ff., 188ff., 205ff.
– push-out   195f.
– reinforced ceramics   175
– reinforced green body   117
– reinforced preform   118
– short fiber   392
– short fiber architecture   387
– supramolecular structure   3
– structure   2ff.
– structure formation   3f.
– type   2
fiber coating   115, 146, 187ff., 205ff., 234f.
– bridging   193
– carbon   193
– cloth   193
– crusting   193
– fabric   193
– fugitive   197
– hermetic   198
– *in-situ*   196
– monazite   193ff.
– preform   193
– strength   194
– thickness   191
fiber matrix interface   115, 178, 189f.
– crack deflection   189ff.
– porous matrix   195
fiber reinforcement   392
– fabrics   404
– three-dimensional   194
filament damage   42
filament winding   173
filler
– inert   177
filter   220
final heat treatment (HTT)   80ff.
finite element model   238ff.
firing nozzle   227
flat-weft-knitting technology   29
flute   295
fracture
– damage-tolerant   219
– energy   234ff., 378
– mechanism   148
– resistance   378
– toughness   148

friction
- application   338, 385ff.
- coefficient   397
- internal   252
- surface   400
- system   134
frictional property   131
fundamental tolerance IT   294
furnace   97ff.
- industrial   227
fusion reactor   156

g

gas cooled fast reactor (GFR)   354ff., 379
gas ionization detector   275
gas reactor technology   354
gas turbine   155, 206, 219ff.
Geiger counter   275
gelation   210
German Aerospace Center (DLR)   209
glass ceramics   207
GRABER   322
grain
- coarsening   206, 217
- cutting   291
graphitic matrix   95
green body   117f., 194, 394
grey cast iron   391ff.
grind-in behavior   292
grinding   288
GUIPEX preform   343f.

h

handling   39
heat capacity   393
heat shield   184, 206
heating
- frictional   255
helix angle   296
high isostatic pressing (HIP)   332
high modulus (HM)   71
high performance brake   402
high temperature furnace   97ff.
high temperature property   246
high temperature test reactor (HTTR)   380
hot pressing   194ff., 207
hot isostatic pressing (HIPing)   194, 207
hysteresis curve   253

i

IDSS, *see* interfacial debond shear strength
IFS, *see* interfacial friction stress
IHX, *see* intermediate heat exchanger
ILSS, *see* inter-laminar shear stress

IM, *see* intermediate modulus
impregnation   83f., 102, 210
impulse excitation method   372
infiltration   29, 37ff.
- gradient   144
- isothermal   147
- isothermal-isobaric   144
- penetration depth   143f.
- time   146
- vacuum-assisted   210
infrared photography   269
inspection analysis, *see also* non-destructive testing
- optical and haptic   262
integration
- CMC-metal system   309
- technology   304ff.
inter-laminar shear stress (ILSS)   376
interface   49ff., 60ff.
- crack   53ff.
- property   235f.
- strength   66
interfacial debond shear strength (IDSS)   378
interfacial friction stress (IFS)   378
intermediate heat exchanger (IHX)   354ff.
intermediate modulus (IM)   71
interphase   49ff., 65
- transverse strength   66
ion irradiation   363
IT, *see* fundamental tolerance

j

jet engine   133
joining
- CMC-metal system   309
- technology   40, 304ff.
joint
- design   304ff.
- stability   307
- strength   307
journal bearing   156ff.
- liquid oxygen (LOx)   159ff.

k

knitting pattern   30

l

laminate   210
laser communication terminal (LCT)   134
laser extensometer   242
launcher propulsion   180
lifetime prediction   219

liquid
- immiscible 191
liquid oxygen (LOx) 159ff.
liquid phase sintering (LPS) 357
liquid rocket engine (LRE) 329ff., 341
- application 341f.
liquid silicon infiltration (LSI) 113ff., 261, 290, 385
- C/SiC 388
load
- sharing 54f.
- transfer 51ff.
loading
- off-axis 238ff.
- on-axis 238ff.
Lockin thermography 271
loop formation 28
low-expansion structure 134
LPI
- material 123ff.
- PIP 136
- PIP-CMC 127
LPS, *see* liquid phase sintering
LRE, see liquid rocket engine
LSI, *see* liquid silicon infiltration

m

machining 211
- hard 301
magnetron-sputtering 224
mandrel 32
manufacturing 394
material
- brittle rigid 300
- database 378
matrix
- ceramic matrix composite (CMC) 172
- pore 214
- porosity 195, 216
- porous 205ff.
- Si/C 119
matrix conversion 86
matrix crack 49ff., 66, 207ff.
matrix cracking stress 233
matrix interface 115
- fiber 115
matrix precursor 73f.
matrix system 73
mechanical property 44
- change 369
- evaluation 371
mechanism
- quasi-ductilizing 175

melt infiltration (MI) 357
- C/C-SiC 127
- C/SiC 126ff.
- CMC material 114ff., 130
- LSI process 123f., 136
- process 113ff.
- SiC/SiC material 133
- SiSiC material 114
mesostructure 214
metal
- matrix composite (MMC) 391
- thermal treatment 102
metal organic chemical vapor deposition (MO-CVD) 224
methanization 107
methyl-trichloro-silan (MTS) 143f.
micro-composite 62
micro-cracking 214, 267
micro-structural evolution 364
military aeronautical application 346
mini-composite 57ff.
modeling 238ff.
- micromechanical 243
- microstructural 231ff.
molding 94
moment of inertia 403
monazite 191ff.
- precursor 197
monolithic armor ceramics 135
morphology 3
mounted point 291ff.
- galvanic 293
- sintered 293
mullite 207ff., 219ff., 246

n

nano-powder infiltration and transient eutectic (NITE) process 357ff.
- NITE-SiC 361
- NITE-SiC/SiC 361f.
nanotube 66
Naxeco process 333
near-net-shape 23
- manufacturing 395
Nextel 246
- 610 191ff.
- 720 197
NITE process, *see* nano-powder infiltration and transient eutectic process
non-destructive testing/evaluation/inspection (NDT/NDE/NDI) technique 261ff.
- echo-pulse analysis 265
- infrared photography 269

Index | 415

– optical and haptic analysis 262
– radiography (X-ray analysis) 272ff.
– thermal imaging 269
– thermography 269ff.
– transmission ultrasonic analysis 264
– ultrasonic analysis 262f.
non-oxide ceramic filament fiber 11
non-oxide fiber 172
non-oxide precursor 171
nonwovens 24f.
notch sensitivity 244
Novoltex 330ff.
nozzle 335
– throat 336
nuclear application 353ff.

o

OPS, *see* oxidation
optimization 290ff.
outer glass sealing layer 90
outer multilayer coating 88
overbraiding 32
oxidation 59, 248
– inhibitor 313
– protection system (OPS) 248ff.
– stability 1
oxide 60
– coating 61
– fiber 172
oxide ceramic fiber 206
oxide ceramic filament fiber 10
oxide matrix
– dense 194
oxide/oxide CMC 189ff., 205ff.
– alumina sol 194
– ceramic precursor 194
– dense matrix 190
– fiber preform 194
– interlaminar 189
– matrix crack 190
– notch sensitivity 197
– off-fiber-axis strength 189
– porous matrix 191
oxide/oxide composite 187ff.
– environmental stability 187
– interlaminar property 187
– matrix-dominated property 187
– microcracking 187
– porous matrix 187
– toughness 187

p

pad material 404
PAN (poly acrylic nitrile) 394
– precursor 15, 331

parameter
– cutting speed 294
– feed 294
– tool rotation speed 294
pattern 25
PCCB, *see* Porsche ceramic composite brake
PCD, *see* polycrystalline diamond
PDC, *see* polymer
pebble bed reactor (PBR) 354
permeability 209
photographic plate 275
photovoltaic cell 105
physical vapor deposition (PVD) 224
– electron-beam (EB-PVD) 224
PIP (polymer infiltration and
  pyrolysis) 113ff., 136, 165ff., 303, 332
– application 165ff., 180ff.
– ceramic matrix composite (CMC) 175
– fiber-coated PIP-CMC 175
– LPI 136
– PIP-C/SiC 181f.
– precursor property 165ff.
– process 171ff.
PIRAC (powder immersion reaction assisted
  coating) 87
pitch
– impregnation 332
– matrix process 334
– precursor 17
polycrystalline diamond (PCD) 291
polymer
– derived ceramics (PDC) 165
– impregnation 115
– infiltration and pyrolysis, *see* PIP
– precursor 9, 170
– shrinkage mechanism 178
polysilicon 105
pore agglomeration 219
porous structure 43
Porsche ceramic composite brake
  (PCCB) 392ff.
pre-pregging 194
precursor 389
– ceramic 194
– cross-linking behavior 167
– non-oxide 171
– property 165ff.
– pyrolysis behavior 169
– system 165ff.
preform 210, 328
– manufacturing 173
preforming 38
– multi-step 38
– one-step 38

prepreg 38, 173
prismatic modular reactor (PMR) 354
process
– multi-step 23
– one-step 23
– strategy 295
processing
– preform 38
– prepreg 38
property
– mechanical 393
– thermophysical 393
proportional limit strength (PLS) 361
pseudo-boehmite 211
push-in 62ff.
push-out 62ff.
pyrolysis 210, 394
– behavior 169
pyrolytic carbon (pyC) 175

q
quality assurance 396
quality management 42

r
radial overbraiding machine 33
radiation damage 362
radiography 272ff.
– C/SiC composite 276
– limitation 278
re-infiltration 216
reaction-bonding 208, 224
redensification/recarbonization cycle 79
reinforcement, *see also* carbon fiber . . . 343
– n-dimensional 328
– preform 329
– three-dimensional 329
relic process 7
resin
– infusion (RI) method 104
– matrix process 334
– transfer moulding (RTM) 117
rotor 388
roving 23

s
S-curve 253
sandwich design 37
sapphire 197
– fiber 194
satellite propulsion 182
Scheelite 197
self-healing layer 89
self-healing matrix 343ff.

SEM 193
SEPCARB 336ff.
SEPCARBINOX 341f.
– A500 344ff.
– application 345
– lifetime duration in oxidative
   environment 345
sewing technology 40
– one-sided 40
shape weaving process 31
shear testing method 376
short fiber reinforced C/SiC 393
short fiber reinforcement 404
SICARBON 180
silicate
– rare earth 223
silicon
– alloy 314
– based precursor 165
– hydroxide 223
– infiltration 119
– molten 119
– Si/SiC 284
– SiC 113ff.
– SiC composite 385
– SiC fiber 116ff.
– SiC matrix 114ff.
– SiC/Al$_2$O$_3$ 246, 306
– SiC/C preform 118
– SiC/Si-B-C composite 348f.
– SiSiC 113ff.
silicon carbide fiber reinforced silicon
   carbide matrix composite (SiC/
   SiC) 113ff., 128ff., 246ff., 304ff., 339ff.,
   353ff., 368
– composite material 379
– NITE process 359ff.
silicon treatment process 86
siliconizing 394
single edge notch beam test (SENB) 244f.
sintering
– pressureless 194
sliding element 95
small specimen test technique (SSTT)
   372
Soft-Cera 198
solar energy market 105f.
solid rocket motor (SRM) 329
– nozzle 335
solid state array 275
solution/sol/slurry precursor 191
space application 131, 327ff., 340ff.
space transportation 206
space vehicle 227

spacecraft   226
spacer fabrics
  – contour warp knitted   38
  – warp-knitted   37
spinning dope   7
spinning process
  – core   24
  – dry   4
  – melt   4
  – wet   4
square-braiding   34
stability
  – high temperature   1
stator   388
stiffness   253
strain   44
strength   50ff., 65
  – interfacial   61
  – strengthening   57
  – transverse   53
  – ultimate   55f.
stress   61
  – concentration   61
  – cyclic   253
  – interface shear   56
  – residual   50ff., 307
  – shear   50
  – state   306
  – tensile   50
subcomponent   318
surface roughness   306
swelling   364
  – point-defect   364ff.

t
TAC, *see* total accumulated cycle
tailored fiber placement (TFP)   40, 104
TEM   193
tensile
  – high (HT)   71
  – loading   233
  – modulus   6
  – strength   6, 41, 62, 195, 216
  – ultimate tensile strength (UTS)   128
test
  – bending   248
  – brake   401f.
  – cantilever   42
  – cyclic   250
  – dynamometer   402
  – fading   402
  – pin-on-disc   160
  – ring-on-disc   157

  – single edge notch beam (SENB)   244f.
  – small specimen test technique (SSTT)   372
testing, *see also* non-destructive testing
  – bulk strength   373
  – composite   374
  – issue   304f.
  – shear testing method   376
textile
  – architecture   23
  – two dimensional   23
  – three dimensional   23ff.
textile preform   21
textile reinforcement structure   21ff.
thermal conductivity   130, 221ff., 366ff.
thermal diffusivity   220
thermal fatigue   220
thermal gradient   187, 220
thermal imaging   269
thermal insulator   220
thermal protection system (TPS)   327
thermal shock   187, 220
thermal stability   218
thermal treatment
  – metal   102
thermography   269ff.
  – Lockin   271
  – ultrasonic (US) induced   272
thermoplastic   74
thermosetting resin   73
thixotropy   312
thrust vector control (TVC)   132
titanium alloy   359
tool   290ff.
  – diamond   395
total accumulated cycle (TAC)   348
toughness   54ff.
tow
  – transverse   50
transmission ultrasonic analysis   264
trepanning   301
tribological behavior   386ff.
tribology
  – data   158
  – friction coefficient   160
  – liquid oxygen (LOx)   159
  – pin-on-disc test   160
  – ring-on-disc test   157
  – system   157ff.
turbine engine   226

u
UD-braid   32
ultimate tensile strength (UTS)   128

u

ultrasonic analysis   262ff.
  – CMC   267
  – physical principle   263

v

vacuum bagging/autoclaving   194
vacuum plasma spraying   224
ventilated brake disk   394
ventilated ceramic brake disk   401
very high temperature reactor (VHTR)   354ff., 380
volatilization   223

w

warm press technique   394
warp-knit   28
  – spacer fabrics   37
water jet cutting   404
water vapor   223
wear   293
  – behavior   398
  – insert   400
  – rate   388
  – resistance   209, 385
weak interface composite (WIC)   231ff., 243ff.
  – CVI derived   232
weak matrix composite (WMC)   231ff., 244ff.
weaving
  – three dimensional   173
weft insertion   29
weft-knit   28
  – multilayer   37
Weibull modulus   373
wettability   305
WHIPOX   211, 224
winding pattern   211
work of fracture   234
woven fabrics   25f.
  – multilayer   31
  – non-crimp   25
  – three dimensional   30
  – tubular   31
woven spacer fabrics   31
wrinkling   42

x

X-ray   272ff.
  – analysis   272
  – C/SiC composite   276
  – computed tomography   278ff.
  – detection   274
  – figure intensifier   275
  – film   275
  – limitation   278
xenotime   195

y

yarn
  – architecture   21ff.
  – commingling   24
  – guiding element   36
Young's modulus determination   372
yttrium   223
yttrium aluminum garnet (YAG)   207, 224

z

zirconia   224
zirconia silica   197